CARBON NANOTUBES

Functionalization and Potential Applications

AAP Research Notes on Nanoscience and Nanotechnology

CARBON NANOTUBES

Functionalization and Potential Applications

Edited by

Ann Rose Abraham, PhD
Soney C. George, PhD
A. K. Haghi, PhD

AAP APPLE ACADEMIC PRESS

First edition published 2023

Apple Academic Press Inc.
1265 Goldenrod Circle, NE,
Palm Bay, FL 32905 USA
760 Laurentian Drive, Unit 19,
Burlington, ON L7N 0A4, CANADA

CRC Press
6000 Broken Sound Parkway NW,
Suite 300, Boca Raton, FL 33487-2742 USA
4 Park Square, Milton Park,
Abingdon, Oxon, OX14 4RN UK

Library and Archives Canada Cataloguing in Publication

Title: Carbon nanotubes : functionalization and potential applications / edited by Ann Rose Abraham, PhD, Soney C. George, PhD, A.K. Haghi, PhD.
Names: Abraham, Ann Rose, editor. | George, Soney C., editor. | Haghi, A. K., editor.
Series: AAP research notes on nanoscience & nanotechnology.
Description: First edition. | Series statement: AAP research notes on nanoscience & nanotechnology | Includes bibliographical references and index.
Identifiers: Canadiana (print) 20220200386 | Canadiana (ebook) 20220200394 | ISBN 9781774638576 (hardcover) | ISBN 9781774638583 (softcover) | ISBN 9781003277194 (ebook)
Subjects: LCSH: Carbon nanotubes. | LCSH: Carbon nanotubes—Industrial applications.
Classification: LCC TA455.C3 C37 2023 | DDC 620.1/93—dc23

Library of Congress Cataloging-in-Publication Data

CIP data on file with US Library of Congress

ISBN: 978-1-77463-857-6 (hbk)
ISBN: 978-1-77463-858-3 (pbk)
ISBN: 978-1-00327-719-4 (ebk)

ABOUT THE AAP RESEARCH NOTES ON NANOSCIENCE & NANOTECHNOLOGY BOOK SERIES

AAP Research Notes on Nanoscience & Nanotechnology reports on research development in the field of nanoscience and nanotechnology for academic institutes and industrial sectors interested in advanced research.

BOOKS IN THE AAP RESEARCH NOTES ON NANOSCIENCE & NANOTECHNOLOGY BOOK SERIES

- **Nanostructure, Nanosystems and Nanostructured Materials: Theory, Production, and Development**
  Editors: P. M. Sivakumar, PhD, Vladimir I. Kodolov, DSc, Gennady E. Zaikov, DSc, A. K. Haghi, PhD
- **Nanostructures, Nanomaterials, and Nanotechnologies to Nanoindustry**
  Editors: Vladimir I. Kodolov, DSc, Gennady E. Zaikov, DSc, and A. K. Haghi, PhD
- **Foundations of Nanotechnology: Volume 1: Pore Size in Carbon-Based Nano-Adsorbents**
  A. K. Haghi, PhD, Sabu Thomas, PhD, and Moein MehdiPour MirMahaleh
- **Foundations of Nanotechnology: Volume 2: Nanoelements Formation and Interaction**
  Sabu Thomas, PhD, Saeedeh Rafiei, Shima Maghsoodlou, and Arezo Afzali
- **Foundations of Nanotechnology: Volume 3: Mechanics of Carbon Nanotubes**
  Saeedeh Rafiei
- **Engineered Carbon Nanotubes and Nanofibrous Material: Integrating Theory and Technique**
  Editors: A. K. Haghi, PhD, Praveen K. M., and Sabu Thomas, PhD
- **Carbon Nanotubes and Nanoparticles: Current and Potential Applications**
  Editors: Alexander V. Vakhrushev, DSc, V. I. Kodolov, DSc, A. K. Haghi, PhD, and Suresh C. Ameta, PhD
- **Advances in Nanotechnology and the Environmental Sciences: Applications, Innovations, and Visions for the Future**
  Editors: Alexander V. Vakhrushev, DSc, Suresh C. Ameta, PhD, Heru Susanto, PhD, and A. K. Haghi, PhD
- **Chemical Nanoscience and Nanotechnology: New Materials and Modern Techniques**
  Editors: Francisco Torrens, PhD, A. K. Haghi, PhD, and Tanmoy Chakraborty, PhD

- **Nanomechanics and Micromechanics: Generalized Models and Nonclassical Engineering Approaches**
  Editors: Satya Bir Singh, PhD, Alexander V. Vakhrushev, DSc, and A. K. Haghi, PhD
- **Carbon Nanotubes: Functionalization and Potential Applications**
  Editors: Ann Rose Abraham, Soney C. George, PhD, and A. K. Haghi, PhD
- **Carbon Nanotubes for a Green Environment: Balancing the Risks and Rewards**
  Editors: Shrikaant Kulkarni, PhD, Iuliana Stoica, PhD, and A. K. Haghi, PhD
- **Nanostructured Carbon for Energy Generation, Storage, and Conversion**
  Editors: V. I. Kodolov, DSc, Omari Mukbaniani, DSc, Ann Rose Abraham, PhD, and A. K. Haghi, PhD

ABOUT THE EDITORS

Ann Rose Abraham, PhD
Assistant Professor, Sacred Heart College (autonomous), Thevara

Ann Rose Abraham, PhD, is currently Assistant Professor of Physics at Sacred Heart College, Thevara, Cochin, India. She is the author of ten book chapters and editor of two books. She has a number of publications to her credit in many peer-reviewed high impact journals of international repute, such as the *Journal of Physical Chemistry C, Physical Chemistry Chemical Physics, New Journal of Chemistry, Philosophical Magazine,* etc. She has research experience at various national institutes, including Bose Institute; SAHA Institute of Nuclear Physics; UGC-DAE CSR Centre, Kolkata; and has collaborated with various international laboratories, such as the University of Johannesburg, South Africa; Institute of Physics Belgrade; etc. She is the recipient of young researcher award in the area of physics, a prestigious forum to showcase intellectual capability. Dr. Abraham received her MSc, MPhil, and PhD degrees in Physics from the School of Pure and Applied Physics, Mahatma Gandhi University, Kerala, India. She has expertise in the field of materials science, nanomagnetic materials, multiferroics, and polymeric nanocomposites, etc.

Soney C. George, PhD
Dean of Research and Director, Amal Jyothi Centre for Nanoscience and Technology, Kerala, India

Soney C. George, PhD, is the Dean of Research and Director of the Amal Jyothi Centre for Nanoscience and Technology, Kerala, India. He is a fellow of the Royal Society of Chemistry, London, and a recipient of "best researcher of the year" award in 2018 from APJ Abdul Kalam Technological University, Thiruvananthapuram, India. He has also received awards such as best faculty award from the Indian Society for Technical Education, best citation award from the International Journal of Hydrogen Energy, a fast-track award for young scientists by the Department of

Science & Technology, India, and an Indian young scientist award instituted by the Indian Science Congress Association. He has published and presented almost 200 publications in journals and at conferences. His major research fields are polymer nanocomposites, polymer membranes, polymer tribology, pervaporation, and supercapacitors. He has guided eight PhD scholars and 95 student projects.

He did his postdoctoral studies at the University of Blaise Pascal, France, and Inha University, South Korea

A. K. Haghi, PhD
Professor Emeritus of Engineering Sciences, Former Editor-in-Chief, International Journal of Chemoinformatics and Chemical Engineering and Polymers Research Journal; Member, Canadian Research and Development Center of Sciences and Culture

A. K. Haghi, PhD., is the author and editor of over 200 books, as well as over 1000 published papers in various journals and conference proceedings. Dr. Haghi has received several grants, consulted for a number of major corporations, and is a frequent speaker to national and international audiences. Since 1983, he has served as professor at several universities. He is the former Editor-in-Chief of the *International Journal of Chemoinformatics and Chemical Engineering* and *Polymers Research Journal* and is on the editorial boards of many international journals. He is also a member of the Canadian Research and Development Center of Sciences and Cultures. He holds a BSc. degree in Urban and Environmental Engineering from the University of North Carolina (USA), an MSc. degree in Mechanical Engineering from North Carolina A&T State University (USA), a DEA in applied mechanics, acoustics, and materials from the Université de Technologie de Compiègne (France), and a PhD degree in Engineering Sciences from Université de Franche-Comté (France).

CONTENTS

CONTRIBUTORS

Jiji Abraham
Department of Chemistry, Vimala College, Thrissur, Kerala, India;
E-mail: jijiabraham02@gmail.com

V. N. Anjana
Sree Sankara Vidyapeetom College, Perumbavoor 683556, Kerala, India

V. N. Archana
Mar Athanasius College, Kothamangalam, India

Reyha Benedict
St Teresa's College, Ernakulam 682011, Kerala, India

K. B. Bhavitha
Department of Physics, St Teresas's College, Ernakulam 682011, Kerala, India
International and Inter University Centre for Nanoscience and Nanotechnology,
Mahatma Gandhi University, Kottayam 686560, Kerala, India; E-mail: bhavithakb@gmail.com

Namitha Binu
Pushpagiri Institute of Medical Sciences and Research Centre, Tiruvalla 689101, Kerala, India

Avinash V. Borgaonkar
Department of Mechanical Engineering, National Institute of Technology Warangal,
Warangal 506004, Telangana, India; E-mail: avi.borgaonkar@gmail.com

Yogesh B. Dalvi
Pushpagiri Institute of Medical Sciences and Research Centre, Tiruvalla 689101, Kerala, India;
E-mail: yogesh.botany@gmail.com

K. S. Devaky
School of Chemical Sciences, Mahatma Gandhi University, Kottayam, Kerala, India;
E-mail: aquaregiazone@gmail.com

N. G. Divya
Cochin University of Science and Technology, Cochin 682022, Kerala, India

Soney C. George
Centre for Nanoscience and Nanotechnology, Amal Jyothi College of Engineering, Kottayam,
Kerala, India

Vidyalakshmi I. S.
SB College, Changanacherry, Kottayam, Kerala, India

Sam John
St Berchmans College, Changanassery, Kottayam, Kerala, India;
E-mail: samjohnthanicken@gmail.com

Nimitha K. C.
Department of Chemistry, Vimala College, Thrissur, Kerala, India

Vidya L.
School of Pure & Applied Physics, Mahatma Gandhi University, Kottayam, Kerala, India

Jaya Lakkakula
Amity Institute of Biotechnology, Amity University, Mumbai, Pune Expressway, Bhatan Post, Somathne, Panvel, Mumbai 410206, Maharashtra, India

Beena Mathew
School of Chemical Sciences, Mahatma Gandhi University, Kottayam, Kerala, India;
E-mail: beenamscs@gmail.com

Sneha Mathew
School of Chemical Sciences, Mahatma Gandhi University, Kottayam, Kerala, India

Jahanvee Mitra
Amity Institute of Biotechnology, Amity University, Mumbai, Pune Expressway, Bhatan Post, Somathne, Panvel, Mumbai 410206, Maharashtra, India

Shital B. Potdar
Department of Chemical Engineering, National Institute of Technology Warangal, Warangal 506004, Telangana, India

Mamatha Susan Punnoose
School of Chemical Sciences, Mahatma Gandhi University, Kottayam, India

Divya Radha
School of Chemical Sciences, Mahatma Gandhi University, Kottayam, Kerala, India

Aparna Raj
School of Pure & Applied Physics, Mahatma Gandhi University, Kottayam, Kerala, India

N. Rakesh
School of Chemical Sciences, Mahatma Gandhi University, Kottayam, Kerala, India

Arunima Reghunadhan
TKM College of Engineering, Karicode, Kollam, Kerala-691005, India;
E-mail: arunimarenjith02@gmail.com

Neelima S.
School of Pure & Applied Physics, Mahatma Gandhi University, Kottayam, Kerala, India

M. V. Santhosh
School of Chemical Sciences, Mahatma Gandhi University, Kottayam, Kerala, India

Aruni Shajkumar
Laboratory for Advanced Research in Polymer Materials, Bhubaneswar, India

G. K. P. Srilekha
Amity Institute of Biotechnology, Amity University, Mumbai, Pune Expressway, Bhatan Post, Somathne, Panvel, Mumbai 410206, Maharashtra, India

C. Swetha
Department of Chemistry, St. Thomas College, Ranni, India; E-mail: swethachengat@gmail.com

Sabu Thomas
International and Inter University Centre for Nanoscience and Nanotechnology, Mahatma Gandhi University, P.D. Hills, Kottayam 686560, Kerala, India

Riju K. Thomas
Bharata Mata College, Thrikkakara, Ernakulam, Kerala, India; E-mail: rijukthomas@gmail.com

Ruby Varghese
Pushpagiri Institute of Medical Sciences and Research Centre, Tiruvalla 689101, Kerala, India

Nilesh Wagh
Amity Institute of Biotechnology, Amity University, Mumbai, Pune Expressway, Bhatan Post, Somathne, Panvel, Mumbai 410206, Maharashtra, India

Srinivasarao Yaragalla
Smart Materials, Istituto Italiano di Tecnologia, Via Morego 30, Genova 16163, Italy; E-mail: yaragallasrinu@gmail.com

ABBREVIATIONS

AC	activated carbon
AFM	atomic force microscope
APC	ammonium perchlorates
APPI	amphiphilic polypropylene imine dendrimers
ARC	antireflection coating
ART	artemisinin
AW	antiwear
BPE	bipolar electrode
BT	bone-targeting
CA	contact angle
CCVD	catalytic chemical vapor deposition
CEA	carcinoembryonic antigen
CFMEs	carbon fiber microelectrodes
CFRP	carbon fiber-reinforced polymer
CLC	CNT-liposomes conjugate
CMWCNTs	carboxylated multiwalled carbon nanotubes
CNPs	carbon nanopearls
CNTs	carbon nanotubes
CNT/DEN	amido amine dendirmers
CNTFET	carbon nanotube field effect transistor
CS	chitosan
CS	co-electrodeposition/stripping
CTAB	hexadecyltrimethylammonium bromide
CVD	chemical vapor deposition
DDS	drug delivery system
DFT	density functional theory
DGU	density-gradient ultracentrifugation
DLS	dynamic light scattering
DMFC	direct methanol fuel cells
DOP	degree of polarization
DOS	density of states
DOX	doxorubicin
DSC	differential scanning calorimetry

DSSC	dye-sensitized solar cells
DWCNT	double-walled carbon nanotube
EAPC	enzyme adsorption, precipitation, and crosslinking
EFC	enzymatic biofuel cell
EG	ethylene glycol
EOC	emerging organic contaminants
EP	emerging pollutants
EP	extreme pressure
FA	folic acid
FED	field emission displays
FESEM	field emission scanning electron microscope
FET	field effect transistors
FF	fill factor
FFS	fringe-free switching
FG	fluoro-graphene
FI	fluorescein isothiocyanate
FITC	fluorescein isothiocyanate
FNM	formononetin
FR	folate receptor
FT	Fischer-Tropsh
GA	gambogic acid
GBR	guided bone regeneration
GHGs	greenhouse gases
GOx	glucose oxidase
GPC	gel permeation chromatography
GTR	guided tissue regeneration
HA	hyaluronic acid
HAS	higher alcohol synthesis
HAS	human serum albumin
HCPUA	heat capacity per unit area
HRC	heat release capacity
HUVEC	human umbilical vein endothelial cells
IBAM	iso-butyramide
IF	inorganic fullerene-like
IS-ILs	imidazolium-salt-based ionic liquids
ITO	indium tin oxide
LBL	layer-by-layer
LCDs	liquid crystal displays

LSP	lightning strike protector
mAb	monoclonal antibody
MB	methylene blue
MC	merocyanine
MDR	multidrug resistance
MFC	microbial fuel cell
MG	methylene green
MGP	modified graphene platelets
MLG-Cu	multilayer graphene impregnated with copper
MO	methyl orange
MoDDP	molybdenum dialkyldithiophosphate
MOSFET	metal oxide semiconductor field effect transistor
MSCs	mesenchymal stem cells
MS	mesoporous silica
MTO	mitoxantrone
MWCNTs	multiwalled carbon nanotubes
NAD	nonabsorbent
NCNTs	nitrogen-doped CNTs
ND	nanodiamond
NIR	near-infrared
NMR	nuclear magnetic resonance spectroscopy
NPs	nanoparticles
OEG	octa ethylene glycol
OLC	onion-like carbon
OLED	organic light-emitting diode
PAMAM	poly(amidoamine)
PAN	polyacrylonitrile
PAO	polyalphaolefin
PCBM	phenyl-C61-butyric acid methyl ester
PCE	photoconversion efficiency
PCE	power conversion efficiency
PCL	polycaprolactone
PCU	polycarbonate urethane
PDADMAC	polyelectrolyte, poly(diallyldimethylammonium) chloride
PDDA	poly(diallyldimethylammonium chloride)
PDT	photodynamic therapeutic
PEDOT	poly(3,4-ethylenedioxythiophene)

PEG	polyethylene-glycol
PEI	polyethyleneimine
PEI/CNT	polyetherimide/CNT
PEM	polymer electrolyte membrane
PEMFCs	proton exchange membrane fuel cells
PEO	polyethylene oxide
PET	polyethylene terephthalate
PFPE	perfluoropolyether
PLA	polylactic acid
PLA-PEG	poly(lactide)-poly (ethylene glycol)
PMMA	poly (methyl) methyl acrylated
PMT	polyampholyte
PP	polypropylene
PPS	polyphenylene sulfide
PS	polystyrene
PSA	prostate specific antigen
PSCA	prostate stem cell antigen
PSS	poly(sodium 4-styrenesulfonate)
PTX	paclitaxel
PVDF	poly(vinyledene fluoride)
PVP	polyvinylpyrrolidone
QD-CNTs	quantum dots-CNT conjugates
RAM	radar absorbing materials
RB	rhodamine
RhB	rhodamine B
RLV	reusable launch vehicles
RPAS	remotely piloted aircraft system
RuPOP	ruthenium polypyridyl
SB	silibinin
SD	standard deviation
SDBS	sodium dodecyl benzene sulfonate
SDS	sodium dodecyl sulfate
SEED	substrate-enhanced electroless deposition
SEM	scanning electron microscopy
SERS	surface-enhanced Raman scattering
SiC	silicon carbide
siRNA	small interfering RNAs

SMCNP	sesame-cookie topography–modified carbon nanotube paper
STS	scanning tunneling spectroscopy
SWCNTs	single-walled carbon nanotubes
TAM	tamoxifen
TEM	transmission electron microscopy
TFT	thin film transistor
TOF	turn over frequency
TON	turnover number
UAVs	unmanned aerial vehicles
US DOE	US Department of Energy
VACNT	vertically aligned carbon nanotube
VOC	open-circuit voltage
WGP	wire-grid polarizer
WSD	wear scar diameter

PREFACE

This book covers many aspects of the science of carbon nanotubes, which is a rapidly advancing field. This book also deals with advanced topics on carbon nanotubes—their extraordinary properties, structure, design, fabrication, development, engineering, functionalization, carbon nanotube-enabled nanocomposites, characterization, and moreover their utility in various aspects. The amazing potential of advanced CNT composites in automotive, aeronautics, spacecrafts, CNT transistors replacing Si electronics, energy, purification, hydrogen storage, tissue regeneration, electrochemical supercapacitor, medical, sensing, biomedical applications, agriculture, energy, and technical applications is outlined. This book covers the overall review of the research advancements on all aspects of carbon nanotube.

This book will prove to be highly useful for graduate and postgraduate students as well as for beginning and senior researchers for a strong foundation on the topic.

The volume:

- Deals with upscaling of CNTs from laboratory stage to fab stage;
- Is a comprehensive guide to theory and practical applications of CNTs and covers extensive spectrum of CNT-based applications;
- Illustrates the lab-to-fab roadmap of CNTs in EMI shielding, energy applications, biomedical applications, agricultural applications, etc.
- Highlights significant and noteworthy properties, applications, and recent developments in the role of nanotechnology in effectively combating various environmental pollution, soil pollution using CNTs.

CHAPTER 1

ENVIRONMENTAL APPLICATIONS OF CARBON NANOTUBES

VIDYALAKSHMI I. S.[1], APARNA RAJ[2], NEELIMA S.[2], VIDYA L.[2], and RIJU K. THOMAS[3*]

[1]*SB College, Changanacherry, Kottayam, Kerala, India*

[2]*School of Pure & Applied Physics, Mahatma Gandhi University, Kottayam, Kerala, India*

[3]*Bharata Mata College, Thrikkakara, Ernakulam, Kerala, India*

[*]*Corresponding author. E-mail: rijukthomas@gmail.com*

ABSTRACT

Carbon nanotubes (CNTs) are the most incredible material in the fullerene family. The graphene sheets with a quasi-one-dimensional structure of nanometer scale diameter in the rolled-up form are called CNTs. Also they are often called tubular fullerene. Due to their exceptional physical, chemical, mechanical, electronic, and thermal properties can make wonders in the various research areas of nanoscience. The significant aspects of CNTs such as their lightweight, diminutive size with a high aspect ratio, fine tensile strength, and good conducting characteristics make them a good candidate for various applications like biomedical, materials in electronics, sensors, field emission devices, energy storage, and conversion, etc. The most interesting fact is that they are environment-friendly materials and this makes them to be the most beloved materials for the environmental applications. In this book chapter, we are presenting a general readership about the environmental applications of CNTs.

1.1 INTRODUCTION

The word "Nano" in nanotechnology is a Greek word and it means dwarf, scientific dealing at the nano level (atomic level) by using particular scientific instruments is acknowledged as nanotechnology and the term Nanotechnology was introduced by a Japanese scientist named Norio Taniguchi in 1974.[1] The arrival of nanotechnology has brought a huge revolution in the area of the research. Generally, materials having dimensions between 1 and 100 nm is called nonmaterial. More conveniently, we can say that a sheet of paper will be about 10^5 nm thick while a single gold atom is about 1/3 nm in diameter.[2] Hence, we can say that nanomaterials are having a size larger than the single atoms or molecules but they are smaller than the bulk materials. Their characteristic properties lie in between classical and quantum physics. The dimensionality of nanoparticles is classified into zero, one, and two. Since they process a low dimensionality, they have a large surface-to-volume ratio and better electrical and optical properties when compared with a bulk sample of the same material. All these properties take nanoparticles to a large area of research. Among all the elements, Carbon has got that much impression in the field of nanotechnology and there vast varieties of carbon-based nanomaterials.

Carbon, the first element of the 14th group in the periodic table is an exceptional and essential element in our world. It is known to be the sixth most common element in the universe, and the 4th most common element in our solar system, and the 17th most common element in the earth's crust.[3] The relative abundance of carbon has been estimated at about between 180 and 270 ppm.[4] Moreover, it is known to be the second most common element in the human body which consists of about 18% of a human's body weight.[5] Carbon possesses a broad variety of metastable phases that can be produced near ambient circumstances and their roomy fields of kinetic stability and this is one of the stupendous descriptions of carbon.[6]

Even though the amount of elemental carbon present on the earth's crust is of about 0.2% of the total mass of the earth, it exhibits implausible essential functions like bonding with other light elements.[6] This has paved the way to revolutionary research fields in subjects like chemistry and biology. The developments in the carbon science lead to the innovation of many such low dimensional allotropes of carbon nanostructures such as graphite, activated carbon (AC), carbon nanotubes (CNTs), and the C_{60}

family of buckyballs, polyaromatic molecules, and graphene in various fields of nanoscience, material science, and engineering technology.[6] The significant application of nanotechnology is that it can produce new materials with elite properties.

1.2 A BRIEF HISTORICAL CHEMICAL BACKGROUND OF SOME ALLOTROPES OF CARBON

For a very long period of time, there were only two natural crystalline allotropes of carbon were discovered and they are the graphite and the diamond. However, the both structures possess extremely different chemistry especially in their structure and properties.[7] The covalent bonding between two carbon atoms may results in the formation of many new other allotropes of carbon.[6] The elements that are identical in their chemical properties and differ in their structural as well as physical properties are known as allotropes. There are several allotropes of carbon have been discovered and they are graphene, Buckminster fullerene, and CNTs and thus made carbon as an element that contains large number of allotropes.[8–10]

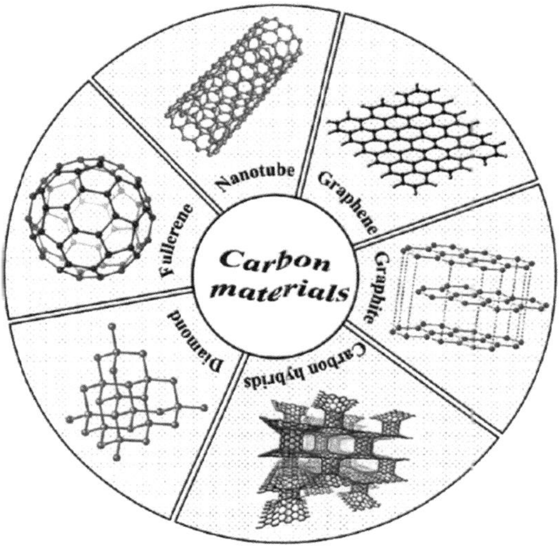

FIGURE 1.1 Structure of carbon nanomaterials: diamond, graphite, C_{60}, CNTs, graphene, and 3D graphene-CNT hybrid materials.

Source: Reprinted with permission from Ref. 10. © 2018 American Chemical Society.

- **Graphite**

A layered structure consisting of six carbon atoms to form a ring that arranged in widely spaced horizontal sheets is known as graphite. Since it has excellent electrical conductivity, it is mainly used for making electrodes in an electrical arc lamp. In graphite, the delocalization of π-electrons is responsible for the ability to conduct electricity. This phenomenon is absent in diamond. In addition to this, Graphite is known as the most stable form of carbon in standard conditions[6].

- **Diamond**

The most peculiar property of diamond is the its highest thermal conductivity when compared to any other material. This is determinedly endorsed to low phonon scattering and strong covalent bonding that held together with the atoms. The natural diamond possesses a thermal conductivity of approximately 2200 W/Mk and it is about five times that of the copper.[11] Due to the high thermal conductivity, diamond is broadly used in the semiconductor engineering to prevent Silicon and other semi-conducting materials from overheating.[6] The other important properties of diamond are good carrier mobility, saturated carrier velocities, and electrical field breakdown strength.[12] According to its physiochemical properties, Diamond is considered to be chemically and physically robust and radiation hard.[12]

- **Graphene and Graphene-derived materials**

Graphene and graphene-derived materials are commonly called as Graphenoids. It was discovered by Andre Geim and Novoselov in 2004. Graphene is the primary building block of the most of the structure of carbon materials.[13] Arrangement of piled-up graphene gives rise to graphite and a rolled-up graphene gives rise to CNTs.[8] The presence of defects, impurities, and structural disorders can cause an adverse effect on the electrical and optical properties of graphene.[8] As a result, a single crystalline graphene thin film with high electrical and thermal conductivities is preferred for the electronic applications.[6]

- **Activated Carbon**

Activated carbon or AC comes under the category of amorphous carbon due to its structure and properties. It has a microporous structure.[6] The materials that are composed of AC exhibit an excellent adsorption capacity of various chemical specious. The significant property of AC is its high surface area-to-volume ratio and, therefore, it is commonly used as an adsorbent in poisoning, reducing the cholesterol level plummeting internal gas flatulence, and many other useful applications.[14,15]

1.3 CARBON NANOTUBES

Fullerenes are often called as molecular forms of carbon or carbon molecules. In 1985, three scientists named H. W. Kroto, R. F. Curl, and R. E. Smalley were discovered fullerenes and awarded with Nobel Prize in chemistry in later. The family of fullerene consists of a number of atomic C_n clusters (n > 20), composed of carbon atoms on a spherical surface and the carbon atoms are typically positioned on the vertices of the pentagons and hexagons.[16] Carbon atoms in fullerenes are in the sp^2-hybrid state and linked together by covalent bonds. The most common and the best-investigated fullerene is the C_{60} fullerene in which the spherical molecule is highly symmetric and it consists of 60 carbon atoms situated at the vertices of 20 hexagons and 12 pentagons.[16] These closed C_{60} fullerenes are informally called *Buckyballs*.

Another member in this fullerene which has a wide range of applications is the CNTs. Being a member of fullerene family, they are often called as *buckytubes*. As the word "nanotube" demands in CNTs, the material consists of only two coaxial cylinders.[6] The discovery of CNTs thus made a revolution in the field of nanotechnology. CNTs were discovered independently by Iijima in 1991 and Benthune et al. in 1993. They are flawless cylinder-shaped macromolecules having a radius of a few nanometers and up to some micrometers in length. The walls of these tubes are made up of a hexagonal lattice of carbon atoms and capped by fullerene-like structures. According to their structure, they can be divided into two, multi-walled CNTs (MWCNTs) and single-walled CNTs (SWCNTs). Coaxial arrangement of two or more concentric cylindrical shells of graphene sheets around a central hollow area having a spacing between the layers will form a MWCNT and they have a large diameter lies between 2 and 100 nm.[2] Whereas a single cylinder graphite sheet held together by Van der Waals bonds will form a SWCNT and have a diameter less than 2 nm.[10] Due to the narrow diameter, they exhibit excellent material properties like high aspect ratio and large surface area. As a result, they can be approximated into one- dimensional nanostructures. The helicity and diameter of a CNT determine the electrical properties.[2] Their changing hybridization states and sensitivity of the structure to perturbations in synthesis conditions exploits their unique physical, chemical, and electronic properties[17]. This leads to innovations in new technologies and applications. Besides, these unique and fine properties offer impending advances in environmental systems from proactive (prevention of environmental degradation, optimizing energy efficiency) to retroactive (wastewater reuse, pollutant transformation).[17]

1.3.1 PROPERTIES OF CARBON NANOTUBES

It is reported that CNTs possess large surface area, high aspect ratios, and remarkably high mechanical strengths.[1] They have a tensile strength which is 100 times greater than that of steel and their electrical and thermal conductivities approaches that of copper.[18]

Electronic Properties of CNTs

The electrical properties of CNTs are in a chiral form.[1] The each carbon atom in CNTs is covalently bonded to each other with the sp^2 hybridization.[19] So in each unit, the fourth valance electron remains free. Hence, the delocalization of these free electrons over all atoms will occur and this motion will help to attain the electrical nature of CNTs. Depending on the type of chirality, CNTs can be either conducting or semiconducting materials. Semiconducting SWNTs will be most likely to be of p-type semiconductors.[20] This electronic nature of CNTs can be applied for the transistors and other switching applications and the most recent application of CNT was an emitter.[1]

Mechanical Properties of CNTs

By considering their mechanical properties, CNTs were considered strongest materials in the nature. They have a Young's modulus ranges from 270 to 950 GPa.[1] They also possess very high tensile strength ranges from 11 to 63 GPa. The covalent carbon bonds present in graphite are found to be the strongest one in the nature and hence CNTs are considered to be the stiffest and toughest structure ever synthesized by scientists.[1]

Thermal Properties of CNTs

Since the CNTs are very small in size, the quantum effects are very significant. Hence, the low temperature specific heat and thermal conductivity show direct evidence of the one-dimensional quantization of the phonon band structure in CNTs.[21] The factors affecting the thermal properties of CNTs are number of phonon-active modes, the length of the free path for the phonons and boundary surface scattering.[22] These properties may also depend on the atomic arrangement, the length, and diameter of the tubes, the number of structural defects and the morphology.[23]

All these excellent properties of CNTs offer great promises in their environmental applications. They are used as sensors and sorbents for the detection and treatment of existing contaminants and thereby preventing the pollution.[24] They are used for the storage of energy such as in solar cell, used for the storage of hydrogen in the hydrogen fuel cells and environment friendly energy storage as in the form of biofuels. Their water treatment applications include the treatment of wastewater, removing the heavy metals from the contaminated water and they are also used for water transportation. Another important application is they are used as sensor for the detection of the toxic gases and thereby they help to control the air pollution. Another environment-friendly application is that they are used for the CO_2 capture. All these peculiarities and their excellent properties make CNTs as the most interesting nanomaterials to the researchers.

1.4 ENVIRONMENTAL APPLICATIONS OF CARBON NANOTUBES

Nowadays, modern society is facing diverse public health implications as a result of environmental pollution. According to WHO, the global ambient air quality database (update 2018) estimates that 7 million people die every year due to exposure to fine particles in polluted air leading to diseases such as respiratory infections, heart disease, stroke, lung cancer, and chronic obstructive pulmonary diseases. Chemicals (natural/synthetic) that have adverse ecological effects in the environment and not commonly monitored are called Emerging pollutants (EP).[25] Emerging pollutants mainly include drugs, illicit drugs, pesticides, furans, dioxins, microplastics, perfluoroalkyl substances, polycyclic aromatic hydrocarbons, and so on. They reach aquatic systems through discharge from wastewater treatment plants, sewage, through run off, etc. It can be bioaccumulated and biomagnified causing adverse effects on ecosystem. The occurrence of EP in aquatic system has not been widely studied.[26] It may be due to the presence of organic compounds, lipids, and proteins that make them complex. Therefore, there is an urgent need for efficient materials for environmental pollution management.[27] Graphene-/CNTs-based materials are promising candidates for the environmental applications from sensing and monitoring to remediation.[28] New nanomaterials for the extraction of emerging organic contaminants (EOC) are CNT's, graphene-based materials, and metallic nanoparticles.[25]

1.4.1 ENERGY APPLICATIONS

1.4.1.1 HYDROGEN GAS STORAGE IN CNTs

The idea of hydrogen as a fuel has been a drastic increase in the production rate and utilization technologies for the gas. The reason is that Hydrogen provides more energies than either gasoline or natural gas on the basis of its weight.[29] Among many ways of storing hydrogen, the most effective one is the use of CNTs. They can store hydrogen even in the room temperatures. But the condition is that the capacities do not exceed ~0 wt%. All in all the reported capacities ranges from 0 to 60 wt%.[1] These are categorized into three. (1) having an H/C ratio of 2(~14 wt% H) and are considered to be un-physical, (2) consistent with expectations based on findings for activated and other, conventional high-surface area carbons(~0 wt%), (3) those are in the intermediate range (1–14 wt%).[1] According to the experimental results, the SWCNT structures can store about of 8 wt% of hydrogen.

Small pores in low pressure enhance the adsorption in pores and thus result in high storage[24] and the reverse effect will occur for high pressures. The excellent conductive properties, low mass densities, and strong physical properties of CNTs make them ideal and durable material for fuel cell electrodes. In addition, the macrostructures formed by assembling the nanotubes possess a high surface area and this makes them a suitable substitute for Pt catalysts and hydrogen adsorption.[30] A hydrogen fuel cell with CNT composite electrodes or membranes helps to reduce the use of rare and expensive catalysts.[30] However, further studies are doing in this field to get a perfect hydrogen storing CNT structure.

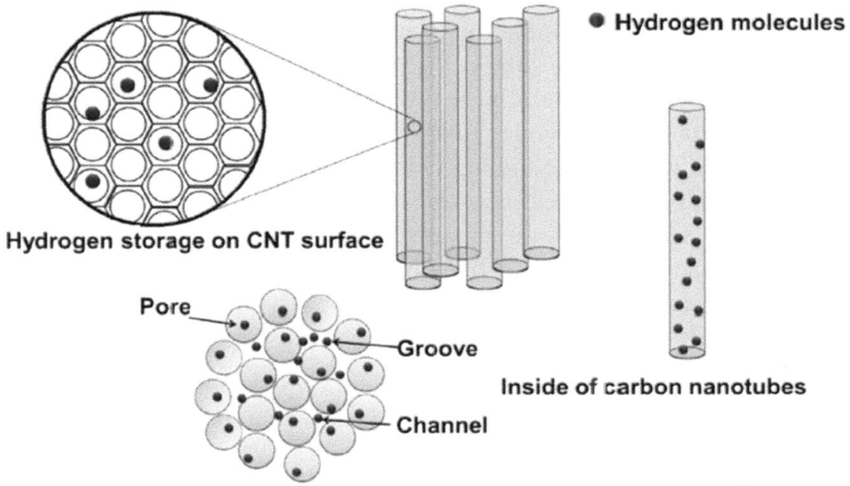

FIGURE 1.2 Different storage sites of CNTs.

Source: Reprinted with permission from Ref. 30a. © 2018 Elsevier.

1.4.1.2 CARBON NANOTUBES IN RENEWABLE ENERGY

Nowadays the energy consumption is increasing across the world. So long-term energy demand has become a challenge. Therefore, we should be aware of the requirement for advance renewable energy technologies to protect the environmental balance. The application of CNTs in photovoltaic devices leads to use them in solar cell was a major breakthrough. The device which absorbs photons from sunlight and converts them to electrical energy is called a photovoltaic device. The affordability and remarkable energy conversion of CNTs make them as an important material in various solar cell structures especially in silicon-based solar cells.[31] Their extremely small surface area enhances massive absorption of photons for harvesting solar energy while the mobility of charge transfer is facilitated by the delocalized π electron system.[32] Apart from silicon-based solar cells, there is another one known as organic solar cell.[30] It consists of a conductive organic polymer like poly(3-octylthiophene) (P3OT), poly(3- hexylthiophene) (P3HT), or [6,6]-phenyl-C61-butyric acid methyl ester (PCBM) for light absorption and charge transfer.[24]

The present research shows an increase in the efficiency upon the incorporation of CNTs in the top electrode, the photoactive layer and the back electrode of organic solar cells.[34–36] The role of CNT as a photoactive material in the photoactive layer provides the efficient electron or hole transport at the CNT/polymer interface and thereby optimize the performance of the cell. Due to the high electron transport capability of CNTs, the photoactive component constructed with P3OT/CNT shows a higher open-circuit voltage.[37] In the top and back electrodes, the CNTs provide a large surface area for the high optical transmittance and low sheet resistance to reduce the power loss.[30] As a result, the rate of photocurrent produced will be increased for the design of a top electrode composed of CNT films and Indium Tin oxide (ITO).[38]

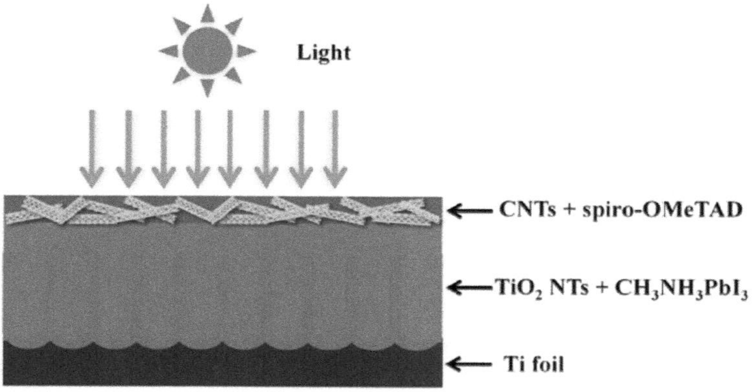

FIGURE 1.3 Schematic of solid-stateperovskite solar cells based on tin foil/TiO_2 nanotubes and CNTs.

Source: Reprinted with permission from Ref. 38a. © 2015 Elsevier.

1.4.1.3 CARBON NANOTUBES IN BIOFUEL CELLS

The uses of cell phones, computers, and other electronic gadgets have become responsible for the large production of lithium batteries in the environment which are absolutely non-environment-friendly chemicals. Hence, the needs for the clean production of electricity have stimulated the origin of new sources of sustainable and renewable energy without the greenhouse gas emission and environment pollution. The increasing interest for environment-friendly technologies has opened doors to the inventions in biotechnology which make specific useful products by using living organisms. A cell that produces electricity by making use of living organisms is known as a biofuel cell. Biofuel cells can be either microbial fuel cell (MFC) or enzymatic biofuel cell (EFC).[38] MFC generates electric power by using its microbial catabolic activities. Taking complex organic waste and the renewable biomass in wastewater as the source materials, MFCs can do better in wastewater treatments.[39] However, the research works on CNT electrodes are under progressing. Due to their high conductivity and large surface area, the CNT electrodes are modifying for increased power production.[40] The high specific surface area of CNT is very useful for the construction of highly porous 3-D CNT electrodes. All these peculiarities have made CNT to become one of the important materials for the designing of electrochemical biosensors and biofuels.[41] In addition to it, by attaching specific redox sites to these nanostructured electrodes through organic functionalization have tuned them very well and this is very suitable for fixing proteins or catalyzing electrochemical reactions with enzymes or coenzymes.[42] This indicates that these CNTs can set up an electrical message with enzymes through an electron transport to enzymes assured by electron hopping between immobilized redox centers. Since they are extremely small in size, CNTs can approach in close propinquity the prosthetic site of enzymes and hence attain a direct electrical writing between the enzymes and the bulk electrode. As a result, modified CNT electrodes have prompted extensive attention in the invention of biofuel cells.[42]

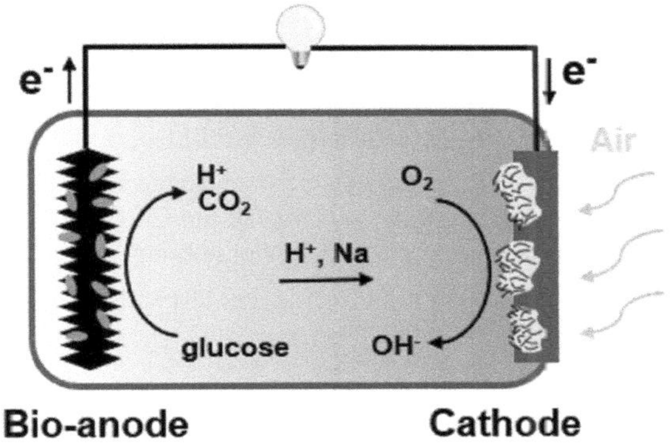

FIGURE 1.4 Schematic set up of microbial fuel cell.

Source: Reprinted with permission from Ref. 42a. © 2019 Elsevier.

1.4.2 WASTEWATER TREATMENT APPLICATION OF CNTS

1.4.2.1 CNT SPONGE

CNT sponge can be developed via chemical vapor decomposition methods using ferrocene as the precursor. A CNT sponge made from this technique possesses a randomly entangled 3-D structure and exhibits high porosity with minimum density. Oil-contaminated water can be cleaned by the CNT sponge. They float on the contaminated surface and will remove the oil by adsorbing it. Since the sponges are hydrophobic, they show a tendency to move to the oil film area, and thereby leading to the exceptional floating and cleaning capacity which is very useful for spill clean-up. This oil-saturated CNT sponge can be regenerated easily through mechanical compression with concurrently the revival of important resources or straight burned in air devoid of destroying the sponge structure.[24]

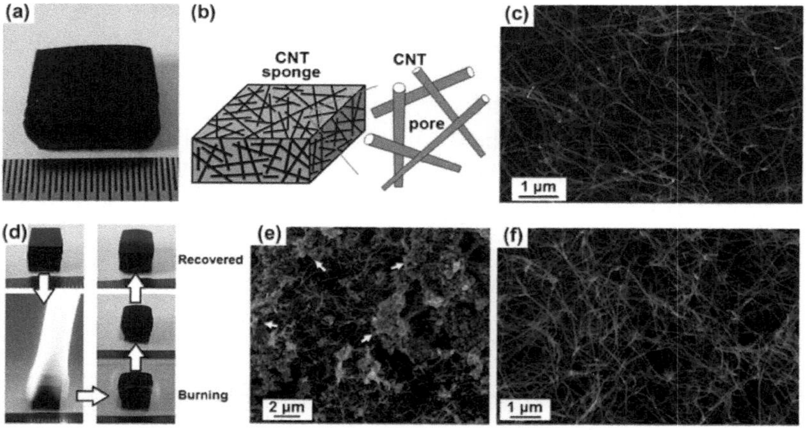

FIGURE 1.5 CNT sponges: (a) photograph; (b) schematic of the pore structure; (c) SEM image of the CNT sponge; (d) burning and reuse of the CNT sponge. SEM images of (e) the surface and (f) the core of the CNT sponge after burning.

Source: Reprinted with permission from Ref. 42b. © 2019 Elsevier.

1.4.2.2 ADSORBENTS

Adsorption is a surface phenomenon used for the elimination of pollutants. It is based on the interaction between absorbable solute and solid adsorbent. It is cost-effective and highly efficient in comparison with other methods. The general techniques for the purification of wastewater are ion exchange, chemical precipitation, reverse osmosis, electrolysis, coagulation/flocculation, ultra-filtration, and adsorption.[43]

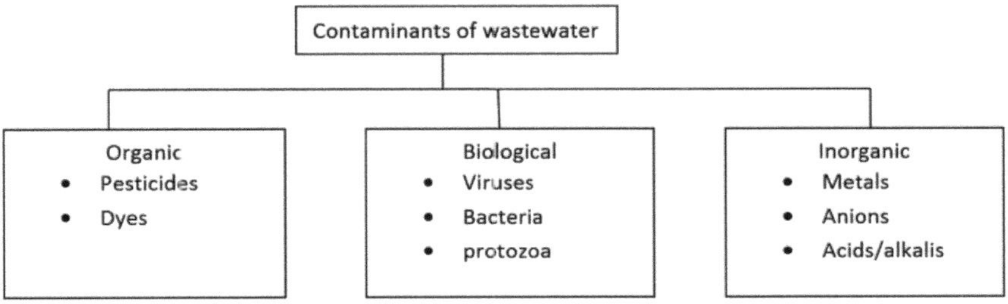

Adsorption can be

- physisorption, caused by Vander Waals forces,
- chemisorption, chemical bonding between adsorbent and adsorbate, and
- Exchange adsorption, caused by charge attraction between adsorbate and the surface.

Importance of non-absorbent comes to play in adsorption process where adsorbent will determine the versatility and efficiency of adsorption process.[44] They have large surface area, multiple sorption sites, tunable pore size, low temperature modification, and surface chemistry that make them suitable candidates for the adsorption process. These nanomaterials can be nanowires, nanotubes, quantum dots, and nanoparticles. CNTs are important for wastewater remediation because of their high adsorption sites and tunable surface chemistry. An ideal nano adsorbent material should be non-toxic, have high sorption capacities and high accuracy for removing the contaminants even in their low concentrations. The surface should be easily reactivated. Recently, eco-friendly synthesized SnO_2-CNT nano hybrid is used as an efficient re-usable material for the onsite water remediation. In this work, oxidative adsorption of As (III) on SnO_2-CNT nanocomposite was presented.[45]

Non-biodegradable wastes have been increasingly released to water directly or indirectly, which causes serious toxic effects in ecosystem.[46]

Removal of such pollutants can be done using sorption, membrane filtration, or chemical precipitation. Effective adsorption of oil from water can be also achieved by new porous adsorbents that possess superhydrophobicity and superleophilicity. Super hydrophobic and super leophilic hybrid foam of CNTs can be fabricated using a two-step chemical vapor deposition technique. They can be selectively used to remove the organic solvents and oils from surface of water which otherwise cause long-term effect of environmental pollution.

Graphene CNT aerogel also shows excellent reusability and stability after adsorption. Organic dyes are non-degradable and have carcinogenic effects. Hence, new development for the removal of dyes is of prime importance. Also, CNTs-graphene has good adsorption capacity for organic dyes. In a work,MWCNT non-absorbent (NAD) called Co-MWCNT NADs is used for removing methylene blue, which is a dyestuff that can cause carcinogenic and mutagenic effects on humans and aquatic life's-MWCNT NAD is fabricated using ultrasonic method.[47]

The physical adsorption of hydrocarbon products (saturated and unsaturated) of Fischer-Tropsh (FT) synthesis on CNT-supported cobalt surface can be investigated using DFT (density functional theory). They found that in FT process (4,4) CNT Co-nanocatalyst had greater efficiency to create heavy hydrocarbons than light types.[48]

In environmental engineering, CNTs are used for various adsorbent application of following species.[49]

- Heavy metals
- Disinfection by products
- Endocrine disrupters
- Biological contaminants
- Natural organic matter and so on

1.4.2.3 ROLE OF CNT IN WASTEWATER TREATMENT

Wastewater coming from industrial, domestic areas, or agricultural sources consists of a huge amount of contaminants like heavy metal ions, some radioactive elements like caesium and strontium, etc.[50] These pollutants are non-degradable, highly toxic, and carcinogenic and hence they badly affect the quality of water as well as our health. So sorbents with good sorption capability and high sorption efficiency can remove these contaminants. Having a high surface active site to volume ratio and controlled pore size distribution, CNTs are very adaptable for the wastewater treatments. Advanced studies discovered that the adsorption capacity of CNTs depends on both the surface functional group and the nature of the sorbate.[38] Generally, CNTs having a pH range of 7–10 are suitable for adsorption.[51] According to current technologies, CNTs are also used as nanofilters to reduce the particle concentration in the wastewater.[52] CNTs have shown an extraordinary performance in transporting water regardless of their hydrophobic nature. Molecular dynamics simulations point out that the hydrophobic nature of CNTs pores builds weak interactions with water molecules, and thereby facilitating a fast and nearly frictionless flow of water.[53] Recent studies show that CNT nanofilters can remove the pathogenic microorganisms such as protozoa, bacteria, viruses present in the wastewater treatment, with microorganisms being reserved on the facade of CNT based on a depth filtration mechanism.[54] It can also remove the *E. Coli* bacteria at low pressure using SWCNTs.[38] Moreover, the antimicrobial activity of CNTs helps to replace chemical disinfectants as a new efficient way for controlling microbial pathogens.[55] Application of CNTs in water disinfectant treatment avoids the creation of harmful disinfection by-products like trihalomethane, haloacetic acids, and aldehydes as they are not strong oxidants and are comparatively static in water.

1.4.3 CNTS FOR CONTROLLING AIR POLLUTION

1.4.3.1 CNTS AS GAS SENSOR

The metallic and semiconductivity of CNT is greatly prejudiced by their one-dimensional cylindrical structure, such as size and chirality although they possess unique and tunable electrical properties. The advantages such as low power consumption, low operating temperature, and high sensitivity will make CNTs dominant over other conventional metal-oxide gas sensors.[56]

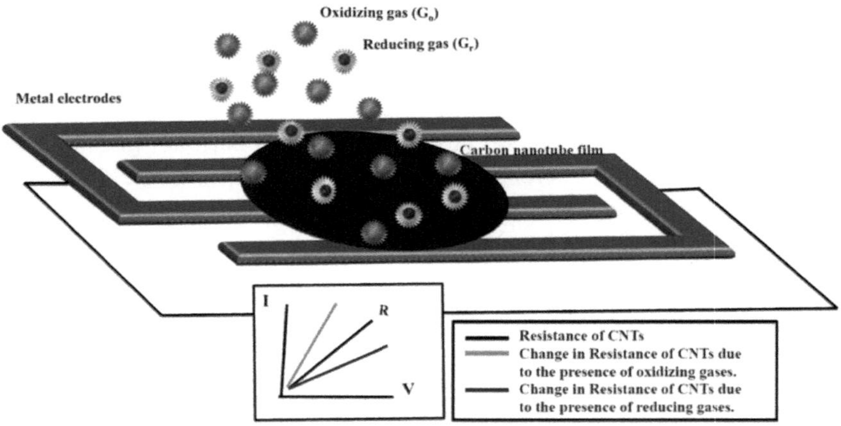

FIGURE 1.6 Schematics of CNT gas sensors for emissions from fossil fuel burning.
Source: Reprinted with permission from Ref. 56b © 2020 Elsevier.

A number of researchers have been used the CNT-based gas sensor for the detection of various poisonous gases like Nitrogen Oxides (NO_x), Nitrogen Dioxide (NO_2), Ammonia (NH_3), and Sulfur Dioxide (SO_2) at room temperature.[57–59]

1.4.3.2 CARBON NANOTUBES AS THE ADSORBENT FOR THE CO2 CAPTURE

The combustion of fossil fuels in power plants is known as one of the most common technology to generate electricity. Due to this combustion, a large amount of CO_2 will be produced. CO_2 is one of the greenhouse gases (GHGs) and cause global warming and this leads to the global climate change. A good remedy for this problem is the capture and storage of CO_2. Impregnation of amines on to solid materials is suitable for adsorption of CO_2. Among them, CNTs are considered as the suitable material for the impregnation of amines. Among the two classes of CNTs, SWCNTs show higher CO_2 adsorption capacity that the MWCNTs.[60]

1.4.4 ACTIVATED CARBON COATED WITH MULTI-WALLED CARBON NANOTUBES FOR REMOVING ANTIBIOTIC MATERIALS

Antibiotics are partly metabolized in the human body and they stay on active after the excretion. These materials can be seen in drinking water, surface water, sewage, and wastewater sources in various scales of amounts. The pharmaceutical wastes from the pharmaceutical industry and hospitals may come into contact with the water sources. This is a threat for both humans and for all the living organisms. This can results in high antibiotic resistance in long term. One of the most common antibiotics is the Ciprofloxacin (CIP). It is present in a large amount in the environment, especially in wastewater. CIP in the aqueous solutions can be removed by AC supported with MWCNTs/AC. The factors affecting the removal of CIPs are the initial concentration of CIP, adsorbent dosage, pH of the solution, temperature, and contact time.[61] They have a removal efficiency of 73% and this made the MWCNT/AC as a potential tool for the antibiotic removal.

1.4.5 CATALYTIC APPLICATIONS OF CNTS

Catalyst is important for accelerating chemical reaction by reducing activation energy. Catalysts can be homogeneous or heterogeneous. An optimal catalyst shows superior performance in mainly four areas: selectivity, activity, durability, and recoverability.[63] The activity of a catalyst expresses the number of molecules converted to product molecules by catalyst per unit time. Turn over frequency (TOF) is used to measure the activity of catalyst. Homogeneous catalysts show high TOF of 0.3 s^{-1} or higher, while heterogeneous catalyst in the range of 0.03 s^{-1} or even lower. The lifetime of a catalyst is a measure of a number of cycles it can undergo without replacement. The turnover number (TON) is the number of products that can be formed by a given amount of catalyst. The recoverability of catalytic active substance is important, in the commercial point of view. It should be able to separate from the reaction mixture and reuse after the termination of reaction. The selectivity, activity, durability, and recoverability are mainly influenced by catalyst size, shape, and surface composition. When the catalyst size comes in nano range, the more active sites are accessible to the substrate. Graphene oxides can be dispersed in CNT and

Pt nanocatalyst. The resulting GOCNT-Pt nano composites show greatly enhanced electrocatalysis and power density.[64] This so prepared enzyme mimic exhibit catalytic performance comparable to protein peroxidase, with higher catalytic efficiency, environmental robustness, stability, and cost-effectiveness.

1.4.5.1 WHY CNT IS SUITABLE FOR CATALYTIC ACTIVITY

Ideally, SWCNT is made of a perfect graphene sheet rolled up to a cylinder and closed at two sides (semi fullerenes). The diameter varies between 0.4 and 2.5 nm and the length can be several millimeters. The metallic particles residue from the production process can be seen inside the MWCNT cavity. Studies show that SWCNT behaves like quantum wires which have electron confinement along the tube axis. Mainly electronic properties are governed by tube diameter and helicity in which the graphene layer is rolled up. Electronic properties are drastically modified by defects in the system.[65]

Alkalized MoO_3 nanocatalyst supported on CNTs is prepared by microemulsion technique. Its activity and selectivity in higher alcohol synthesis (HAS) have been assessed in fixed bed microreactor. In their previous work, they used impregnation method to prepare CNT supported MoO_3-k_2O HAS catalyst. CNT as a catalyst support allows better metal dispersion and controls the metal support interactions. The effect of molybdenum particle size on the catalyst properties, activity, and selectivity was also studied supported MoO_3-k_2o nanocatalyst synthesized using microemulsion technique shows a decreased hydrocarbon selectivity from 32.4 to 24.5%, while the alcohol selectivity increased from 45.3 to 61.4%.

1.5 CONCLUSION

Nowadays, modern society is facing diverse public health implications as a result of environmental pollution. According to WHO, the global ambient air quality database (update 2018) estimates that 7 million people die every year due to exposure to fine particles in polluted air leading to diseases such as respiratory infections, heart disease, stroke, lung cancer, and chronic obstructive pulmonary diseases. These environmental pollutants can be bioaccumulated and biomagnified causing adverse effects on ecosystem. Therefore, there is an urgent need for efficient materials for the environmental pollution management. CNT-based materials are the promising candidates for the environmental applications. With advancements in the practical applications of CNT in environmental engineering, we can broaden the horizon of human life.

ACKNOWLEDGMENT

Ms. Neelima S acknowledged the department of science and technology, government of India (DST-INSPIRE) for the award of JRF.

KEYWORDS

- **carbon nanotubes**
- **pollution**
- **adsorbent**
- **catalyst**
- **wastewater treatment**
- **air pollution**

REFERENCES

1. Ibrahim, K. S. Carbon Nanotubes-Properties and Applications: A Review. *Carbon Lett.* **2013,** *14*, 131–144.
2. Lahiff, E.; Lynam, C.; Gilmartin, N.; O'Kennedy, R.;Diamond, D. The Increasing Importance of Carbon Nanotubes and Nanostructured Conducting Polymers in Biosensors. *Anal. Bioanal. Chem.* **2010,** *398*, 1575–1589.
3. Zhang, Y.;Yin, Q. Z. Carbon and Other Light Element Contents in the Earth's Core Based on First-Principles Molecular Dynamics. *Proc. Natl. Acad. Sci. USA.* **2012,** *109*, 19579–19583.
4. Allègre, C. J.; Poirier, J.-P.; Humler, E.; Hofmann, A. W. The Chemical Composition of the Earth. *Earth Planet. Sci. Lett.* **1995,** *134*, 515–526.
5. Pace, N. R The Universal Nature of Biochemistry. *Proc. Natl. Acad. Sci. USA* **2001,** *98*, 805–808.
6. Nasir, S.; Hussein, M. Z.; Zainal, Z.;Yusof, N. A. Carbon-Based Nanomaterials/ Allotropes A Glimpse of Their Synthesis, Properties and Some Applications. *Materials (Basel)***2018,** 11, 1–24.

7. Dai, L.; Chang, D. W.; Baek, J.-B.;Lu, W. Carbon Nanomaterials for Advanced Energy Conversion and Storage. *Small 8*, 1130–1166.
8. Geim, A. K.; Novoselov, K. S. The Rise of Graphene. *Nat. Mater.* **2007,** *6*, 183–191.
9. Kroto, H. W.; Heath, J. R.; O'Brien, S. C.; Curl, R. F.;Smalley, R. E. C60: Buckminsterfullerene. *Nature* **1985,** *318*, 162–163.
10. Hu, C.; Qu, J.; Xiao, Y.; Zhao, S.; Chen, H.; Dai, L. Carbon Nanomaterials for Energy and Biorelated Catalysis: Recent Advances and Looking Forward. 2018. https://doi.org/10.1021/acscentsci.8b00714.
11. Mildren, R.;Rabeau, J. *Optical Engineering of Diamond. Optical Engineering of Diamond*, 2013. doi:10.1002/9783527648603
12. Pierson, H. O. Handbook of Carbon, Graphite, Diamonds and Fullerenes. *Handb. Carbon Graph. Diamonds Fullerenes* **1993,** 43–69. doi:10.1016/B978-0-8155-1339-1.50008-6
13. Choudhary, N.; Hwang, S.;Choi, W. Carbon Nanomaterials: A Review. **2014,** 709–769. doi:10.1007/978-3-642-31107-9_37
14. Mahmoudi, K.; Hamdi, N.;Srasra, E. Preparation and Characterization of Activated Carbon from Date Pits by Chemical Activation with Zinc Chloride for Methyl Orange Adsorption. 2014.
15. Hernández-Montoya, V.;Bonilla-Petriciolet, A. *Lignocellulosic Precursors Used in the Synthesis of Activated Carbon: Characterization Techniques and Applications in the Wastewater Treatment*. 2012.
16. Zaytseva, O.;Neumann, G. Carbon Nanomaterials: Production, Impact on Plant Development, Agricultural and Environmental Applications. *Chem. Biol. Technol. Agric.* **2016,** *3*, 1–26.
17. Ong, Y. T.; Ahmad, A. L.; Zein, S. H. S.; Tan, S. H. A Review on Carbon Nanotubes in an Environmental Protection and Green Engineering Perspective. *Braz. J. Chem. Eng.* **2010,** *27*, 227–242.
18. Ebbesen, T-W. et al. Electrical Conductivity of Individual Carbon Nanotubes. *Nature* **1996,** *382*, 54–56.
19. Eatemadi, A. et al. Carbon Nanotubes: Properties, Synthesis, Purification, and Medical Applications. *Nanosc. Res. Lett.* **2014,** *9*, 1–13.
20. Amente, C.; Dharamvir, K. Thermally Agitated Self Assembled Carbon Nanotubes and the Scenario of Extrinsic Defects. *World J. Nano Sci. Eng.* **2015,** *5*, 17–25.
21. Hahm, M.-G.; Hashim, D. P.; Vajtai, R.; Ajayan, P. M. A Review: Controlled Synthesis of Vertically Aligned Carbon Nanotubes. *Carbon Lett.* **2011,** *12*, 185–193.
22. Sinnott, S. B.; Shenderova, O. A.; White, C. T.; Brenner, D. W. Mechanical Properties of Nanotubule Fibers and Composites Determined from Theoretical Calculations and Simulations. *Carbon NY* **1998,** *36*, 1–9.
23. Maeda, T.; Horie, C. Phonon Modes in Single-Wall Nanotubes with a Small Diameter. *Phys. B Condens. Matter* **1999,** *263*, 479–481.
24. Bhanuse, U. M.; Babar, A. B.;Ranveer, A. C. Journal of Environmental Science, Computer Science and Engineering & Technology. *Jecet* **2016,** *5*, 10–18.
25. Nouri, N.; Khorram, P.; Duman, O.; Tunç, S.; Sereshti, H. Trends in Environmental Analytical Chemistry Overview of Nanosorbents Used in Solid Phase Extraction Techniques for the Monitoring of Emerging Organic Contaminants in Water and Wastewater Samples. **2020,** 25. https://doi.org/10.1016/j.teac.2020.e00081.

26. Álvarez-ruiz, R.; Picó, Y. Trends in Environmental Analytical Chemistry Analysis of Emerging and Related Pollutants in Aquatic Biota. **2020,** 25. https://doi.org/10.1016/j.teac.2020.e00082.
27. Mauter, M. S.; Elimelech, M. Critical Review Environmental Applications of Carbon-Based Nanomaterials. **2008,** 5843–5859.
28. Fan, W. Springer Briefs in Molecular Science Graphene-Carbon Nanotube Hybrids for Energy and Environmental Applications.
29. Becher, M. et al. Hydrogen Storage in Carbon Nanotubes. *Comptes. Rendus Phys.* **2003,** *4,* 1055–1062.
30. Olusegun, A. et al. We are IntechOpen, the World's Leading Publisher of Open Access Books Built by Scientists, for Scientists TOP 1%. *Intechi* **2012,** 38.
30a Singh, R.; Srivastava, S.; Kumar, P. Nano-Structures & Nano-Objects Structural and Surface Modification of Carbon Nanotubes for Enhanced Hydrogen Storage Density. *Nano-Struct. Nano-Objects* **2018,** *14,* 57–65. https://doi.org/10.1016/j.nanoso.2018.01.005.
31. Schmidt, J.; Peibst, R. Surface Passivation of Crystalline Silicon Solar Cells: Present and Future. 2018. doi:10.1016/j.solmat.2018.06.047
32. Kamat, P. V. Meeting the Clean Energy Demand: Nanostructure Architectures for Solar Energy. **2007,** 2834–2860.
33. Janssen, R. Polymer-Fullerene Bulk Heterojunction Solar Cells. 2005. doi:10.1557/mrs2005.6
34. Pasquier, A. Du, Unalan, H. E.; Kanwal, A.; Miller, S.; Chhowalla, M. Conducting and Transparent Single-Wall Carbon Nanotube Electrodes for Polymer-Fullerene Solar Cells *Appl. Phys. Lett.* **2005,** *87,* 1–3.
35. Jeon, I.; Matsuo, Y.;Maruyama, S. Single-Walled Carbon Nanotubes in Solar Cells. *Top. Curr. Chem.* **2018,** *376,* 2–4.
36. Rowell, M. W. et al. Organic Solar Cells with Carbon Nanotube Network Electrodes. *Appl. Phys. Lett.* **2006,** *88,* 23–25.
37. Cataldo, S., Salice, P.; Menna, E.;Pignataro, B. Carbon Nanotubes and Organic Solar Cells. *Energy Environ. Sci.* **2012,** *5,* 5919–5940.
38. Ong, Y. T.; Ahmad, A. L.; Zein, S. H. S.; Tan, S. H. A Review on Carbon Nanotubes in an Environmental Protection and Green Engineering Perspective. *Braz. J. Chem. Eng.* **2010,** *27,* 227–242.
38a. Wang, X.; Li, Z.; Xu, W.; Kulkarni, S. A.; Batabyal, S. K.; Zhang, S.; Cao, A.; Helena, L. TiO_2 Nanotube Arrays Based Fl Exible Perovskite Solar Cells with Transparent Carbon Nanotube Electrode. *Nano Energy* **2014,** 1–8. https://doi.org/10.1016/j.nanoen.2014.11.042.
39. Goh, K.; Chen, Y. Controlling Water Transport in Carbon Nanotubes. *Nano Today.* 2017. doi:10.1016/j.nantod.2016.12.015
40. Sharma, T.; Mohana Reddy, A. L.; Chandra, T. S.; Ramaprabhu, S. Development of Carbon Nanotubes and Nanofluids Based Microbial Fuel Cell. *Int. J. Hydrogen Energy* **2008,** *33,* 6749–6754.
41. Lahiff, E.; Lynam, C.; Gilmartin, N.; O'Kennedy, R.; Diamond, D. The Increasing Importance of Carbon Nanotubes and Nanostructured Conducting Polymers in Biosensors *Anal. Bioanal. Chem.* **2010,** *398,* 1575–1589.

42. Cosnier, S.; Holzinger, M.; Goff, A. Le. Recent Advances in Carbon Nanotube-Based Enzymatic Fuel Cells. *Front. Bioeng. Biotechnol.* **2014,** *2,* 1–6.
42a. Du, Y.; Ma, F.; Xu, C.; Yu, J.; Li, D.; Feng, Y.; Zhen, L. Nano Energy Nitrogen-Doped Carbon Nanotubes / Reduced Graphene Oxide Nanosheet Hybrids towards Enhanced Cathodic Oxygen Reduction and Power Generation of Microbial Fuel Cells. May **2019,** *61,* 533–539. https://doi.org/10.1016/j.nanoen.2019.05.001.
42b. Gui, X.; Li, H.; Wang, K.; Wei, J.; Jia, Y.; Li, Z.; Fan, L.; Cao, A.; Zhu, H.; Wu, D. Recyclable Carbon Nanotube Sponges for Oil Absorption. *Acta Mater.* **2011,** *59* (12), 4798–4804. https://doi.org/10.1016/j.actamat.2011.04.022.
43. El-sayed, M. E. A. Science of the Total Environment Nanoadsorbents for Water and Wastewater Remediation. *Sci. Total Environ.* **2020,** *739,* 139903. https://doi.org/10.1016/j.scitotenv.2020.139903.
44. Olveira, S.; Forster, S. P.; Seeger, S. Nanocatalysis: Academic Discipline and Industrial Realities. **2014.**
45. Mohanta, D.; Nath, A. Environmentally Benign Fabrication of SnO_2 -CNT Nanohybrids and Their Multifunctional Efficiency as an Adsorbent , Catalyst and Antimicrobial Agent for Water Decontamination. *Sci. Rep.* August **2019,** 1–19. https://doi.org/10.1038/s41598-019-49181-2.
46. Rocha, R. P.; Pereira, M. F. R. Tuning CNT Properties for Metal-Free Environmental Catalytic Applications. 2016. https://doi.org/10.3390/c2030017.
47. Çalımlı, M. H. Magnetic Nanocomposite Cobalt—Multiwalled Carbon Nanotube and Adsorption Kinetics of Methylene Blue Using an Ultrasonic Batch. *Int. J. Environ. Sci. Technol.* **2020,** No. 0123456789. https://doi.org/10.1007/s13762-020-02855-1.
48. Shariatinia, Z.; Abdollahi-moghadam, M. Original Article DFT Computations on Surface Physical Adsorption of Hydrocarbons Produced in the Fischer-Tropsch Synthesis on a CNT/Co Nanocatalyst. *J. Saudi Chem. Soc.* **2018,** *22* (7), 786–808. https://doi.org/10.1016/j.jscs.2018.01.001.
49. Upadhyayula, V. K. K.; Gadhamshetty, V. Appreciating the Role of Carbon Nanotube Composites in Preventing Biofouling and Promoting Bio Fi Lms on Material Surfaces in Environmental Engineering : A Review. **2010,** *28,* 802–816. https://doi.org/10.1016/j.biotechadv.2010.06.006.
50. Wang, Y.; Pan, C.; Chu, W.; Vipin, A. K.; Sun, L. Environmental Remediation Applications of Carbon Nanotubes and Graphene Oxide: Adsorption and Catalysis. *Nanomaterials* **2019,** *9.*
51. Rao, G.; Lu, C.; Su, F. Sorption of Divalent Metal Ions from Aqueous Solution by Carbon Nanotubes: A Review. *Sep. Purif. Technol.* **2007,** *58,* 224–231.
52. Srivastava, A.; Srivastava, O. N.; Talapatra, S.; Vajtai, R.; Ajayan, P. M. Carbon Nanotube Filters. *Nat. Mater.* **2004,** *3,* 610–614.
53. Noy, A. et al. Nanofluidics in Carbon Nanotubes. *Nano Today* **2007,** *2,* 22–29.
54. Bohonak, D. M.; Zydney, A. L. Compaction and Permeability Effects with Virus Filtration Membranes. *J. Memb. Sci.* **2005,** *254,* 71–79.
55. Savage, N.; Diallo, M. S. Nanomaterials and Water Purification: Opportunities and Challenges. *J. Nanopart. Res.* **2005,** *7,* 331–342.
56. Endo, M.; Strano, M.; Ajayan, P. Potential Applications of Carbon Nanotubes. *Topics Appl. Phys.* **2007,** *111,* 13–61.

56a. Mittal, M.; Kumar, A. Accepted Us T. *Sens. Actuat. B. Chem.* 2014. https://doi.org/10.1016/j.snb.2014.05.080.
57. Ueda, T. et al. Development of Carbon Nanotube-Based Gas Sensors for NO $_{x}$ Gas Detection Working at Low Temperature. *Phys. E Low-Dimensional Syst. Nanostruct.* **2008,** *40,* 2272–2277.
58. Wang, Y.; Yeow, J. T. W. A Review of Carbon Nanotubes-Based Gas Sensors. *J. Sensors* **2009, 2009,** 493904.
59. Suehiro, J.; Zhou, G.; Hara, M. Detection of Partial Discharge in SF6 Gas Using a Carbon Nanotube-Based Gas Sensor. *Sens. Actuat. B Chem.* **2005,** *105,* 164–169.
60. Osler, K.; Dheda, D.; Ngoy, J.; Wagner, N.; Daramola, M. O. Synthesis and Evaluation of Carbon Nanotubes Composite Adsorbent for CO_2 Capture: A Comparative Study of CO_2 Adsorption Capacity of Single-Walled and Multi-Walled Carbon Nanotubes. *Int. J. Coal Sci. Technol.* **2017,** *4,* 41–49.
61. Sharifpour, N.; Moghaddam, F. M.; Mardani, G.; Malakootian, M. Evaluation of the Activated Carbon Coated with Multiwalled Carbon Nanotubes in Removal of Ciprofloxacin from Aqueous Solutions. *Appl. Water Sci.* **2020,** *10,* 1–17.
62. Zubizarreta, L. Sustainable Carbon as Efficient Support for Metal-Based Nanocatalyst: Applications in Energy. 2020.
63. Manuscript, A. Nanoscale. https://doi.org/10.1039/b000000x.
64. Serp, P.; Corrias, M.; Kalck, P. Carbon Nanotubes and Nanofibers in Catalysis. **2003,** *253,* 337–358. https://doi.org/10.1016/S0926-860X(03)00549-0.
65. Tavasoli, A.; Karimi, S.; Shoja, M. Synthesis and Application of CNT-Supported MoO_3 -K_2O Nanocatalyst Using Microemulsion Technique: Role of Nanoparticle Size on Catalyst Activity and Selectivity in Higher Alcohol Synthesis. **2013,** 1–11.

CHAPTER 2

CARBON NANOTUBES FOR A GREENER ENVIRONMENT

DIVYA RADHA[1], N. RAKESH[1], M. V. SANTHOSH[1], K. S. DEVAKY[1,*], and SAM JOHN[2,*]

[1]*School of Chemical Sciences, Mahatma Gandhi University, Kottayam, Kerala, India*

[2]*St Berchmans College, Changanassery, Kottayam, Kerala, India*

**Corresponding author. E-mail: aquaregiazone@gmail.com, samjohnthanicken@gmail.com*

ABSTRACT

Reckless human activities have a negative impact on the environment. Hence, there is a need for the development of technologies that help to give a new dimension to the eco-friendly environment. The advancement in the field of nanotechnology provides so many multifunctional materials that help in making the environment more sustainable and greener. Many multifunctional materials like carbon nanotubes (CNTs), quantum dots, graphene augur well for such applications. Recent progress in the frontiers of nanotechnologies has supported benchmarking CNTs as one of the most investigated carbon-based nanomaterials. A plethora of applications of CNTs has been proposed in various fields of research due to their unequaled physical, chemical, optical, and electronic properties. This chapter points out the contribution of CNTs in terms of sustainable environment and eco-friendly technologies perspectives like wastewater treatment, air pollution monitoring, desalination, renewable microbial fuel cells, supercapacitors, and green nanocomposites. Based on the available reports, CNTs proved to be a promising candidate for application in various environmental fields.

2.1 INTRODUCTION

The amelioration in the field of nanotechnology has brought tremendous changes in various fields of research. Nanochemistry has put forth a variety of multifunctional materials. Carbon nanotubes (CNTs) are one such kind. According to Iijima[1] and Bethune,[2] CNTs are seamless cylindrical-shaped tubular structures with length in the range of micrometers and radius in the range of nanometers. CNTs are generally classified into single-walled CNTs (SWCNTs) and multiwalled CNTs (MWCNTs) depending on their unique structures. SWCNTs are made up of single graphite-like sheet held together by weak Van der Waals forces whereas MWCNTs consist of two or more concentric cylinders of graphene sheets coaxially arranged around. The major commercial methods employed to synthesize the CNTs include chemical vapor deposition technique, laser ablation, and electric arc discharge.[3–5]

The properties of CNTs stated in Table 2.1 could be tuned due to their mutable hybridization state and their sensitivity to the conditions of the synthetic method. These lead to the exploitation of their new developments and applications in various fields. CNTs are widely exploited in environmental systems. They are being used in wastewater treatment, Air pollution monitoring, supercapacitors, green catalysis, biotechnologies, green nanocomposites design, renewable energy sources, and so on.[6]

This chapter documents the role of CNTs in making the environment greener and eco-friendly.

TABLE 2.1 Properties of SWCNTs and MWCNTs.

Properties	SWCNTs	MWCNTs
Modulus of elasticity	~1000 GPa	~30–150 GPa
Specific gravity	0.8 g/cm^3	1.8 g/cm^3
Tensile strength	500 GPa	60 GPa
Resistivity	5–50	5–50
Thermal conductivity	~3000 W/m/K	~3000 W/m/K
Thermal stability	>750°C (in air)	>750°C (in air)
Specific surface area	400–900 m^2/g	200–400 m^2/g

2.2 APPLICATIONS OF CARBON NANOTUBES

2.2.1 WASTEWATER TREATMENT

Water contamination has been a global concern. The common sources of contaminants include heavy metal ions, various hydrocarbon compounds, and dyes.[7] These pollutants seem to be nonbiodegradable, toxic, and carcinogenic that pose a serious threat to the aquatic forms of life as well as the human beings. The most celebrated method of removal of these pollutants from water relies on the surface phenomenon called adsorption. Adsorption is greatly affected by the capacity of an adsorbent to adsorb the adsorbates. CNTs are found to be a good adsorbent for this utility as they possess properties like greater surface area, higher sorption capacity and efficiency, and greater pore size distribution. Compared with conventional adsorbents like granular and powder-activated charcoal, CNTs are excellent adsorbents for the removal of pollutants. The adsorption capacity of the CNTs to an extent depends mainly on the functional group present and the nature of the adsorbed species. The reports suggested that the various group functionalized CNTs support the adsorption of polar compounds[8] whereas the nonfunctionalized favors the adsorption of nonpolar compounds like hydrocarbons.[9,10]

Properties of the adsorbed functional groups on to CNTs and metal ions, chemical interaction for polar components, physical interaction for the nonpolar compounds, the range of pH and ionization and competition between ionic species also play a pivotal role in the adsorption behavior of CNTs. The CNTs are excellent adsorbents for the removal of heavy metal ions and dyes. CNTs are generally used for the removal of heavy metal ions like Zn(II), As(III,V), Pb(II), Cu(II), Cr(II), Co(II,III), Hg(II), Cd(II), Ni(II), Sr(II), and dyes like Methylene Blue, Congo red. The most commonly reported mechanisms of adsorption include chemisorption, physisorption, the electrostatic interaction, surface complexation, Van der Waals interactions, soft acid–soft base interaction, ion exchange, hydrophobic interactions, external diffusion, intraparticle diffusion, cation–π interaction, and anion–π interaction between the adsorbent and the metal ions. Among these reported mechanisms, the major mechanism of adsorption is chemical interaction and electrostatic interactions between the surface functional groups of CNTs and the heavy metal ions.[11,12]

The presence of toxic and nonbiodegradable hydrocarbons in water is a serious menace to the environment. Various hydrocarbons like dichlorobenzene, polycyclic hydrocarbons, Triton, ionizable aromatic compounds, and phenolic compounds are examples of such kind. CNTs are used for the removal of 1,2-dichlorobenzene (DCB) due to its rough surface. The CNTs exhibited excellent adsorption capacity for DCB at an optimum pH range of 3–10. When the pH gets increased, the adsorption capacity of CNTs get decreased due to the adsorption of water molecules on to the carboxyl group that ultimately impairs the access of DCB.[13] SWCNTs purified by the sulfuric acid and nitric acid are used for the removal of phenolic compounds,[14] Triton surfactants[15] and polycyclic aromatic hydrocarbons.[16]

CNTs functionalized with 8-hydroxyquinoline is used for the elimination of Pb(II), Cd(II), Cu(II), and Zn(II).[17] PTSA doped polyaniline @ graphene oxide-MWCNT composite were used for the adsorption of the dye Congo red and Cr(VI) ions.[18] CNTs modified with polyamidoamine (PAMAM) dendrimer were effectively used for the adsorption of CU(II) and Pb(II) ions.[19] Acid functionalized chitosan carbon nanosheets were also reported as the adsorbents for the removal of copper ions.[20] Lead, nickel, and copper ions were removed from the aqueous solution using the functionalized CNTs wrapped with polypyrrole nanocomposites.[21] Chitosan/MWCNTs and chitosan beads were excellent absorbing agents for the removal of copper ions from aqueous solution. CS/MWCNTs beads have a maximum adsorption capacity of 454.55 mg/g greater than that of the free chitosan beads.[22] Aminopyrazole functionalized MWCNTs adsorbed cadmium (III) ions with an adsorption capacity of 83.7%.[23] Isocyanate functionalized MWCNTs were used for the adsorption of lead ions from wastewater and also used for the separation of Tl from Pb ions.[24]

Arsenic ion is one of the major contaminants present in industrial effluents. Functionalized CNTs are interesting candidates for its removal. SWCNTs doped with walnut shells were evaluated for their capacity to remove the arsenic ions and found that they were excellent in the heavy metal removal than the unmodified SWCNTs.[25] Arsenic (III,IV) ions could be effectively removed from the contaminated water using the zerovalent iron-doped MWCNTs chelated with EDTA.[26] MWCNTs functionalized with phosphonium based deep eutectic solvents were found to be good adsorbents for the removal of arsenic ions. CNTs functionalized using phosphonium-based solvents like methyl triphenylphosphonium bromide (MTPB) and benzyl triphenyl phosphonium chloride (BTPC) were used to adsorb the arsenic ions.[27] PAMAM/CNT nanocomposite was also found to be an excellent adsorbent for arsenic ions.[28] Ceria nanoparticles (NPs) supported on CNTs were also used to remove arsenic ions. The presence of calcium and magnesium ions in the water improved the adsorption capacity of the ceria-supported adsorbents due to the formation of the ternary surface complex.[29] Another heavy metal ion arsenic was excellently removed from the aqueous solution using the magnetic Fe_3O_4 functionalized nonoxidative graphene/CNT composite.[30]

Stannic molybdophosphate MWCNTs (MWCNT SMP) hybrid materials excellently removed strontium from aqueous solution with a 4.9-fold adsorption capacity than that of the raw MWCNTs.[31] Single pot polycondensation reaction were used to synthesize polyamine/CNT composite by mixing melamine, paraformaldehyde, 1,6-diaminohexane and chlorinated CNTs in the presence of DMF as a solvent. The pure melamine-formaldehyde-diaminihexane (MFDH) Polymer and functionalized composite(MFDH-1, MFDH-2, MFDH-3, and MFDH-4) with 0.01%, 0.02%, 0.05%, and 0.1% weight of chlorinated CNTs were used to remove the lead ions from the effluents. The comparative studies found that MFDH-4 has a greater adsorption capacity for the Pb removal.[32]

The metals like Pb(II), Cu(II), Ni(II), and Zn(II) are effectively get adsorbed on the diamino functionalized mesoporous silica-coated MWCNTs((NN-mSiO_2@MWCNTs) and mesoporous SiO_2-coated MWCNTs (mSiO_2@MWCNTs), oxidized MWCNTs (OMWCNTs), and MWCNTs. The experiments were done to investigate the ability of these functionalized CNTs for these metals removal by stirring 20 mg of adsorbents in 20 mL of 100 mg/L aqueous heavy metal solutions at pH of 6.2 and temperature of 25°C. The experimental results showed that lead ions could be effectively removed by these adsorbents followed by Copper and zinc ions. It is also found that the adsorption capacity of the adsorbents follows the order as NNmSiO_2@MWCNTs > mSiO_2@MWCNTs > OMWCNTs > MWCNTs.[33] Voelcker et al. reported that Thiol derivatized SWCNTs could be used for the removal of mercury ions. These adsorbents were synthesized by reacting decarboxycysteine hydrochloride and oxidized SWCNTs. The study observed that these adsorbents has greater metal adsorption capacity than that of raw MWCNTs and natural adsorbent-activated carbon.[34]

Besides serving as an excellent adsorbent for the heavy metal ions and dyes, CNTs acts as nanofillers in the reduction of particle concentration in contaminated water. Functionalization of these CNT fillers can make miracles in these applications. Computational studies viz. molecular simulation studies reported that the CNTs could effectively transport water molecules even though they are hydrophobic. The easy frictionless flow of water could be explained by the weaker interactions with water or due to the nanoscale confinement of the tubes that ultimately leads to the narrowing of the energy distribution thereby weakening its interaction with water.[35] CNT nanofilters are effectively used in removing the microorganisms like viruses, bacteria, and protozoa. CNTs excellently act as sorbent by retaining these organisms on their surface.[36] SWCNTs filters are reported to use for the filtration of *Escherichia coli* from the water at low pressure.[37] SWCNTs immobilized on the ceramic filter were used to enhance the performance of filtration of *E. coli*. The immobilized CNTs were found to be more robust, thermal resistant, and reusable. Another CNT-based filter prepared by the spray pyrolysis method having a controllable porosity could effectively able to remove the MS2 virus at a low pressure of 8–11 bar.[38] CNTs were widely used to remove pollutant microcystins. Microcystins are the toxins produced by the freshwater blue-green algae. The adsorption capacity of CNTs toward these toxins is due to the fitting of molecular dimensions of toxins with the pore size of the CNTs.[39]

There were so many reports regarding the antimicrobial properties of CNTs which helps to make them excellent substitutes for the existing toxic disinfectants .[43] Since they are not strong oxidants and inert to water, the use of CNTs in water disinfection treatment avoids the formation of toxic disinfection by-products like aldehydes, ketones and haloacetic acid diamines, phenols, formaldehyde, hydrogen peroxide, silver ions, halogens, glutaraldehyde, and acridines.[42,44] Purified CNTs have strong antimicrobial activities against bacterial spores, gram-positive, and gram-negative bacterial strains. They inhibit the growth of these strains by intervening metabolic process of the pathogen and impair the synthesis of cellular constituents like DNA and enzymes needed for the vitality of the growth of pathogens.[42,43]

CNTs also act as scaffolds for the macromolecules and metal oxides with high adsorption capability.[42] Examples for CNTs acting as scaffolds for the pollutant removal include CNT decorated with iron oxide for the removal of europium,[44] CNT decorated with chitosan for the removal of Methyl Orange dye,[45] CNTs decorated with ceria NPs for chromium adsorption[46] and CNT/polyaniline composite for the removal of Malachite

Green.[47] CNTs act as scaffolds for antimicrobial reagents like silver NPs and lysozymes.[48,49]

CNTs also act as additives to the anti-fouling membranes. Fouling of the membranes could be controlled by increasing their hydrophilic character.[50] Even though CNTs have hydrophobicity, the acid treatment can improve its hydrophilic character. The membrane of CNTs blended with polysulfone,[52] polyethersulfone,[53] and isophthaloyl chloride[54] exhibits fouling resistance due to the presence of a carboxylic acid group. The introduction of CNTs with inherent antibacterial properties to the polymer membranes would result in imparting biocidal properties to the membrane. This could ultimately result in excellent resistance to the fouling of the membrane. Apart from this anti-fouling properties, the incorporation of CNTs could impart better tensile strength and water permeability to the polymer membrane.[55]

2.2.2 DESALINATION OF WATER

Desalination of water is a technique that assures the supply of freshwater. It relies on the principle of reverse osmosis. High pressure and energy are used to separate the dissolved salts and fine solids from the saline water. Several reports presented that the aligned CNT membranes could act as excellent candidates for seawater desalination. It is reported that CNTs transport water very efficiently and smoothly than conventional porous membrane filters of comparable size .[56] This greater efficiency is due to the presence of smooth inner hydrophobic walls of CNTs. Other features that promote the applications of CNT in desalination treatments include greater water flux, selectivity, antimicrobial, antifouling, structural and mechanical properties. The easy and frictionless flow of water across the CNTs could be explained by the presence of highly ordered hydrogen bonds in the CNT- water interface that helps in generating a water vapor phase barrier between this interface. This barrier helps in the easy and frictionless flow of water .[57]

Holt et al. synthesized an aligned CNT membrane using the CVD technique. They were then uncapped using the laser etching technique. Despite having miniature pore size, the as-synthesized membrane has excellent water permeability than the conventional commercial polycarbonate membrane .[58] Vertically aligned CNT forest membrane of high density has exclusive ion selectivity which aids in improving the desalination process.[59]

To date, the reported diameter of CNT is about 0.3 nm which is approximately equal to the radius of hydrated sodium ions .[60] According to Corry et al. sodium-like ions have a high energy barrier that hinders the movement of ions through narrower tubes and arm chair type tubes but pass almost freely via wider nanotubes. The elixir of life, water has no such hindrance owing to the presence of hydrogen bond which naturally helps to cross the nanotubes at a very large rate .[61] Thus, the development of CNT membranes with sub-nanometre pore size can shed light on the path of desalination technologies.

2.2.3 AIR POLLUTION MONITORING

The potential applications of CNTs as sensors have been widely explored owing to the excellent electrical, optical, and thermal properties of CNTs. Both types of CNTs, MWCNTs and SWCNTs are generally used in gas sensors. Adsorption of gases in SWCNTs generally occurs at four reactive sites presented outside the bundles, interstitial channels in between the tubes in bundles, in the pores of CNTs and in grooves between the nearby tubes[62,63] while the aggregated forms of MWCNTs due to the inability to form bundle structure have their adsorption sites in their outer and inner pores.[64] It is therefore concluded that SWCNT bundles having interstitial channels are efficient in the adsorption of small molecules such as gases[65] while MWCNTs for the adsorption of biomolecules like bacteria and viruses. In general, CNTs are considered as attractive components for constructing gas sensors involved in environmental monitoring. They are considered the most promising of all nanomaterials developed so far.

There are so many advantages celebrated by the CNT-based sensors over the conventional sensors that include sensitivity, less power consumption, and low working temperature. In a typical CNT-based sensor, CNT act as the anode that generates a high electric field that helps in the ionization of the gas. Detection of change in the resistance or conductance of the CNTs on its contact with the gas helps in the sensing of gases .[66] CNT-based sensors are generally classified into four types based on the mechanism of sensing – sorption, ionization, capacitance, and resonance frequency shift.[67] The most commonly used sensor is based on sorption which relies on the adsorption of gases that results in a change in conductance due to the loss or gain of electrons.

CNT-based sensors are generally employed in the detection of organic gases such as toluene, methyl ethyl ketone, dimethyl methyl phosphonate, hexane, acetone, chloroform, acetonitrile, and methanol, nitrogen oxides (NO*x*),[68] nitrogen dioxide (NO_2),[69–73] ammonia (NH_3),[74–76] ozone,[77] and sulfur dioxide (SO_2)[78] at room temperature. During the sensing of these vapors, the conductivity of CNTs gets altered that is subsequently quantified for detecting its presence.

Even though these sensors gave better results, one of the greatest challenges suffered was time-consuming recovery. Several techniques have been put forth for the recovery of the adsorbed gas from the sensor. Heating the CNT sensor with ultraviolet rays and improving the flux rate of purging gas improved the gas recovery from the sensor.

The sensitivity and affinity of the sensor can be greatly improved by polymer functionalization.[79,80] Modified MWCNTs film-based gas sensor in which CNT act as the microelectrode were generally employed in the detection of chlorine gas. The sensing of gas is confirmed by the formation of red color that ultimately results in the generation and detection of galvanic effect. CNTs functionalized with carboxylic acid group detected the poisonous carbon monoxide with a detection limit of 1 ppm whereas the gas sensor fitted with pristine CNT didn't detect the CO gas. The functionalization using acidic groups increase the reactive sites in the walls of CNTs and improved the detection.[81] Incorporating CNTs in the conducting polymers like Polyaniline and polypyrrole improves the sensitivity of the sensor. CNT-based sensor incorporated with polyethyleneimine has a greater affinity for the detection of NO_2 at a concentration less than 1 ppb without interference from ammonia due to the low affinity of ammonia on electron-rich CNTs. In contrast to the PEI-coated sensor, Nafion-coated CNT-based sensor has a greater affinity toward the NH_3 by blocking the NO_2 sensing .[82] Polymethyl methacrylate/CNT composite based sensor is reported to have higher sensitivity toward the volatile organic compounds like DCM, chloroform, and acetone.[83] CNT-modified using 3-aminopropyl trimethoxy silane was reported to adsorb the carbon dioxide gases over their amine group containing CNT surface. SWCNTs functionalized using NaClO helped in the adsorption of isopropyl vapors by the Van der Waals force.[84] CNTs deposited on quartz filter removed the volatile organic compounds by pi-pi interactions[76] and Si and B doped CNT sensor removed CO and methanol gases by physisorption and chemisorption.[77] It is reported that the CNT-based sensors have strong competition with metal oxide sensors. CNT sensors have greater efficiency in responsive time and sensitivity than conventional metal oxide sensors for the sensing of ammonia and nitrous oxide gas at room temperature.[78] The enhanced performance is due to the easy access of these gases to the nano passages and the variation in conductance.[79–84]

Various forms of CNT-based filters were reported for Particulate Matter purification, such as CNT coating,[85] CNT film, aligned CNT sheet,[86] hierarchical CNT framework, and agglomerated CNT fluidized bed.[87] Among these materials, due to the macroporous nature and structural stability, the hierarchical CNT framework has excellent efficiency as atmospheric filters. Micron-sized fibrous material filters , glass fiber filters,[88] and ceramic filters[89] were used as substrates for the development of CNT to design hierarchical CNT-fitted filters. The results show that these novel CNT filters have features that are much akin to electrospun fabrics, making them a better option for potential filtration applications .[90]

2.2.4 CARBON NANOTUBES IN SUPERCAPACITORS

The rapid escalation in the energy consumption and the environmental impact of conventional energy sources demanded the research strategies on clean and renewable energy storage systems like solar water and wind energy. Since these are regionally limited sources the development of efficient energy storage systems like ultracapacitors or supercapacitors is essential. The supercapacitors are always superior to conventional batteries owing to their properties like their small size, high energy and power density, environment friendly, long life cycle and so on. The efficiency of these materials depends on the structure and constituents of the materials to a greater extent.

The development in the field of nanotechnology opened the frontiers of CNTs based efficient energy storage systems. SWCNTs and MWCNTs are potential candidates for electrochemical electrodes due to their inherent attributes like size in the nanometer range, high surface area, low resistivity, great stability, mechanical strength, and novel structure. The specific capacitance of the CNTs arise due to the electric double layer capacitor(EDLC) mechanism.[91]

Even though CNTs have high surface area and stability, these could not act as electrode material alone due to their low capacitance.[92] Therefore, CNTs are proposed to act as a substrate for the transition metal oxides having high capacitance.[93] CNT-based supercapacitors are reported to have a ten-fold increase in capacitance and energy density than the conventional capacitors.[94] A hybrid supercapacitor with greater surface area, high conductivity, and stability is designed by integrating the carbon nanofiber, graphene and CNT is reported to show a high capacitance and energy density and retained 94.98% retention after many cycles.[95] $KNiF_3$ @ CNTs synthesized by the solvothermal method have the potential to act as the supercapacitor electrodes. The incorporation of CNTs into the system enhanced the specific capacity, conductivity, and electron transport capacity. The features like high specific capacity, capacity retention, long cycling ability, and high energy density altogether support the role of as-synthesized capacitors in the energy storage systems.[96]

There were so many reports regarding the use of CNT supported systems with high specific capacitance, high energy density, and high cycling stability as supercapacitors like CNT-decorated $NiSe_2$ nanosheets,[97] CNTs decorated with Mo_2C and W_2C NPs,[98] MnO_2@CNTs/Graphene/diatomite hybrid material, asymmetric capacitor prepared by MnO2@CNTs/graphene/diatomite and microwave exfoliated graphene oxide,[99] SWCNTs decorated with MnO_2,[100] ferrocene-functionalized CNTs,[101] electrodeposited films of a mixture of graphene and CNTs,[102] and ternary graphene-CNT-carbon nanofibers.[103] A new recyclable supercapacitor based on CNT and organic nanocrystals was designed by Yanshyana et al.[104] Recently, a new type of supercapacitor fabricated using the waste materials spotlight the development in the technology of developing waste sources into energy. Waste MWCNT obtained after the adsorption of methylene blue was used for the fabrication of supercapacitors. MWCNTs/MB/Ppy composites were found to be excellent superconductors. The waste MWCNTs containing the adsorbed dye methylene blue improved the properties of the composite. The presence of Methylene blue increased the dispersion of CNTs in aqueous solvents and the specific capacitance of the MWCNTs and the synthesized system (Table 2.2).[105]

TABLE 2.2 Types of CNT Based Supercapacitors and Their Mechanism of Action.

CNT-based supercapacitors	Electrode/ electrolyte	Mechanism	References
N, P, Co- doped polypyrrole derived carbon nanotubes	Electrode	*Increasing the ion accessible pores to facilitate the flow and hetero atom exposure at the surface of electrode electrolyte improved the specific capacitance	[105]
CNT-decorated NiSe2 nanosheets	Electrode	*Electrode exhibited a long tenure cycling stability *The supercapacitor based on CNT@$NiSe_2$ can light up to four LEDs	[97]
CNTs decorated with MO_2C and W_2C nanoparticles	Electrode	*Electrodes contain abundant pores that facilitate contact with electrolyte and conductivity thereby the electron flow *Exhibited excellent hydrogen evolutions in both acidic and basic media	[98]
CNT incorporated 1-butyl-3-methylimidazolium bromide based polymer gel	Electrolyte	*Quasi solid-state supercapacitor with high charge density and less self-discharge due to the presence of 3D CNT network	[106]
CoPC-CMP/CNT	Electrode	*Flexible and binder-free electrode for supercapacitors *Exhibited a good electrical performance due to the pseudocapacitance and conductivity of CoPC-CMP, and CNTs	[108]
Nanocellulose/CNT hybrid	Electrode	*Flexible and self-healable supercapacitors	[109]
Polypyrrole/CNT	Electrodes	*The flexible supercapacitors did not show any decrease in capacitance performance on bending at an angle of 0°, 90°, and 180° and hence used as free standing energy storage systems	[110]

TABLE 2.2 *(Continued)*

CNT-based supercapacitors	Electrode/ electrolyte	Mechanism	References
Reduced graphene/ Helical CNTs	Electrodes	*CNTs hampered the stacking of graphene and increased the space between them that helps the generation of conducting three-dimensional architectures and hence imparted stability	[111]
CNT functionalized EGaIn (Eutectic Gallium Indium alloy)	Electrode	The presence of oxygen-containing groups on the surface of CNTs aided the strong adhesion between the surfaces; this adhesion supports the formation of stretchable soft electrodes under mechanical stress	[112]

2.2.5 CARBON NANOTUBES IN PHOTOCATALYSIS

The conventional semiconductors that are commonly used as photocatalysts include TiO_2, CdS, Fe_2O_3, and ZnO.[113] These conventional catalysts suffer from various disadvantages. The titanium oxide with a large band gap requires the excitement by high energy UV illumination, thus could not make use of the solar energy. Cadmium sulfide and zinc oxide usually undergo photo corrosion thereby reducing their activity and stability. All these systems have low quantum efficiency due to the recombination of positive and negative ions which limits their applications. But these disadvantages could be minimized to an extent by CNT-based photocatalysts owing to their excellent optical, thermal, and electrical properties. The incorporation of CNTs into the conventional photocatalyst makes miracles. When CNTs come into contact with TiO_2 NPs electron transfer from the conducting band of titanium to the surface of CNT gets improved. Hence, CNTs accept and retain electrons and hinder the recombination of ions. Thus, these electrons get accepted by another electron acceptor species followed by the degradation and mineralize the pollutants.[114] CNT/TiO_2 nanocomposites efficiently photodegraded the pollutants like benzene derivative,[115] carbamazepine,[116] nitrophenols,[117] and dyes like methylene blue, Methyl Orange, Rhodamine B, azodye, and atrazine.[118] Various applications of CNT supported photocatalyst are given in Table 2.3.

TABLE 2.3 Various Applications of CNT Supported Photocatalyst.

CNT-based photocatalyst	Target organic pollutants	Mechanism of action	References
CNT/TiO_2	Azo dye	*CNTs decrease the recombination of holes and electrons * Better adsorption and photocatalytic ability than conventional activated carbon support * Presence of radicals in the system helps in the degradation of azo dyes	[118]
CNT/ mesoporous TiO_2	Azo dye	* Decreased the recombination of electron–hole pairs * Increased the surface area of TiO_2 in the presence of CNTs *Improved photocatalytic activity than conventional supports	[119]
CNT/TiO_2 (Anatase)		*SWCNTs/TiO_2 is a better photocatalyst than MWCNT-based supports * Facilitates the flow of electrons from TiO_2 to CNTs and helps to stabilize the charge separation and hence decreased the recombination of charges	[120]
CNT/TiO_2 heterojunction Arrays	Phenol	* Better photocatalyst than the pristine TiO_2 nanomaterial * By altering the deposition time, the thickness of the TiO_2 layer can be tuned	[121]
MWCNT/TiO_2	DNPC	The composite catalyst was found to be efficient even after five cycles	[122]

TABLE 2.3 *(Continued)*

CNT-based photocatalyst	Target organic pollutants	Mechanism of action	References
MWCNT/TiO_2	DNP	*Photocatalytic degradation of pollutants depends on the temperature *Effective in the decolorization and reduction of COD	[123]
CNT/TiO_2	Benzene derivatives like aniline, benzoic acid, and nitro benzene	*The photoefficiency of the catalyst greatly depends on the ring reactivity properties of the substrate *The oxygen-containing surface functional groups contribute to the photocatalytic activity of the composite	[124]
CNT/TiO_2	Benzene and Methyl Orange	CNT acts as an electron carrier and hinders the recombination of charges. CNTs also act as dispersing support to control the morphology of TiO_2	[125]
CNT/Titanium silicate	4-Nitrophenol and Rhodamine B	* Presence of inorganic materials in CNT improved the catalytic performance. *CNT hybrid exhibits a good photocatalytic activity than their corresponding nanocomposites	[126]
MWCNT/TiO_2	Herbicide Atrazine	*Microwave irradiation greatly enhanced the photocatalytic degradation of herbicide *CNTs strongly absorbed the microwave irradiation	[127]
CNT/ZnS	Methylene Blue	*Post refluxing treatment and the order in which reactants are added played a pivotal role in the improvement of interaction between ZnS and CNTs * Close contact between CNT and ZnS crystals facilitates the electron transfer and reduces the charge recombination	[128]

TABLE 2.3 *(Continued)*

CNT-based photocatalyst	Target organic pollutants	Mechanism of action	References
CNT/ZnS	Methyl Orange	* Microwave radiation improved the dispersion of ZnS	[129]
CNT/Cds	Azo dye	* CNTs impeded the photocorrosion of CdS	[130]
CNT/WO3	Rhodamine B	* Optimal mass ratio of CNT: WO3 is found	[131]
CNT/ZnO	Methylene Blue	*Photocatalytic activity of the system increased with the concentration of carbon nanotubes *On hybridizing with CNTs, bandgap energy of the ZnO photocatalyst was blue shifted	[132]
TiO_2/CNT/PMMA	Methylene Blue, Rhodamine B, Sulforhodamine B, and Methyl Orange	*Incorporation of electrospun PMMA nanofibers and CNTs improved the photocatalytic activity, stability, and recyclability	[133]
ZnO/Ag/CNT	Methylene Blue	*The catalytic efficiency was related to the synergistic influence between Ag nanoparticles and CNTs which facilitates the charge splitting via the generation of Schottky barrier at the surface and electron-accepting properties of CNT	[134]
MWCNT/BiOI	Antipyrene	* Photogenerated hole and superoxide radicals contribute to the photodegradation of antipyrene * 3-dimensional micro mesoporous morphology of the hybrid facilitates the diffusion of electro–hole pairs	[135]
MWCNTs/ Ag_3PO_4/Bi_2WO_6 Nanocomposite	Norfloxacin	The ternary composite exhibited an excellent photocatalytic activity due to the narrow bandgap, strong visible light absorbance, high surface area, and charge separation efficiency	[136]

TABLE 2.3 *(Continued)*

CNT-based photocatalyst	Target organic pollutants	Mechanism of action	References
BiSI/BiOI/CNTs	Malachite Green	*Nanocomposite with 2% Wt. of CNT has the highest photocatalytic activity with 93.1% degradation of dye in 210 minutes	[137]
$GdFeO_3$/CNT NPs &H_2O_2	Rhodamine B	H_2O_2 played a crucial role in generating the hydroxyl radical scavengers that help in the degradation of dye	–
F-MWCNT/ Co-Ti oxide nanocomposite	Rhodamine B	Functionalization via acid treatment improved the dispersion properties and interaction with metal nanoparticles*Rate of degradation is higher in basic medium	[138]

2.2.6 CARBON NANOTUBES IN CATALYTIC WET AIR OXIDATION

Wet air oxidation is an innovative and environmentally benign way of treating hazardous and non-biodegradable effluents in wastewater. It is the method of oxidizing the waste effluents using oxygen as the oxidizer at a very high temperature (200°C–300°C) and pressure (20–200 bar). Oxidizer helps in the evolution of the active oxygen radical species like hydroxyl (OH) and hydroperoxyl (HOO) which ultimately helps in the oxidation of the organic compounds into nonhazardous and biodegradable by-products.[164] Majority of the organic compounds are prone to wet air oxidation except for carboxylic acids and polychlorinated biphenyls. But during the oxidation, these organic compounds get degraded into unfavorable carboxylic acids that is one of the major disadvantages suffered by the WAO. Wet air oxidation is generally used in the treatment of industrial sludge, distillery waste, pharmaceutical wastewater, black liquor treatment, sewage sludge, hazardous waste, kinetic hydrate inhibitors, cyanide wastewater, ammonium sulfate crystallizer mother liquor, oxy desulfurization of coal, and so on.

Catalysts can be used to improve the oxidation efficiency, destruction of COD, and reduce the costs. Both homogeneous and heterogeneous catalysts have been used. By using these catalyst systems, the efficiency of the WAO can be enhanced by decreasing the operational temperature (130–250) and pressure (5–50 bar). Homogeneous catalysts are the dissolved transition metals whereas the heterogeneous one is usually the mixed metal catalyst. But both systems suffer some disadvantages in spite of their good performances. The high price, easy deactivation due to the deposition of carbonaceous materials, and leaching of the active sites, which induced the secondary pollution altogether hampers the application of these catalysts in wet air oxidation. Wet air oxidation has so many potential benefits over the other conventional wastewater treatments such as formation of nontoxic by-products, requirement of low reaction temperature, and residence time. These advantages shed light on the research in the field of wet air oxidation. The bottleneck of these processes is the lack of efficient, active, and eco-friendly catalysts.

Recently, the CNTs received greater attraction as a catalyst in wet air oxidation owing to their high surface area, resistance to corrosion, environment friendliness, tunability of its texture, and stability. Several investigations used CNT-supported catalysts in the WAO of toxic wastewater. The mesoporous nature of CNTs supported the diffusion of pollutants over the surface of the catalyst. CNT-supported transition metals like Pd, Pt, Rh, and Ru have been widely used in the wet air oxidation of phenols and nitrogen-containing pollutants like aniline and azodyes.[139–146] CNTs without any supported metals themselves can act as a catalyst for wet air oxidation. MWCNTs display the catalytic activity for the wet air oxidation of phenol. Under milder conditions, purified CNTs catalyzed the oxidation of P-Coumaric acid, the biphenolic compound present in the olive oil processing water and degraded the total organic carbon.[147] CNTs with parallel and herringbone walls were reported as the catalyst for the WAO of phenols. CNTs with herring bone walls have greater catalytic activity due to the synergistic effects of edge carbons and active sites of carboxylic groups. These synergistic effects accelerate the activation of oxygen molecule.[148] Modification of MWCNTs with urea, nitric acid, and gas phase treatments can increase the basic character thereby the catalytic activity. MWCNTs modified using nitric acid, urea, and gas phase thermal treatment efficiently helps in the WAO of oxalic acid.[149]

The catalytic activity of the CNTs for the wet air oxidation relies on the presence of an active site that may be acidic or basic. The catalytic activity of CNTs can be enhanced by increasing the number of basic sites. In general, the catalytic activity of CNT increase with the increase in basicity but the presence of certain acidic groups like sulfur-containing functional groups also increase its activity.[150]

2.2.7 CN ELECTRODES IN MICROBIAL FUEL CELL

Due to the depleting fossil fuel energies, it is appreciable that the researchers consider looking for new substitutes for energy. Biofuel cells are new sources of energy that are eco-friendly and don't cause any pollution. These environment-friendly cells are found to be a promising candidate in the sustainable way of energy generation. Sustainable energy generation from different substrates such as organic wastes has intensified the research in this area.

The microbial fuel cell is an eco-friendly method for treating the water effluents along with the electricity generation using microorganisms. It is an electrochemical device that converts chemical energy present in the organic or inorganic substance to electrical energy using microorganisms along with waste treatment of water. The degradation of chemical oxygen demand and electricity generation by MFC depends on various factors like pH, design of the reactor, nature of electrode material, and concentration of microbial species. MFC consist of an anode in which the organic matter is oxidized by the transfer of electrons with the help of microorganisms like exoelectrogens. The commonly used exoelectrogens include *Escherichia coli, Saccharomyces, Shewanella putrefaciens, Geobacteraceae sulferreducens and Rhodoferax ferrireducens*.[151–155] The anode material should possess a large specific area for microbe colony formation, high conductivity, stability, and catalytic activity. The most commonly employed materials are carbon-based materials like carbon paper, carbon cloth, carbon foam, etc. Carbon-based nanomaterials due to their high conductivity, high volume to surface ratio, microbial adhesion, cheapness, and presence of the surplus amount of pi electrons represent a group of promising catalysts in the cathode reduction. But bare carbonaceous materials are found to be very unsatisfying candidates due to the presence of inactive pi electrons, which impedes its application in the field of microbial reactions.[156]

CNTs and graphene-based materials can be used as oxidation-reduction reaction (ORR) catalysts in MFCs. However, CNTs could not be used for direct purposes due to their toxicity that ultimately results in the inhibition of cell proliferation and finally the death of microorganisms. Doping of heteroatoms into the sites of these carbon supports can activate the pi electrons by increasing the number of sites and hence increase its oxidation-reduction reaction efficiency. But the heteroatom atom doped CNT and graphene-based materials don't provide satisfactory performance as ORR catalyst. This can be due to the stacking ability of graphene and nanotubes which would results in the blocking of active sites. In 2019, Du et al. found a solution to this problem by designing a mixed dimensional hybrid of N Doped CNT and reduced graphene oxide nanosheet. The mixed dimensional architecture of the as synthesized systems prevents the stacking and enhance the surface area, activate the catalytic sites, and thereby facilitate the charge transfer of electron from catalytic active CNT to rGo nanosheets.[157]

CNTs coated with conductive materials like polyaniline[158] and polypyrrole[159] are efficient anode materials with high surface charge area and charge efficiency. A three-dimensional CNT anode supported on macroporous materials like sponge and textile[160,161] interacts strongly with microbial film and facilitates the electron transfer from microbes to the surface of the electrode. This electron transfer results in the decrease of resistance of charge transfer and thereby improving the efficiency of the cell. Microbes like bacteria possess redox proteins which help in generating oxygen via reductions. The CNTs due to their structural features can make a contact with these redox active centers and help in the flow of electrons and improves the reduction of oxygen.[162,163] CNTs are also found to be good support for redox catalyst like MnO_2.

A chemically and thermally stable MFC based on carbon felt nano molybdenum carbide synthesized using microwave irradiation is reported to be an excellent anode material.[164] Species like tungsten and molybdenum carbide with almost the same electrocatalytic activity can be used as low-cost alternatives to platinum electro catalyst. The electrotransfer occurs owing to the presence of biocompatible and highly conducting nanowires of bacteria. The carbon felt Mo_2C/CNT is also reported to have an excellent power density of about 170 mWm.

Another MFC anodic system synthesized by coating nanocomposites of MWCNT and SnO_2 on a glassy carbon electrode was found to have a greater power density and charge transfer ability. The addition of CNTs to the conductive polymer-based anode greatly enhanced its power output.[165] The MFC based on the Fe_3O_4/CNT nanocomposites has been reported to be of high power density due to the inclusion of iron oxide particles. The magnetic properties of these particles assisted the attachment of CNT to the surface of the anode, growth of bacteria, and hence facilitate the electron transfer.[160]

Microsized MFCs are reported to generate high power and electricity. The advantage of microsized fuel cells is that they can avoid the use of chemicals and power sources by making on-chip power sources or in lab on-chip uses. Justin E mink and his coworkers designed a microsized MFC with an anode of vertically grown forest type MWCNTs and nickel silicide as the contact base. The as-synthesized MFCs are found be more biocompatible and conductive.[166]

As mentioned above, various carbon-based materials have been used for electrode materials, such as carbon felt, carbon cloth, and carbon rods. In spite of their low porosity and biocompatibility, CNT-based nanocomposites outperformed these materials as an electrode in MFCS. In gist, all the tunable, meritorious structural and electrical properties of CNTs make them promising candidates in MFC electrodes.

2.2.8 CARBON NANOTUBES IN DESIGN OF GREEN NANOCOMPOSITE

Waste generation is a global concern across the world. The wastes produced from various industrial processes cause a negative impact on the environment. Thus there is an urgent need to minimize the waste generation. According to the suggestion by European community management of waste can be done mainly via two strategies: minimizing the waste by improving the design of product synthesized and emphasizing on the recycling and reuse of the produced waste.[167] The principle of green nanocomposites is directly connected with the trend of reducing and recycling of waste. Even though biodegradable polymers are an excellent alternative to the nonbiodegradable and nonrenewable sources of energy like coal and petroleum product, there are certain disadvantages like poor mechanical properties and heat distortion temperature that impedes their applications in wide areas. But the introduction of CNTs as reinforcement materials for those degradable polymers improved their thermal and mechanical strength as well as their durability. The premium of these CNT incorporated composites depends greatly on the alignment of CNTs, their adhesion with the polymer, and their dispersion in the polymer matrix.[168–170]

The incorporation of the CNTs into the poly thermoplastic aliphatic biodegradable polymer resin, polybutylene succinate improved its thermal stability[171] whereas the CNTs reinforcement with polypropylene fumarate, and biodegradable polyester enhanced the mechanical strength. Besides, CNT/PPF green nanocomposites exhibited an excellent biocompatibility that supports its use as scaffolds in bone tissue engineering applications.[172]

A nanocomposite designed from magnetically synthesized SWCNTs, chitosan, and nanocrystalline hydroxyl apatite have an excellent osteoblast density which hailed the synthesized material as a promising candidate in bone regeneration applications.[173] The *in vitro* and *in vivo* biocompatibility shown by the MWCNT/ Poly 3-hydroxybutyrate-*co*-3-hydroxyvalerate (PHBV) nanocomposites are reported to have the capability in repairing bone defects.[174]

Another desirable characteristics of the green composite are its ability to recycle the reinforced CNTs due to the degradability of the polymer. Degradability of the polymer is generally attained via enzymatic or microbial pathways. The CNTs leftover after degradation could be reused for the synthesis of new composites. This principle of reuse and recycling the CNTs minimized the waste disposal and simultaneously be cost effective for the processing.

2.3 CONCLUSIONS

The chapter spotlights the applications of CNTs in making the environment greener. The major proactive to retroactive uses of CNTs have been widely discussed in the chapter. The excellent optical, electrical, structural, chemical, and physical properties of CNTs make them promising tools in waste water management, air pollution monitoring, desalination of water, and energy sources. CNTs act as adsorbents, antimicrobial agents, antifouling membranes, filters to remove organic, inorganic pollutants, and pathogenic microorganisms like bacteria and viruses. CNTs are also prompt gas sensors in sensing the toxic and polluting gases like NO_2, NH_3, etc.

In addition, CNTs are also presented as one of the main challenges in producing "green and sustainable" energy, which involves safe and clean combustion. The generation of electricity by sustainable energy technologies based on biofuel cells, solar energy-based photovoltaic cells and hydrogen-based fuel cell assure a clean combustion by preventing the discharge of nonpoisonous gases to the atmosphere. Apart from this, the high power density provided by CNT-based supercapacitors has been viewed as a second option to replace traditional ones. Supercapacitors with prolonged usage time could minimize the quantity of waste disposed to the environment over time. Progress in this field is still on going in order to meet the long-term environmental protection challenge. Among green materials, the role of CNTs in green nanocomposite design strictly follows the concept of life cycle assessment which emphasizes reduction, reuse, and recyclability of raw materials.

It can be concluded that CNT-based nanomaterials celebrate boons over traditional materials in environmental applications. The evolution of cost-effective and highly efficient manufacturing strategies augurs well for integrating CNTs into various environmental applications. Functionalization of CNTs via surface modification and control at a macroscopic level of CNTs are often fruitful methods to fully take advantage of CNTs' unequaled physical, chemical, and electrical properties. Despite the elation, the potential menace of CNTs to the environment and humans should be considered before its large-scale applications. Major drawbacks include the manufacturing cost of CNT and their optimization in terms of alignment, dispersion, functionalization, and opening of tip. Since CNTs are cytotoxic, the risk associated with the leakage of CNT in applications especially water treatment needs to be properly assessed. It is hopeful that the application of CNTs in environmental applications brings a new revolution and enhances the current technologies. Considering this threat, the best way of minimizing the exposure of CNTs toward the environment and humans is source reduction.

KEYWORDS

- **carbon nanotubes**
- **environment**
- **wastewater treatment**
- **air pollution monitoring**
- **desalination**
- **microbial fuel cell**
- **photocatalysis**
- **supercapacitors**
- **green nanocomposites design**

REFERENCES

1. Iijima, S. Helical Microtubules of Graphitic Carbon. *Nature* **1991,** *354* (6348), 56–58.
2. Bethune, D. S.; Kiang, Ch H.; De Vries, M. S.; Gorman, G.; Savoy, R.; Vazquez, J.; Beyers, R. Cobalt-Catalysed Growth of Carbon Nanotubes with Single-Atomic-Layer Walls. *Nature* **1993,** *363* (6430), 605–607.
3. Awasthi, K.; Srivastava, A.; Srivastava, O. N. Synthesis of Carbon Nanotubes. *J. Nanosci. Nanotechnol.* **2005,** *5* (10), 1616–1636.
4. Che, G.; Lakshmi, B. B.; Martin, C. R.; Fisher, E. R.; Ruoff, R. S. Chemical Vapor Deposition Based Synthesis of Carbon Nanotubes and Nanofibers Using a Template Method. *Chem. Mater.* **1998,** *10* (1), 260–267.
5. Dai, H. Carbon Nanotubes: Synthesis, Integration, and Properties. *Acc. Chem. Res.* **2002,** *35* (12), 1035–1044.
6. Endo, M.; Strano, M.; Ajayan, P. PotentialApplications of Carbon Nanotubes. *Carbon Nanotubes* **2008,** 13.
7. Fagan, S. B.; Santos, E. J. G.; Souza Filho, A. G.; Mendes Filho, J.; Fazzio, A. Ab Initio Study of 2, 3, 7, 8-Tetrachlorinated Dibenzo-p-Dioxin Adsorption on Single Wall Carbon Nanotubes. *Chem. Phys. Lett.* **2007,** *437* (1–3), 79–82.
8. Rao, G. P.; Lu, C.; Su, F. Sorption of Divalent Metal Ions from Aqueous Solution by Carbon Nanotubes: A Review. *Sep. Purif. Technol.* **2007,** *58* (1), 224–231.
9. Wang, X.; Lu, J.; Xing, B. Sorption of Organic Contaminants by Carbon Nanotubes: Influence of Adsorbed Organic Matter. *Environ. Science Technol.* **2008,** *42* (9), 3207–3212.
10. Xu, D.; Tan, X.; Chen, C.; Wang, X. Removal of Pb (II) from Aqueous Solution by Oxidized Multiwalled Carbon Nanotubes. *J. Hazard. Materials* **2008,** *154* (1–3), 407–416.
11. Stafiej, A.; Pyrzynska, K. Solid Phase Extraction of Metal Ions Using Carbon Nanotubes. *Microchem. J.* **2008,** *89* (1), 29–33.

12. Pan, B.; Xing, B. Adsorption Mechanisms of Organic Chemicals on Carbon Nanotubes. *Environ. Sci. Technol.* **2008,** *42* (24), 9005–9013.
13. Peng, X. J.; Li, Y. H.; Luan, Z. K.; Di, Z. C.; Wang, H. Y.; Tian, B. H.; Jia, Z. P. Adsorption of 1, 2 -Dichlorobenzene from Water to Carbon Nanotubes. *Chem. Phys. Lett.* **2003,** *376* (1–2), 154–158.
14. Lin, D. H.; Xing, B. S. Adsorption of Phenolic Compounds by Carbon Nanotubes: Role of Aromaticity and Substitution of Hydroxyl Groups. *Environ. Sci. Technol.* **2008,** *42* (19), 7254–7259.
15. Bai, Y. C.; Lin, D. H.; Wu, F. C.; Wang, Z. Y.; Xing, B. S. Adsorption of Triton X-Series Surfactants and Its Role in Stabilizing Multi-Walled Carbon Nanotube Suspensions. *Chemosphere* **2010,** *79* (4), 362–367.
16. Yang, K.; Zhu, L. Z.; Xing, B. S. Adsorption of Polycyclic Aromatic Hydrocarbons by Carbon Nanomaterials. *Environ. Sci. Technol.* **2006,** *40* (6), 1855–1861.
17. Kosa, S. A.; Al-Zhrani, G.; Salam, M. A. Removal of Heavy Metals from Aqueous Solutions by Multi-Walled Carbon Nanotubes Modified with 8-Hydroxyquinoline. *Chem. Eng. J.* **2012,** *181*, 159–168.
18. Ansari, M. O.; Kumar, R.; Ansari, S. A.; Ansari, S. P.; Barakat, M. A.; Alshahrie, A.; Cho, M. H. Anion Selective pTSA Doped Polyaniline@ Graphene Oxide-Multiwalled Carbon Nanotube Composite for Cr (VI) and Congo Red Adsorption. *J. Colloid Interf. Sci.* **2017,** *496*, 407–415.
19. Hayati, B.; Maleki, A.; Najafi, F.; Daraei, H.; Gharibi, F.; McKay, G. Super High Removal Capacities of Heavy Metals (Pb2+ and Cu2+) Using CNT Dendrimer. *J. Hazard. Mater.* **2017,** *336*, 146–157.
20. Tofighy, M. A.; Mohammadi, T. Copper Ions Removal from Aqueous Solutions Using Acid-Chitosan Functionalized Carbon Nanotubes Sheets. *Desalination Water Treatment* **2016,** *57* (33), 15384–15396.
21. Kanthapazham, R.; Ayyavu, C.; Mahendiradas, D. Removal of Pb2+, Ni2+ and Cd2+ Ions in Aqueous Media Using Functionalized MWCNT Wrapped Polypyrrole Nanocomposite. *Desalination and Water Treatment* **2016,** *57* (36), 16871–16885.
22. Popuri, S. R.; Frederick, R.; Chang, C-Y.; Fang, S-S.; Wang, C-C.; Lee, L-C. Removal of Copper (II) Ions from Aqueous Solutions onto Chitosan/Carbon Nanotubes Composite Sorbent. *Desalination Water Treatment* **2014,** *52* (4–6), 691–701.
23. Alimohammady, M.; Jahangiri, M.; Kiani, F.; Tahermansouri, H. A New Modified MWCNTs with 3-Aminopyrazole as a Nanoadsorbent for Cd (II) Removal from Aqueous Solutions. *J. Environ. Chem. Eng.* **2017,** *5* (4), 3405–3417.
24. Ouni, L.; Mirzaei, M.; Ashtari, P.; Ramazani, A.; Rahimi, M.; Bolourinovin, F. Isocyanate Functionalized Multiwalled Carbon Nanotubes for Separation of Lead from Cyclotron Production of Thallium-201. *J. Radioanaly. Nucl. Chem.* **2016,** *310* (2), 633–643.
25. Saadat, S.; Karimi-Jashni, A.; Doroodmand, M. M. Synthesis and Characterization of Novel Single-Walled Carbon Nanotubes-Doped Walnut Shell Composite and Its Adsorption Performance for Lead in Aqueous Solutions. *J. Environ. Chem. Eng.* **2014,** *2* (4), 2059–2067.
26. Sankararamakrishnan, N.; Gupta, A.; Vidyarthi, S. R. Enhanced Arsenic Removal at Neutral pH Using Functionalized Multiwalled Carbon Nanotubes. *J. Environ. Chem. Eng.* **2014,** *2* (2), 802–810.

27. AlOmar, M. K.; Alsaadi, M. A.; Hayyan, M.; Akib, S.; Hashim, M. A. Functionalization of CNTs Surface with Phosphonuim Based Deep Eutectic Solvents for Arsenic Removal from Water. *Appl. Surf. Sci.* **2016,** *389*, 216–226.
28. Hayati, B.; Maleki, A.; Najafi, F.; Daraei, H.; Gharibi, F.; McKay, G. Synthesis and Characterization of PAMAM/CNT Nanocomposite as a Super-Capacity Adsorbent for Heavy Metal (Ni2+, Zn2+, As3+, Co2+) Removal from Wastewater. *J. Mol. Liquids* **2016,** *224*, 1032–1040.
29. Peng, X. J.; Luan, Z. K.; Ding, J.; Di, Z. H.; Li, Y. H.; Tian, B. H. Ceria Nanoparticles Supported on Carbon Nanotubes for the Removal of Arsenate from Water. *Mater. Lett.* **2005,** *59* (4), 399–403.
30. Park, W. K.; Yoon, Y.; Kim, S.; Yoo, S.; Do, Y.; Kang, J-W.; Yoon, D. H.; Yang, W. S. Feasible Water Flow Filter with Facilely Functionalized Fe_3O_4-Non-Oxidative Graphene/CNT Composites for Arsenic Removal. *J. Environ. Chem. Eng.* **2016,** *4* (3), 3246–3252.
31. Asadollahi, N.; Yavari, R.; Ghanadzadeh, H. Preparation, Characterization and Analytical Application of Stannic Molybdophosphate Immobilized on Multiwalled Carbon Nanotubes as a New Adsorbent for the Removal of Strontium from Wastewater. *J. Radioanalyt. Nucl. Chem.* **2015,** *303* (3), 2445–2455.
32. Yang, W.; Ding, P.; Zhou, L.; Yu, J.; Chen, X.; Jiao, F. Preparation of Diamine Modified Mesoporous Silica on Multi-Walled Carbon Nanotubes for the Adsorption of Heavy Metals in Aqueous Solution. *Appl. Surf. Sci.* **2013,** *282*, 38–45.
33. Fornasiero, F.; Park, H. G.; Holt, J. K.; Stadermann, M.; Grigoropoulos, C. P.; Noy, A.; Bakajin, O. Ion Exclusion by Sub-2-nm Carbon Nanotube Pores. *Proc. Natl. Acad. Sci.* **2008,** *105* (45), 17250–17255.
34. Noy, A.; Park, H. G.; Fornasiero, F.; Holt, J. K.; Grigoropoulos, C. P.; Bakajin, O. Nanofluidics in Carbon Nanotubes. *Nano Today* **2007,** *2* (6), 22–29.
35. Hummer, G.; Rasaiah, J. C.; Noworyta, J. P. Water Conduction through the Hydrophobic Channel of a Carbon Nanotube. *Nature* **2001,** *414* (6860), 188–190.
36. Bohonak, D. M.; Zydney, A. L. Compaction and Permeability Effects with Virus Filtration Membranes. *J. Membr. Sci.* **2005,** *254* (1–2), 71–79.
37. Brady-Estévez, A. S.; Kang, S.; Elimelech, M. A Single-Walled-Carbon-Nanotube Filter for Removal of Viral and Bacterial Pathogens. *Small* **2008,** *4* (4), 481–484.
38. Mostafavi, S. T.; Mehrnia, M. R.; Rashidi, A. M. Preparation of Nanofilter from Carbon Nanotubes for Application in Virus Removal from Water. *Desalination* **2009,** *238* (1–3), 271–280.
39. Yan, H.; Gong, A.; He, H.; Zhou, J.; Wei, Y.; Lv, L. Adsorption of Microcystins by Carbon Nanotubes. *Chemosphere* **2006,** *62* (1), 142–148.
40. Savage, N.; Diallo, M. S. Nanomaterials and Water Purification: Opportunities and Challenges. *J. Nanopart. Res.* **2005,** *7* (4–5), 331–342.
41. Cortes, P.; Deng, S.; Smith, G. B. The Adsorption Properties of Bacillus Atrophaeus Spores on Single-Wall Carbon Nanotubes. *J. Sens.* **2009,** *2009*.
42. Kang, S.; Pinault, M.; Pfefferle, L. D.; Elimelech, M. Single-Walled Carbon Nanotubes Exhibit Strong Antimicrobial Activity. *Langmuir* **2007,** *23* (17), 8670–8673.
43. Li, Q.; Mahendra, S.; Lyon, D. Y.; Brunet, L.; Liga, M. V.; Li, D.; Alvarez, P. J. J. Antimicrobial Nanomaterials for Water Disinfection and Microbial Control: Potential Applications and Implications. *Water Res.* **2008,** *42* (18), 4591–4602.

44. Chen, C. L.; Wang, X. K.; Nagatsu, M. Europium Adsorption on Multiwall Carbon Nanotube/Iron Oxide Magnetic Composite in the Presence of Polyacrylic Acid. *Environ. Sci. Technol.* **2009,** *43* (7), 2362–2367.
45. Zhu, H. Y.; Jiang, R.; Xiao, L.; Zeng, G. M. Preparation, Characterization, Adsorption Kinetics and Thermodynamics of Novel Magnetic Chitosan Enwrapping Nanosized γ-Fe_2O_3 and Multi-Walled Carbon Nanotubes with Enhanced Adsorption Properties for Methyl Orange. *Bioresour. Technol.* **2010,** *101* (14), 5063–5069.
46. Di, Z-C.; Ding, J.; Peng, X-J.; Li, Y-H.; Luan, Z-K.; Liang, J. Chromium Adsorption by Aligned Carbon Nanotubes Supported Ceria Nanoparticles. *Chemosphere* **2006,** *62* (5), 861–865.
47. Zeng, Y.; Zhao, L.; Wu, W.; Lu, G.; Xu, F.; Tong, Y.; Liu, W.; Du, J. Enhanced Adsorption of Malachite Green Onto Carbon Nanotube/Polyaniline Composites. *J. Appl. Polym. Sci.* **2013,** *127* (4), 2475–2482.
48. Morones, J. R.; Elechiguerra, J. L.; Camacho, A.; Holt, K.; Kouri, J. B.; Ramírez, J. T.; Yacaman, M. J. The Bactericidal Effect of Silver Nanoparticles. *Nanotechnology* **2005,** *16* (10), 2346.
49. Yuan, W.; Jiang, G.; Che, J.; Qi, X.; Xu, R.; Chang, M. W.; Chen, Y.; Lim, S. Y.; Dai, J.; Chan-Park, M. B. Deposition of Silver Nanoparticles on Multiwalled Carbon Nanotubes Grafted with Hyperbranched Poly (Amidoamine) and Their Antimicrobial Effects. *The J. Phys. Chem.* **2008,** *C 112* (48), 18754–18759.
50. Nepal, D.; Balasubramanian, S.; Simonian, A. L.; Davis, V. A. Strong Antimicrobial Coatings: Single-Walled Carbon Nanotubes Armored with Biopolymers. *Nano Lett.* **2008,** *8* (7), 1896–1901.
51. Choi, J-H.; Jegal, J.; Kim, W-N. Fabrication and Characterization of Multi-Walled Carbon Nanotubes/Polymer Blend Membranes. *J. Membr. Sci.* **2006,** *284* (1–2), 406–415.
52. Celik, E.; Park, H.; Choi, H.; Choi, H. Carbon Nanotube Blended Polyethersulfone Membranes for Fouling Control in Water Treatment. *Water Res.* **2011,** *45* (1), 274–282.
53. Qiu, S.; Wu, L.; Pan, X.; Zhang, L.; Chen, H.; Gao, C. Preparation and Properties of Functionalized Carbon Nanotube/PSF Blend Ultrafiltration Membranes. *J. Membr. Sci.* **2009,** *342* (1–2), 165–172.
54. Gunawan, P.; Guan, C.; Song, X.; Zhang, Q.; Leong, S. S. J.; Tang, C.; Chen, Y. et al. Hollow Fiber Membrane Decorated with Ag/MWNTs: Toward Effective Water Disinfection and Biofouling Control. *ACS Nano* **2011,** *5* (12), 10033–10040.
55. Majeed, S.; Fierro, D.; Buhr, K.; Wind, J.; Du, B.; Boschetti-de-Fierro, A.; Abetz, V. Multi-Walled Carbon Nanotubes (MWCNTs) Mixed Polyacrylonitrile (PAN) Ultrafiltration Membranes. *J. Membr. Sci.* **2012,** *403*, 101–109.
56. Wang, R.; Chen, D.; Wang, Q.; Ying, Y.; Gao, W.; Xie, L. Recent Advances in Applications of Carbon Nanotubes for Desalination: A Review. *Nanomaterials* **2020,** *10* (6), 1203.
57. Hummer, G.; Rasaiah, J. C.; Noworyta, J. P. Water Conduction through The Hydrophobic Channel of a Carbon Nanotube. *Nature* **2001,** *414* (6860), 188–190.
58. Holt, J. K.; Park, H. G.; Wang, Y. M.; Stadermann, M.; Artyukhin, A. B.; Grigoropoulos, C. P.; Noy, A.; Bakajin, O. Fast Mass Transport through Sub-2-Nanometercarbon Nanotubes. *Science* **2006,** *312* (5776), 1034–1037.

59. Yu, M.; Funke, H. H.; Falconer, J. L.; Noble, R. D. High Density, Vertically-Aligned Carbon Nanotube Membranes. *Nano Letters* **2009,** *9* (1), 225–229.
60. Zhao, X.; Liu, Y.; Inoue, S.; Suzuki, T.; Jones, R. O.; Ando, Y. Smallest Carbon Nanotube Is 3 Angstrom in Diameter. *Phys. Rev. Lett.* **2004,** *92* (12).
61. Corry, B. Designing Carbon Nanotube Membranes for Efficient Water Desalination. *J. Phys. Chem. B* **2008,** *112* (5), 1427–1434.
62. Hoa, N. D.; Quy, N. V.; Cho, Y. S.; Kim, D. Nanocomposite of SWNTs and SnO_2 Fabricated by Soldering Process for Ammonia Gas Sensor Application. *Phys. Status Solidi (A) App. Mater.* **2007,** *204* (6), 1820.
63. Wei, B. Y.; Hsu, M. C.; Su, P. G.; Lin, H. M.; Wu, R. J.; Lai, H. J. A Novel SnO_2 Gas Sensor Doped with Carbon Nanotubes Operating at Room Temperature. *Sens. Actuat. B: Chem.* **2004,** *101* (1–2), 81.
64. Agnihotri, S.; Mota, J. P. B.; Rostam-Abadi, M.; Rood, M. J. Structural Characterization of Single-Walled Carbon Nanotube Bundles by Experiment and Molecular Simulation. *Langmuir* **2005,** *21* (3), 896–904.
65. Upadhyayula, V. K. K.; Deng, S.; Mitchell, M. C.; Smith, G. B. Application of Carbon Nanotube Technology for Removal of Contaminants in Drinking Water: A Review. *Sci. Total Environ.* **2009,** *408* (1), 1–13.
66. Bogue, R. W. Nanotechnology: What Are the Prospects for Sensors? *Sens. Rev.* **2004,** *24* (3), 253.
67. Zaporotskova, I. V.; Boroznina, N. P.; Parkhomenko, Y. N.; Kozhitov, L. V. Carbon Nanotubes: Sensor Properties. A Review. *Mod. Electron. Mater.* **2016,** *2*, 95–105.
68. Ueda, T.; Bhuiyan, M. M. H.; Norimatsu, H.; Katsuki, S.; Ikegami, T.; Mitsugi, F. Development of Carbon Nanotube-Based Gas Sensors for NOx Gas Detection Working at Low Temperature. *Phys. E: Low-Dimensional Syst. Nanostruct.* **2008,** *40* (7), 2272.
69. Kong, J.; Franklin, N. R.; Zhou, C.; Chapline, M. G.; Peng, S.; Cho, K. and Dai, H. Nanotube Molecular Wires as Chemical Sensors. *Science* **2000,** *287* (5453), 622.
70. Valentini, L.; Cantalini, C.; Armentano, I.; Kenny, J. M.; Lozzi, L.; Santucci, S. Highly Sensitive and Selective Sensors Based on Carbon Nanotubes Thin Films for Molecular Detection. *Diamond Relat. Mater.* **2004,** *13* (4–)8, 1301.
71. Cho, W. S.; Moon, S. I.; Paek, K. K.; Lee, Y. H.; Park, J. H.; Ju, B. K. Patterned Multiwall Carbon Nanotube Films as Materials of NO_2 Gas Sensors. *Sens. Actuat. B: Chemical* **2006,** *119* (1), 180.
72. Cantalini, C.; Valentini, L.; Lozzi, L.; Armentano, I.; Kenny, J. M.; Santucci, S. NO_2 Gas Sensitivity of Carbon Nanotubes Obtained by Plasma Enhanced Chemical Vapor Deposition. *Sens. Actuat. B Chem.* **2003,** *93* (1–3), 333.
73. Moon, S. I.; Paek, K. K.; Lee, Y. H.; Park, H. K.; Kim, J. K.; Kim, S. W.; Ju, B. K. Bias Heating Recovery of MWCNT Gas Sensor. *Mater. Lett.* **2008,** *62* (16), 2422.
74. Nguyen, H. Q.; Huh, J. S. Behavior of Singlewalled carbon Nanotube-Based Gas Sensors at Various Temperatures of Treatment and Operation. *Sens. Actuat. B: Chem.* **2006,** *117* (2), 426.
75. Nguyen, L. H.; Phi, T. V.; Phan, P. Q.; Vu, H. N.; Nguyen-Duc, C.; Fossard, F.; Synthesis of Multi-Walled Carbon Nanotubes for NH3 Gas Detection. *Phys. E: Low-Dimensional Syst. Nanostruct.* **2007,** *37* (1–2), 54.

76. Quang, N. H.; Van Trinh, M.; Lee, B. H. and Huh, J. S. Effect of NH3 Gas on the Electrical Properties of Single-Walled Carbon Nanotube Bundles. *Sens. Actuat., B Chem.* **2006,** *113* (1), 341.
77. Azam, M. A.; Alias, F. M.; Tack, L. W.; Seman, R. N. A. R.; Taib, M. F. M. Electronic Properties and Gas Adsorption Behaviour of Pristine, Silicon-, and Boron-Doped (8, 0) Single-Walled Carbon Nanotube: A First Principles Study. *J. Mol. Graphics Modell.* **2017,** *75*, 85–93.
78. Suehiro, J.; Zhou, G.; Hara, M. Detection of Partial Discharge in SF6 Gas Using a Carbon Nanotube-Based Gas Sensor. *Sens. Actuat. B Chem.* **2005,** *105*, 2, 164.
79. Wei, C.; Dai, L.; Roy, A.; Tolle, T. B. Multifunctional Chemical Vapor Sensors of Aligned Carbon Nanotube and Polymer Composites. *J Am. Chem. Soc.* **2006,** *128* (5), 1412.
80. Lu, J.; Kumar, B.; Castro, M.; Feller, J. F. Vapour Sensing with Conductive Polymer Nanocomposites (CPC): Polycarbonate-Carbon Nanotubes Transducers with Hierarchical Structure Processed by Spray Layer by Layer. *Sens. Actuat. B Chem.* **2009,** *140* (2), 451.
81. Fu, D.; Lim, H.; Shi, Y.; Dong, X.; Mhaisalkar, S. G.; Chen, Y. Moochhala, S.; Li, L-J. Differentiation of Gas Molecules Using Flexible and All-Carbon Nanotube Devices. *J. Phys. Chem. C* **2008,** *112* (3), 650–653.
82. Qi, P.; Vermesh, O.; Grecu, M.; Javey, A.; Wang, Q.; Dai, H.; Peng, S.; Cho, K. J. Toward Large Arrays of Multiplex Functionalized Carbon Nanotube Sensors for Highly Sensitive and Selective Molecular Detection. *Nano Lett.* **2003,** *3* (3), 347.
83. Abraham, J. K.; Philip, B.; Witchurch, A.; Varadan, V. K.; Reddy, C. C. A Compact Wireless Gas Sensor Using a Carbon Nanotube/PMMA Thin Film Chemi Resistor. *Smart Mater. Struct.* **2004,** *13* (5), 1045.
84. Gong, J.; Sun, J.; Chen, Q. Micro Machined Sol Gel Carbon Nanotube/SnO_2 Nano Composite Hydrogen Sensor. *Sens. Actuat. B Chem.* **2008,** *130* (2), 829.
85. Soares, S. F.; Fernandes, T.; Trindade, T.; Daniel-da-Silva, A. L. Trimethyl Chitosan/ Siloxane-Hybrid Coated Fe_3O_4 Nanoparticles for the Uptake of Sulfamethoxazole from Water. *Molecules* **2019,** *24* (10), 1958.
86. Nasibulin, A. G.; Kaskela, A.; Mustonen, K.; Anisimov, A. S.; Ruiz, V.; Kivisto, S.; Rackauskas, S. et al. Multifunctional Free-Standing Single-Walled Carbon Nanotube Films. *ACS Nano* **2011,** *5* (4), 3214–3221.
87. Wang, C.; Li, P.; Zong, Y.; Zhang, Y.; Li, S.; Wei, F. A High Efficiency Particulate Air Filter Based on Agglomerated Carbon Nanotube Fluidized Bed. *Carbon* **2014,** *79*, 424–431.
88. Park, K-T.; Hwang, J. Filtration and Inactivation of Aerosolized Bacteriophage MS2 by a CNT Air Filter Fabricated Using Electro-Aerodynamic Deposition. *Carbon* **2014,** *75*, 401–410.
89. Parham, H.; Bates, S.; Xia, Y.; Zhu, Y. A Highly Efficient and Versatile Carbon Nanotube/Ceramic Composite Filter. *Carbon* **2013,** *54*, 215–223.
90. Yildiz, O., Bradford, P. D. Aligned Carbon Nanotube Sheet High Efficiency Particulate Air Filters. *Carbon* **2013,** *64*, 295–304.
91. Simon, P.; Gogotsi, Y. Materials for Electrochemical Capacitors. *Nat. Mater.* **2008,** *7* (11), 845.

92. Liu, X.; Huber, T. A.; Kopac, M. C.; Pickup, P. G. Ru Oxide/Carbon Nanotube Composites for Supercapacitors Prepared by Spontaneous Reduction of Ru(VI) and Ru(VII). *Electrochimica Acta* **2009,** *54* (27), 7141.
93. Yan, J.; Fan, Z.; Wei, T.; Cheng, J.; Shao, B.; Wang,K.; Song, L.; Zhang, M. Carbon Nanotube/MnO_2 Composites Synthesized by Microwave-Assisted Method for Supercapacitors with High Power and Energy Densities. *J. Power Sourc.* **2009,** *194* (2), 1202.
94. Hadjipaschalis, I.; Poullikkas, A.; Efthimiou, V. Overview of Current and Future Energy Storage Technologies for Electric Power Applications. *Renew. Sustain. Energy Rev.* **2009,** *13* (6–7), 1513.
95. Kshetri, T.; Tran, D. T.; Nguyen, D. C.; Kim, N. H.; Lau, K-t.; Lee, J. H. Ternary Graphene-Carbon Nanofibers-Carbon Nanotubes Structure for Hybrid Supercapacitor. *Chem. Eng. J.* **2020,** *380*, 122543.
96. Fan, H.; Zhang, X.; Wang, Y.; Lang, J.; Gao, R. Highly Conductive KNiF3@ Carbon Nanotubes Composite Materials with Cross-Linked Structure for High Performance Supercapacitor. *J. Power Sour.* **2020,** *474*, 228603.
97. Zheng, Y.; Tian, Y.; Sarwar, S.; Luo, J.; Zhang, X. Carbon Nanotubes Decorated $NiSe_2$ Nanosheets for High-Performance Supercapacitors. *J. Power Sourc.* **2020,** *452*, 227793.
98. Hussain, S.; Rabani, I.; Vikraman, D.; Feroze, A.; Karuppasamy, K.; Haq, Z.; Seo, Y-S.; Chun, S-H.; Kim, H-S.; Jung, J. Hybrid Design Using Carbon Nanotubes Decorated with Mo_2C and W_2C Nanoparticles for Supercapacitors and Hydrogen Evolution Reactions. *ACS Sustain. Chem. Eng.* **2020,** *8* (32), 12248–12259.
99. Hu, Z.; Ma, K.; Tian, W.; Wang, F.; Zhang, H.; He, J.; Deng, K.; Zhang, Y. X.; Yue, H.; Ji, J. Manganese Dioxide Anchored on Hierarchical Carbon Nanotubes/ Graphene/Diatomite Conductive Architecture for High Performance Asymmetric Supercapacitor. *Appl. Surf. Sci.* **2020,** *508*, 144777.
100. Gong, W.; Fugetsu, B.; Li, Q.; Vipin, A. K.; Konishi, T.; Ueki, T.; Sakata, I. et al. Improved Supercapacitors by Implanting Ultra-Long Single-Walled Carbon Nanotubes into Manganese Oxide Domains. *J. Power Sourc.* **2020,** 479, 228795.
101. Ali, G. A.; Megiel, E.; Cieciórski, P.; Thalji, M. R.; Romański, J.; Algarni, H.; Chong, K. F. Ferrocene Functionalized Multi-Walled Carbon Nanotubes as Supercapacitor Electrodes. *J. Mol. Liquids* **2020,** *318*, 114064.
102. Bardi, N.; Giannakopoulou, T.; Vavouliotis, A.; Trapalis, C. Electrodeposited Films of Graphene, Carbon Nanotubes, and their Mixtures for Supercapacitor Applications. *ACS Appl. Nano Mater.* **2020.**
103. Kshetri, T.; Tran, D. T.; Nguyen, D. C.; Kim, N. H.; Lau, K-t.; Lee, J. H. Ternary Graphene-Carbon Nanofibers-Carbon Nanotubes Structure for Hybrid Supercapacitor. *Chem. Eng. J.* **2020,** *380*, 122543.
104. Yanshyna, O.; Weissman, H.; Rybtchinski, B. Recyclable Electrochemical Supercapacitors Based on Carbon Nanotubes and Organic Nanocrystals. *Nanoscale* **2020,** *12* (16), 8909–8914.
105. Chen, Y.; Xu, C.; Zhao, J.; Huang, J.; Xu, H.; Gou, G. Creating a New Electrode Material of Supercapacitors from the Waste Multi-Walled Carbon Nanotubes. *Electrochim. Acta 330*, 135237.

106. Luo, X.; Yang, Q.; Dong, Y.; Huang, X.; Kong, D.; Wang, B.; Liu, H.; Xiao, Z.; Zhi, L. Maximizing Pore and Heteroatom Utilization within N, P-co-Doped Polypyrrole-Derived Carbon Nanotubes for High-Performance Supercapacitors. *J. Mater. Chem. A* **2020,** *8* (34), 17558–17567.
107. Fan, L-Q.; Tu, Q-M.; Geng, C-L.; Huang, J-L.; Gu, Y.; Lin, J-M.; Huang, Y-F.; Wu, J-H. High Energy Density and Low Self-Discharge of a Quasi-Solid-State Supercapacitor with Carbon Nanotubes Incorporated Redox-Active Ionic Liquid-Based Gel Polymer Electrolyte. *Electrochimica Acta* **2020,** *331*, 135425.
108. Mei, L.; Cui, X.; Duan, Q.; Li, Y.; Lv, X.; Wang, H-g. Metal Phthalocyanine-Linked Conjugated Microporous Polymer Hybridized with Carbon Nanotubes as a High-Performance Flexible Electrode for Supercapacitors. *Int. J. Hydrogen Energy* **2020,** *45* (43), 22950–22958.
109. Wang, H.; Biswas, S. K.; Zhu, S.; Lu, Y.; Yue, Y.; Han, J.; Xu, X.; Wu, Q.; Xiao, H. Self-Healable Electro-Conductive Hydrogels Based on Core-Shell Structured Nanocellulose/Carbon Nanotubes Hybrids for Use as Flexible Supercapacitors. *Nanomaterials* **2020,** *10* (1), 112.
110. Wang, S.; Liang, Y.; Zhuo, W.; Lei, H.; Javed, M. S.; Liu, B.; Wang, Z.; Mai, W. Freestanding Polypyrrole/Carbon Nanotubes Electrode with High Mass Loading for Robust Flexible Supercapacitors. *Mater. Chem. Front.* 2020.
111. Xu, D.; Xuan, C.; Li, X.; Luo, Z.; Wang, Z.; Tang, T.; Wen, J.; Li, M.; Xiao, J. Novel Helical Carbon Nanotubes-Embedded Reduced Graphene Oxide in Three-Dimensional Architecture for High-Performance Flexible Supercapacitors. *Electrochim. Acta* **2020,** *339*, 135912.
112. Kim, M. G.; Lee, B.; Li, M.; Noda, S.; Kim, C.; Kim, J.; Song, W. J.; Lee, S. W.; Brand, O. All-Soft Supercapacitors Based on Liquid Metal Electrodes with Integrated Functionalized Carbon Nanotubes. *ACS Nano* **2020.**
113. Hoffmann, M. R.; Martin, S. T.; Choi, W.; Bahnemann, D. W. Environmental Applications of Semiconductor Photocatalysis. *Chem. Rev.* **1995,** *95* (1), 69–96.
114. Kongkanand, A.; Kamat, P. V. Electron Storage in Single Wall Carbon Nanotubes. Fermi Level Equilibration in Semiconductor-SWCNT Suspensions. *ACS Nano* **2007,** *1* (1), 13–21.
115. Silva, C. G. Faria, J. L. Photocatalytic Oxidation of Benzene Derivatives in Aqueous Suspensions: Synergic Effect Induced by the Introduction of Carbon Nanotubes in a TiO_2 Matrix. *Appl. Cataly. B: Environ.* **2010,** *101* (1–2), 81–89.
116. Martínez, C. M. C. L.; Fernández, M. I.; Santaballa, J. A.; Faria, J. Kinetics and Mechanism of Aqueous Degradation of Carbamazepine by Heterogeneous Photocatalysis Using Nanocrystalline TiO_2, ZnO and Multi-Walled Carbon Nanotubes–Anatase Composites. *Appl. Catalys. B: Environ.* **2011,** *102* (3–4), 563–571.
117. Wang, H.; Wang, H. L.; Jiang, W. F.; Li, Z. Q. Photocatalytic Degradation of 2,4-Dinitrophenol (DNP) by Multi-Walled Carbon Nanotubes (MWCNTs)/TiO_2 Composite in Aqueous Solution under Solar Irradiation. *Water Res.* **2009,** *43* (1), 204–210.
118. Yu, Y.; Jimmy, C. Y.; Chan, C-Y.; Che, Y-K.; Zhao, J-C.; Ding, L.; Ge, W-K.; Wong, P-K. Enhancement of Adsorption and Photocatalytic Activity of TiO_2 by Using

Carbon Nanotubes for the Treatment of Azo Dye. *Appl. Cataly. B Environ.* **2005,** *61* (1–2), 1–11.

119. Yu, Y.; Jimmy, C. Y.; Yu, J-G.; Kwok, Y-C.; Che, Y-K.; Zhao, J-C.; Ding, L.; Ge, W-K.; Wong, P-K. Enhancement of Photocatalytic Activity of Mesoporous TiO_2 by Using Carbon Nanotubes. *Appl. Cataly. A Gen.* **2005,** *289* (2), 186–196.
120. Yao, Y.; Li, G.; Ciston, S.; Lueptow, R. M.; Gray, K. A. Photoreactive TiO_2/Carbon Nanotube Composites: Synthesis and Reactivity. *Environ. Sci. Technol.* **2008,** *42* (13), 4952–4957.
121. Yu, H.; Quan, X.; Chen, S.; Zhao, H.; Zhang, Y. TiO_2–Carbon Nanotube Heterojunction Arrays with a Controllable Thickness of TiO_2 Layer and Their First Application in Photocatalysis. *J. Photochem. Photobiol. A Chem.* **2008,** *200* (2–3), 301–306.
122. Wang, H.; Wang, H-L.; Jiang, W-F. Solar Photocatalytic Degradation of 2, 6-Dinitro-p-Cresol (DNPC) Using Multi-Walled Carbon Nanotubes (MWCNTs)–TiO_2 Composite Photocatalysts. *Chemosphere* **2009,** *75* (8), 1105–1111.
123. Wang, H.; Wang, H-L.; Jiang, W-F.; Li, Z-Q. Photocatalytic Degradation of 2, 4-Dinitrophenol (DNP) by Multi-Walled Carbon Nanotubes (MWCNTs)/TiO_2 Composite in Aqueous Solution under Solar Irradiation. *Water Res.* **2009,** *43* (1), 204–210.
124. Silva, C. G.; Faria, J. L. Photocatalytic Oxidation of Benzene Derivatives in Aqueous Suspensions: Synergic Effect Induced by the Introduction of Carbon Nanotubes in a TiO_2 Matrix. *Appl. Cataly. B Environ.* **2010,** *101* (1–2), 81–89.
125. Xu, Y-J.; Zhuang, Y.; Fu, X. New Insight for Enhanced Photocatalytic Activity of TiO_2 by Doping Carbon Nanotubes: A Case Study on Degradation of Benzene and Methyl Orange. *J. Phys. Chem. C* **2010,** *114* (6), 2669-2676.
126. Krissanasaeranee, M.; Wongkasemjit, S.; Cheetham, A. K.; Eder, D. Complex Carbon Nanotube-Inorganic Hybrid Materials as Next-Generation Photocatalysts. *Chem. Phys. Lett.* **2010,** *496* (1–3), 133–138.
127. Chen, H.; Yang, S.; Yu, K.; Ju, Y.; Sun, C. Effective Photocatalytic Degradation of Atrazine Over Titania-Coated Carbon Nanotubes (CNTs) Coupled with Microwave Energy. *J. Phys. Chem. A* **2011,** *115* (14), 3034–3041.
128. Feng, S.-a.; Zhao, J-h.; Zhu, Z-p. The Manufacture of Carbon Nanotubes Decorated with ZnS to Enhance the ZnS Photocatalytic Activity. *N. Carbon Mater.* **2008,** *23* (3), 228–234.
129. Liu, X.; Zhang, S.; Pan, B. Potential of Carbon Nanotubes in Water Treatment. *Recent Progr. Carbon Nanotube Res.* **2012,** *201110* (5772), 51332.
130. Ma, L-L.; Sun, H-Z.; Zhang, Y-G.; Lin, Y-L.; Li, J-L.; Wang, E-k.; Yu, Y.; Tan, M.; Wang, J-B. Preparation, Characterization and Photocatalytic Properties of CdS Nanoparticles Dotted on the Surface of Carbon Nanotubes. *Nanotechnology* **2008,** *19* (11), 115709.
131. Wang, S.; Shi, X.; Shao, G.; Duan, X.; Yang, H.; Wang, T. Preparation, Characterization and Photocatalytic Activity of Multi-Walled Carbon Nanotube-Supported Tungsten Trioxide Composites. *J. Phys. Chem. Solids* **2008,** *69* (10), 2396–2400.
132. Phin, H-Y.; Ong, Y-T.; Sin, J-C. Effect of Carbon Nanotubes Loading on the Photocatalytic Activity of Zinc Oxide/Carbon Nanotubes Photocatalyst Synthesized via a Modified Sol-Gel Method. *J. Environ. Chem. Eng.* **2020,** *8* (3), 103222.

133. Li, Y.; Wang, Z.; Zhao, H.; Yang, M. Composite of TiO_2 Nanoparticles and Carbon Nanotubes Loaded on Poly (Methyl Methacrylate) Nanofibers: Preparation and Photocatalytic Performance. *Synth. Metals* **2020,** *269*, 116529.
134. Mohammed, M. K. A. Carbon Nanotubes Loaded ZnO/Ag Ternary Nanohybrid with Improved Visible Light Photocatalytic Activity and Stability. *Optik* **2020,** 164867.
135. Gao, P.; Yin, Z.; Feng, L.; Liu, Y.; Du, Z.; Duan, Z.; Zhang, L. Solvothermal Synthesis of Multiwall Carbon Nanotubes/BiOI Photocatalysts for the Efficient Degradation of Antipyrine under Visible Light. *Environ. Res.* **2020,** 109468.
136. Zhu, Z.; Wang, C.; Liang, L.; Yu, D.; Sun, J.; Zhang, L.; Zhong, S.; Liu, B. Synthesis of Novel Ternary Photocatalyst Ag_3PO_4/Bi_2WO_6/Multi-Walled Carbon Nanotubes and Its Enhanced Visible-Light Photoactivity for Photodegradation of Norfloxacin. *J. Nanosci. Nanotechnol.* **2020,** *20* (4), 2247–2258.
137. Bargozideh, S.; Tasviri, M.; Ghabraei, M. Effect of Carbon Nanotubes Loading on the Photocatalytic Activity of BiSI/BiOI as a Novel Photocatalyst. *Environ. Sci. Pollut. Res.* **2020,** *27* (29), 36754–36764.
138. Zada, N.; Saeed, K.; Khan, I. Decolorization of Rhodamine B Dye by Using Multiwalled Carbon Nanotubes/Co–Ti Oxides Nanocomposite and Co–Ti Oxides as Photocatalysts. *Appl. Water Sci.* **2020,** *10* (1), 40.
139. Taboada, C. D.; Batista, J.; Pintar, A.; Levec, J. Preparation, Characterization and Catalytic Properties of Carbon Nanofiber-Supported Pt, Pd, Ru Monometallic Particles in Aqueous-Phase Reactions. *Appl. Cataly. B Environ.* **2009,** *89* (3–4), 375–382.
140. Gomes, H T.; Samant, P. V.; Serp, Ph; Kalck, Ph; Figueiredo, J. L.; Faria, J. L. Carbon Nanotubes and Xerogels as Supports of Well-Dispersed Pt Catalysts for Environmental Applications. *Appl. Cataly. B Environ.* **2004,** *54* (3), 175–182.
141. Garcia, J.; Gomes, H. T.; Serp, P.; Kalck, P.; Figueiredo, J. L.; Faria, J. L. Platinum Catalysts Supported on MWNT for Catalytic Wet Air Oxidation of Nitrogen Containing Compounds. *Cataly. Today* **2005,** *102*, 101–109.
142. Garcia, J.; Gomes, H. T.; Serf, P.; Kalck, P.; Figueiredo, J. L.; Faria, J. L. Carbon Nanotube Supported Ruthenium Catalysts for the Treatment of High Strength Wastewater with Aniline Using Wet Air Oxidation. *Carbon* **2006,** *44* (12), 2384–2391.
143. Yang, S.; Zhu, W.; Li, X.; Wang, J.; Zhou, Y. Multi-Walled Carbon Nanotubes (MWNTs) as an Efficient Catalyst for Catalytic Wet Air Oxidation of Phenol. *Cataly. Commun.* **2007,** *8* (12), 2059–2063.
144. Yang, S.; Li, X.; Zhu, W.; Wang, J.; Descorme, C. Catalytic Activity, Stability and Structure of Multi-Walled Carbon Nanotubes in the Wet Air Oxidation of Phenol. *Carbon* **2008,** *46* (3), 445–452.
145. Barge, A. S.; Vaidya, P. D. Ruthenium-Decorated Carbon Nanotubes as Catalyst for Wet Air Oxidation. *J. Environ. Chem. Eng.* **2019,** *7* (1), 102914.
146. Yang, S.; Sun, Y.; Yang, H.; Wan, J. Catalytic Wet Air Oxidation of Phenol, Nitrobenzene and Aniline over the Multi-Walled Carbon Nanotubes (MWCNTs) as Catalysts. *Front. Environ. Sci. Eng.* **2015,** *9* (3), 436–443.
147. Milone, C.; Hameed, A. R. S.; Piperopoulos, E.; Santangelo, S.; Lanza, M.; Galvagno, S. Catalytic Wet Air Oxidation of p-Coumaric Acid over Carbon Nanotubes and Activated Carbon. *Indust. Eng. Chem. Res.* **2011,** *50* (15), 9043–9053.

148. Cao, Y.; Li, B.; Zhong, G.; Li, Y.; Wang, H.; Yu, H.; Peng, F. Catalytic Wet Air Oxidation of Phenol over Carbon Nanotubes: Synergistic Effect of Carboxyl Groups and Edge Carbons. *Carbon* **2018,** *133*, 464–473.
149. Rocha, R. P.; Sousa, J. P. S.; Silva, A. M. T.; Pereira, M. F. R.; Figueiredo, J. L. Catalytic Activity and Stability of Multiwalled Carbon Nanotubes in Catalytic Wet Air Oxidation of Oxalic Acid: The Role of the Basic Nature Induced by the Surface Chemistry. *Appl. Cataly. Environ.* **2011,** *104* (3–4), 330–336.
150. Rocha, R. P.; Silva, A. M. T.; Romero, S. M. M.; Pereira, M. F. R.; Figueiredo, J. L. The Role of O-and S-Containing Surface Groups on Carbon Nanotubes for the Elimination of Organic Pollutants by Catalytic Wet Air Oxidation. *Appl. Cataly. B Environ.* **2014,** *147*, 314–321.
151. Bretschger, O. et al. Microbial Fuel Cells and Microbial Ecology: Applications in Ruminant Health and Production Research. *Microb. Ecol.* **2010,** *59* (3), 415–427.
152. Potter, M. C.; Waller, A. D. Electrical Effects Accompanying the Decomposition of Organic Compounds. *Proc. R. Soc. Lond.—Ser. B Contain Pap a Biol. Character* **1911,** *84* (571), 260–276.
153. Park, D.; Zeikus, J. Impact of Electrode Composition on Electricity Generation in a Single-Compartment Fuel Cell Using Shewanella Putrefaciens. *Appl. Microbiol. Biotechnol.* **2002,** *59* (1), 58–61.
154. Bond, D. R.; Lovley, D. R. Electricity Production by Geobacter Sulfurreducens Attached to Electrodes. *Appl. Environ. Microbiol.* **2003,** *69* (3), 1548.
155. Li, F. et al. Microbial Fuel Cells: The Effects of Configurations, Electrolyte Solutions, and Electrode Materials on Power Generation. *Appl. Biochem. Biotechnol.* **2009,** *160* (1), 168.
156. Kaur, R.; Marwaha, A.; Chhabra, V. A.; Kim, K-H.; Tripathi, S. K. Recent Developments on Functional Nanomaterial-Based Electrodes for Microbial Fuel Cells. *Renew. Sustain. Energy Rev.* **2020,** *119*, 109551.
157. Du, Y.; Ma, F-M.; Xu, C-Y.; Yu, J.; Li, D.; Feng, Y.; Zhen, L. Nitrogen-Doped Carbon Nanotubes/Reduced Graphene Oxide Nanosheet Hybrids towards Enhanced Cathodic Oxygen Reduction and Power Generation of Microbial Fuel Cells. *Nano Energy* **2019,** *61*, 533–539.
158. Qiao, Y.; Li, C. M.; Bao, S. J.; Bao, Q. L. Carbon Nanotube/Polyaniline Composite as Anode Material for Microbial Fuel Cells. *J. Power Sourc.* **2007,** *170* (1), 79–84.
159. Zou, Y. J.; Xiang, C. L.; Yang, L. N.; Sun, L. X.; Xu, F.; Cao, Z. A Mediatorless Microbial Fuel Cell Using Polypyrrole Coated Carbon Nanotubes Composite Asanode Material. *Int. J. Hydrogen Energy* **2008,** *33* (18), 4856–4862.
160. Xie, X.; Hu, L. B.; Pasta, M.; Wells, G. F.; Kong, D. S.; Criddle, C. S.; Cui, Y. Three-Dimensional Carbon Nanotube-Textile Anode for High-Performance Microbial Fuel Cells. *Nano Lett.* **2011,** *11* (1), 291–296.
161. Xie, X.; Ye, M.; Hu, L. B.; Liu, N.; McDonough, J. R.; Chen, W.; Alshareef, H. N.; Criddle, C. S.; Cui, Y. Carbon Nanotube-Coated Macroporous Sponge for Microbial Fuel Cell Electrodes. *Energy Environ. Sci.* **2012,** *5* (1), 5265–5270.
162. Liu, X. W.; Sun, X. F.; Huang, Y. X.; Sheng, G. P.; Wang, S. G.; Yu, H. Q. Carbon Nanotube/Chitosan Nanocomposite as a Biocompatible Biocathode Material to Enhance the Electricity Generation of a Microbial Fuel Cell. *Energy Environ. Sci.* **2011,** *4* (4), 1422–1427.

163. He, Y.; Liu, Z.; Xing, X-h.; Li, B.; Zhang, Y.; Shen, R.; Zhu, Z.; Duan, N. Carbon Nanotubes Simultaneously as the Anode and Microbial Carrier for Up-Flow Fixed-Bed Microbial Fuel Cell. *Biochem. Eng. J.* **2015,** *94*, 39–44.
164. Liu, D.; Chang, Q.; Gao, Y.; Huang, W.; Sun, Z.; Yan, M.; Guo, C. High Performance of Microbial Fuel Cell Afforded by Metallic Tungsten Carbide Decorated Carbon Cloth Anode. *Electrochim. Acta* **2020,** *330*, 135243. Reduction and Power Generation of Microbial Fuel Cells. *Nano Energy* **2019,** *61*, 533–539.
165. Mehdinia, A.; Ziaei, E.; Jabbari, A. Multi-Walled Carbon Nanotube/SnO_2 Nanocomposite: A Novel Anode Material for Microbial Fuel Cells. *Electrochim. Acta* **2014,** *130*, 512–518.
166. Wang, Y. et al. Nano-Molybdenum Carbide/Carbon Nanotubes Composite as Bifunctional Anode Catalyst for High-Performance *Escherichia coli*-Based Microbial Fuel Cell. *Biosens. Bioelectron.* **2014,** *51*, 349–355.
167. Delord, B.; Neri, W.; Bertaux, K.; Derre, A.; Ly, I.; Mano, N.; Poulin, P. Carbon Nanotube Fiber Mats for Microbial Fuel Cell Electrodes. *Bioresour. Technol.* **2017,** *243*, 1227–1231.
168. Baillie, C., Ed. *Green Composites: Polymer Composites and the Environment*; CRC Press, 2005.
169. Grossiord, N.; Loos, J.; Koning, C. E. Strategies for Dispersing Carbon Nanotubes in Highly Viscous Polymers. *J. Mater. Chem.* **2005,** *15* (24), 2349–2352.
170. Ray, S. S.; Vaudreuil, S.; Maazouz, A.; Bousmina, M. Dispersion of Multi-Walled Carbon Nanotubes in Biodegradable Poly (Butylene Succinate) Matrix. *J. Nanosci. Nanotechnol.* **2006,** *6* (7), 2191–2195.
171. Vaudreuil, S.; Labzour, A.; Sinha-Ray, S.; El Mabrouk, K.; Bousmina, M. Dispersion Characteristics and Properties of Poly (Methyl Methacrylate)/Multi-Walled Carbon Nanotubes Nanocomposites. *J. Nanosci. Nanotechnol.* **2007,** *7* (7), 2349–2355.
172. Song, L.; Qiu, Z. Crystallization Behavior and Thermal Property of Biodegradable Poly (Butylene Succinate)/Functional Multi-Walled Carbon Nanotubes Nanocomposite. *Polym. Degrad. Stab.* **2009,** *94* (4), 632–637.
173. Sitharaman, B.; Shi, X.; Walboomers, X. F.; Liao, H.; Cuijpers, V.; Wilson, L. J.; Mikos, A. G.; Jansen, J. A. In Vivo Biocompatibility of Ultra-Short Single-Walled Carbon Nanotube/Biodegradable Polymer Nanocomposites for Bone Tissue Engineering. *Bone* **2008,** *43* (2), 362–370.
174. Im, O.; Li, J.; Wang, M.; Zhang, L. G.; Keidar, M. Biomimetic Three-Dimensional Nanocrystalline Hydroxyapatite and Magnetically Synthesized Single-Walled Carbon Nanotube Chitosan Nanocomposite for Bone Regeneration. *Int. J. Nanomed.* **2012,** *7*, 2087.
175. Li, J-L.; Yang, Z.; Loo, W. T. Y.; Xiao, X.; Zhang, D.; Cheung, M. N. B.; Tsang, W. W. N.; Ng, E. L. Y.; Wang, M. In Vitro and in Vivo Biocompatibility of Multi-Walled Carbon Nanotube/Biodegradable Polymer Nanocomposite for Bone Defects Repair. *J. Bioact. Compatible Polym.* **2014,** *29* (4), 350–367.

CHAPTER 3

BIOMEDICAL APPLICATIONS OF CARBON NANOTUBES

RUBY VARGHESE[1], NAMITHA BINU[2] and YOGESH B. DALVI[2*]

[1]*School of Sciences, Department of Chemistry, Jain Deemed to be University, Bengaluru, Karnataka*

[2]*Pushpagiri Research Centre, Pushpagiri Institute of Medical Sciences and Research Centre, Tiruvalla, Kerala*

[*]*Corresponding author. E-mail: yogesh.botany@gmail.com*

ABSTRACT

Nanomaterials are the foundation of the convergence of Nano and Biotechnology that made remarkable advances in improving human well being by enabling their scientific and technical knowledge. Carbon nanotubes (CNTs) are tube-like materials, which are allotropes of carbon, made up of graphite, and with diameter measured in nanoscale. CNTs are fabricated using three different methods, that is, chemical vapor deposition, electrical method, and laser deposition method. Depending on the number of carbon layers, CNTs can be single-walled, double-walled, and multi-walled with an impressive high electric and thermal conductivity, mechanical potency, and structural properties.

The surface of CNTs can be functionalized by the binding of active molecules, making them compatible with biological systems. Functionalizations of CNTs surface enable them to be diversely used for biomedical applications such as drug delivery, biosensing, tissue engineering, dental applications, and extraction and analysis of drugs and pollutants. This review focuses on providing a brief outline on the pharmacokinetics, toxicity, and metabolism of CNTs in living system and discusses its application in biomedical field.

3.1 INTRODUCTION

Nanoparticles, either organic or inorganic, are the most common nanotechnology-based products that are diversely used in biomedical and environmental applications. These materials have dimensions below 100 nm that enable access to develop functionalized and improvised material of one's choice with wider variety of magnetic, electronic, mechanical, or optical properties.[54,86–88]

Carbon nanotubes (CNTs) are one of the most promising nanomaterials due to its diversified nature like biologically active, spatially designed and closely mimic the native extracellular matrix.[14]

CNTs constitute of one or several concentric graphite layer with a diameter ranging from 0.4 to 10 and length 100 nm.[47] It is classified into single, double, triple, and multiwalled CNTs depending on the number of grapheme layers (Table 3.1). Most CNTs are synthesized by three methods:

1. Arc Discharge Method: In this method, a high temperature (>3000°C) is used to convert carbon atoms into a plasma and form SWCNTs (single-walled CNTs) and MWCNTs (multi-walled CNTs). Catalytic agents like Cobalt (Co), Yttrium (Y), Nickel (Ni), and Iron (Fe) are required to form individual SWCNT and not required for the synthesis of MWCNT.
2. Laser Ablation Method: In this method, graphite is vaporized in an electrical furnace (1200°C) to ensure a high level purity and high ratio of resulting products.
3. Chemical Vapor Deposition Method: It uses hydrocarbon sources such as Co, Methane, and ethylene for the development of CNTs.[4]

CNTs exhibit excellent physiochemical properties such as high tensile strength, ultra-light weight, increased surface area, biocompatibility, and biodegradability without affecting the biological living system. This review focuses on the biomedical applications of CNTs in various fields like dentistry, as bio-sensing/bio-imaging agent with promising role in cancer diagnosis, as immunosensors, in drug delivery and also in tissue engineering and regenerative medicine.

TABLE 3.1 Types of Carbon Nanotubes.

Types of carbon nanotubes	Diameter	Length	Layers
Single-walled carbon nanotube (SWCNT)	0.4–3 nm	100–1,000 nm	Single graphite sheet
Double-walled carbon nanotube (DWCNT)	1–4 nm	100–1,000 nm	Two concentric cylindrical graphite layers
Multi-walled carbon nanotube (MWCNT)	1.4–100 nm	50 μm	several concentric graphite layers (2–10)

3.2 BIOMEDICAL APPLICATION OF CNTS

3.2.1 APPLICATIONS OF CNTS IN DENTISTRY

Utilization of CNT in the dentistry has been introduced in early 1990s and its application in the dental field can be categorized into following areas:

A. Dental Restorative Materials

CNT dental composite resin is tooth colored, resin-based matrix made up of substances like bisphenol-A-glycidyl methacrylate and silica as inorganic filler that provides improved mechanical property and translucency with wear resistance ability. To this dental composite SWCNT are applied to improve its tensile strength and Young's modulus for better restorative property, flexural strength, and longevity in oral cavity.[86–88]. CNTs have been also used at the interface of dentin and composite resin to prevent micro leakage development thereby preventing bacterial colonization.[3]

Another application of CNT is in the development of dental base, currently available base is made up of Poly (methyl) methyl acrylate (PMMA) and is of low-cost, low density with excellent esthetics and is repairable. Though denture base has low fracture strength which makes it vulnerable to crack during any external pressure. Hence, MWCNTs together with PMMA improved flexural strength and fracture toughness thereby showing an improved mechanical properties.[10]

B. Guided Bone Regeneration and Replacement of Bone Defects

CNTs have emerged as a new biomaterial that presents itself as a multifaceted composite with a potential role in oral regenerative medicine and to support periodontal and implant regenerative procedures.[48] Guided Tissue Regeneration (GTR) corrects periodontal defects by combining bone grafts to a membrane barrier[30] as well as it is useful in restoring deficient alveolar sites and in such cases it's called as Guided bone Regeneration (GBR). For periodontal or bone regeneration, synthetic polymer membranes are used that are either polyester or tissue-derived collagen[14] but their utility is limited by structural, functional, and mechanical utility.

Mendes et al.[51] and Sa et al.[68] showed that the utilization of hyaluronate together with SWCNT promoted bone repair/regeneration of tooth sockets of rats by preventing bacterial colonization[23] and it improved the low stability of hyaluronate to sustain in the aqueous environment of oral cavity.[52]

Tissue engineering with CNT alone or in combination may be an optimal regenerative option in implantology because it provides better alveolar ridge preservation and accelerates osteointegration process by attaching to the receptor sites of dental implants thereby reducing post-operative problems. An important factor for the successful bone defect replacement therapy is the availability of blood supply at grafted site, immobilized, and accurate grafts (w.r.t dimensions) to carry out a proper healing. Now, when comes to a larger bony defects, it does not have a supporting natured bone for immobilization as well as no adequate blood supply as compared to small bony defects. This criteria is easily fulfilled by ultra-short SWCNT[70] which is highly porous and it mimics natural trabecular bone structure and also by PLGA—nanofibrins-MWCNT yarns[21,22] with a uniform cell distribution and spanning of scaffold can be use for bone replacement.

CNT-PCL composite fabricated layer-by-layer and SWCNT also facilitates osteogenic cell proliferation by utilizing CNT's electric properties.

3.2.2 BIO-SENSING/BIO-IMAGING AND CANCER DIAGNOSIS

Due to small size, large size-volume, and inherent physiochemical properties, CNTs act as transducer nanomaterials with higher sensitivity and lower detection limits.[74] By virtue of unrivalled properties like electronic, optical, electrochemical, and mechanical properties, CNTs are extremely direction dependent and tunable. Hence, it is a promising candidate for detection purposes. CNT offers several assets for the detection as described in the following:

1. By adding bioactive molecules CNTs act as functionalized CNT
2. Due to its ability to conduct electricity CNTs can generate electrical signals upon target recognition
3. CNTs are stronger than steel and its thermal conductivity is more than that of diamond.
4. CNTs can cross biological membranes, hence can be employed for the photoacoustic imaging.

Depending on CNTs target recognition capacity and transduction, CNT-based biosensors are broadly categorized into following:

A. Electronic and Electrochemical CNT Biosensors

Till date biosensors developed are electrochemical-natured and are preferred due to its low-cost, relatively fast-response time, easy to use and small size. Biosensors generally contain a reference electrode, a working electrode with target analyzer immobilized on it and a counter electrode. In case of enzyme-coupled when an enzyme is recognized by an analyte on working electrode, an enzyme–substrate complex reaction takes place that causes electron transfer and causes production of a current that is displayed in signal processor (amplifier processor displayer). Due to electrical and electrochemical properties, CNTs act as a promising material for enhancing electron transfer a suitable for integration into electrochemical biosensor.[8,12,28,37,75,80]

Various electrochemical CNT biosensors have been developed and used for biomedical applications and they are as follows:

1. CNT–Glucose biosensor based on conjugation of glucose-oxidase.[39,63]
2. MWCNT—Cholesterol biosensor for quantification of total cholesterol in the blood.[42]
3. SWCNT—RNA aptasensor for detection of disease related glycoprotein in the blood.[84]
4. SWCNT-based detection of single-molecule H_2O_2 signaling from epidermal growth factor receptor of human epidermal carcinoma cells.[34]
5. CNT-biosensor arrayed on a chip together with pneumatic micropumps for the detection of several biomarkers in the blood.[76]

6. Disposable paper-based bipolar electrode (BPE) with MWCNT for the detection of the Prostate Specific Antigen (PSA).[25] Similarly SWCNTs and MWCNTs functionalized with DNA strand helped in early stage detection of PSA in the blood.[72]
7. SWCNT biosensors showed noticeable sensitivities toward volatile organic compounds in human breath to diagnose lung cancer.[60]
8. Field effect—CNT was developed to detect osteopontin—a possible biomarker for prostate cancer.[38]
9. CNT biosensors have emerged as potential sensing platform for detection and quantification of clinically relevant material such as glucose, cholesterol, glutamate, lactate etc. and also for the detection of cancer biomarkers such as alpha-fetoprotein, carcinoembryonic antigen (CEA), cancer antigens etc.[19]

In conclusion, CNT exhibit these promising and relevant diagnostic approach for biosensing applications relying on enzymatic reactions due to its small size, large surface area, high electrochemical properties (conductivity, chemical stability, and sensitivity),[90] fast electron-transfer rate, and high electrocatalytic effect.[42]

3.3 IMMUNOSENSORS

Antigen recognition by antibody which is immobilized on to a substrate and becomes a part of the receptor moiety of the biosensor is known as the immunosensor. In recent year with emergence in Nanotechnology, recombinant antibodies have been immobilized onto the nano-immunosensors like CNTs which in turn increases binding capacity and scientific threshold compared to traditional biosensors. CNT-based immunosensors have been developed to probe cancer biomarker.[36,77]

A. CNT-based Immunosensors in Cancer Diagnosis

1. SWCNT-based immunosensors are developed to detect very low and elevated levels of IL-6 in squamous cell carcinoma of head and neck.[46]
2. Screen—printed CNT based electrochemical biosensors are developed for detection of both PSA and IL-8.[78]

3. Modified gold nanoparticle decorated multiwall carbon nanotube-ionic liquid electrode enhances the stability, immunoactivity and immobilization of HER2 Ab Herceptin for the detection of low concentration of HER2 in serum samples of breast cancer patients.[5]

B. Optical CNT Biosensors

Detection of an analyte or a biomolecule by a biosensor through changes in emission of light (light—UV, Visible, IR) is known as optic biosensors. SWCNT contains inherent photoluminescence between 650 and 1400 nm that ensures deep penetration and imaging of the biological tissue.[73] SWCNTs are well-suited for fluorescence-based sensing application as it is affected significantly by surface functionalization, environmental changes, or interaction with biomolecules.[11,15,32,55] Moreover, they remain stable to photo bleaching which is again an add-on for bioimaging.[9]

Optical Biosensors for Cancer Applications

1. SWCNT—Indocyanine green sensors conjugated with cyclic Arg-Gly-Asp peptides achieved an improved sub-nanomolar sensitivity and higher photoacoustic contrast (300 times better than previously reported SWCNT) to target α (V) β (3) integrins in tumor- bearing mice.[85]
2. SWCNT-biosensor detects nanomolar concentration of cyclin A – a common cell cycle regulator that is overexpressed in various human cancer.[79]

C. Challenges and Limitations

1. Better control of physical and chemical properties is required.
2. CNTs like pristine cannot be directly used in diagnostic due to their metallic nature and insolubility in water.
3. Despite the advantage of accuracy and precision, there is always concern related to high-purity throughput in the development of CNT-based biosensors.

3.4 DRUG DELIVERY

Due to exceptionally outstanding and unique characteristics, CNTs are ideal candidates for drug delivery but its utilization is limited due to poor dispersibility.[50] Hence, functionalization of CNTs has emerged as an intriguing nanotechnology tool for drug delivery due to following reasons:

1. Increased solubility of CNTs
2. Biocompatibility
3. Ultra-lightweight
4. Increased surface area
5. Improved surface chemistry
6. Negligible immunogenicity
7. Photoluminescence
8. Biliary excretion
9. Rapid uptake by cell due to anisotropic morphology or needle-like morphology
10. Functionalized CNT act as nanomatrix containing drug molecules with controlled drug release mechanism

CNT–Hybrid Conjugates for Drug Delivery and Targeting

CNT–hybrid conjugates have attained an applauding attention for their tremendous role in pharmaceutical, biotechnological, and biomedical areas as CNTs provide improved safety and efficacy of pharmaceutical product with minimal toxicity.

A. Dendrimer–CNT Conjugate

Dendrimers are well-known nanotool for solubilizing the insoluble substances due to its multifaceted properties such as monodispersity, it is deriviatisable, can be compartmentalize, is highly branched, contains a lot of functional groups, and so on.[35] Dendrimer conjugation with CNT is a unique drug delivery agent due to increased biocompatibility, lower viscosity with ease in assistance to manipulation, modification, and development of nanoconjugates.

Following section will provide a brief biomedical use of Dendrimer–CNT conjugate:

1. PAMAM conjugated with MWCNT increased DNA uptake with low cytotoxicity and hence considered as novel gene vector.[43]
2. PAMAM-DEN-CNT mediated anti-survivin oligonucleotides effectively reduced cell proliferation of MCF-7 cells.[61]

3. Dendrimer-modified MWCNTs provided sites in CNTs/Ag nanohybrid and exhibited a strong antimicrobial activity than dendrimer-grafted nanotubes.[83]
4. PAMAM-CNTs exhibited efficient gene delivery while 5-G PAMAM dendrimer-CNT exhibited improved transfection efficiency and inhibited cell proliferation in tumor cells.[62,82]
5. FITC-dye and Folic acid A- PAMAM-CNTs use for cancer cell targeting and imaging while 5-G PAMAM-CNT-FITC-labeled anti sense c-myc oligonucleotides exhibited maximum transfection efficiency and minimum tumor proliferation.[62]
6. MWCNT functionalized—amphiphilic polypropylene imine dendrimers (APPI) and Ag nanoparticle-deposited MWCNT exhibited increased permeability to the bacterial surface and antimicrobial activity against both gram-positive bacteria (*Bacillus subtilis* and *Staphylococcus aureus*) and gram-negative bacteria (*Escherichia coli*). These activities were exhibited due to adsorption of positively charged bacterial cells, facilitate cross the cell membrane due to decreased generation of quaternary NH_4^+ of dendritic moieties and finally disruption of bacteria and cytoplasmic constituents thereby significantly high percentage of bacterial cells.[56]

B. Liposome–CNT Conjugate

Drug-loaded liposomes covalently attached to CNT to form a CNT–liposomes conjugate (CLC). This approach can deliver drugs into cells, with minimal or negligible adverse systemic effects of CNT when administered at high doses. CLC provide versatile and controlled means of enhanced drug delivery of one or more agents stably associated with the liposomes.[26,67]

C. Nanoparticle–CNT Conjugates

Attachment of nanoparticles with CNTs shed light into the new avenue such as therapeutic delivery. Fan et al.[24] synthesized a grapheme nanosheet-CNT-Fe nanoparticle hybrid deliver anticancer drug (5 fluorouracil). Folic acid and Fe nanoparticle with difunctionalized MWCNTs loaded with DOX exhibits six times higher toxicity than DOX itself against Hela cells.[40] Due to their multifunctional properties such as high functional density, size, shape, etc., these nanohybrid composites have wide spread biomedical applications such as in the development of biosensing nanodevices and nanoproducts, drug delivery, and targeting and biological imaging.[50]

D. Quantum Dots–CNT Conjugates (QD-CNTs)

Conjugation of QD-CNTs provide both imaging and inhibition or total destruction of cancer cells, that is, imaging of cancerous cells via QD and CNTs kill tumor cells via photothermal radiation. Hence, this hybrid can function as promising nanohybrid tool in imaging and controlled drug delivery. It has the ability to be less likely recognized by macrophages thus minimizing phagocytsis and increased time in blood circulation.

3.5 TISSUE ENGINEERING

CNTs can be called as nano innovative tool and its addition to polymeric (synthetic) scaffolds have revolutionized the field of regenerative medicine. Incorporation of CNTs to the polymer scaffolds enhances structural and chemical properties such as scaffold strength and flexibility, better biocompatibility, induction of cytotoxicity against cancerous cells, promote neovascularization during wound healing, reduced thrombosis and interference with gene expression in newly developed cells without affecting the degradation rate of synthetic scaffolds.[29]

A. Mechanical and Structural Properties

Incorporation of CNTs into polymers like polycarbonate urethane (PCU), Polycaprolactone (PCL), and Polystyrene (PS) showed a 36–42% increase in elastic modulus and 25% increase in tensile strength,[66] 5% decrease in polymer crystallinity and 12% increase in tensile strength with reduced rate of reduction;[45] finally CNTs have improved mechanical stability[33] without accelerating the degradation in polymer scaffold[45] thereby supporting cell growth, angiogenesis, and cytokine interaction with cells which in turn helps in tissue formation, that is, it is structurally and chemically independent of the original scaffold.[6,18]

B. Role of CNTs in Biological System and Tissue Engineering

1. CNTs effectively adsorb on extracellular and serum protein which in turn make scaffolds more bioactive and biocompatible[41,44,89]

2. Adsorption of CNTs effectively improves the cell adhesion, growth, and differentiation which depend on the size and arrangements of the CNTs.[57]
3. Orientation of CNTs influences cell proliferation, differentiation, and activation of cell adhesion mediators.
4. SWCNT aligned on glass substrate increases mesenchymal stem cells (MSCs) proliferation increases differentiation of osteocyte lineage as compared to randomly oriented SWCNT.[58]
5. Vascular endothelial cells grown on MWCNTs up-regulated cell adhesion mediators RhoA, ROCK2, and FAK as well as proliferated cells and enhanced collagen type IV secretion.[53]
6. Higher elastic moduli and hydrophobicity also plays a pivotal role in attachment of cells to the scaffold which was only possible due to the presence of MWCNTs that provides thermal stability and enhanced cell adhesive properties.[16]
7. CNTs exhibit excellent conductive properties that is useful in the design of scaffold used in bone, cardiac, and neural tissue engineering thus allowing scaffold to mimic electrical properties of nerve and myocytes and electrically stimulate development of cells and tissue[13] (Table 3.2).

TABLE 3.2 Role of CNTs in Tissue Engineering.

Neural tissue engineering			
CNT	**Cells**	**Function**	**References**
SWCNT multilayer films	Neuronal cells	Nerve repair	[27]
4-Hydroxynonenal-MWCNTs	Neuronal cells	Embryonic rat neuronal growth, long term viability, doubles axon length, a 3 times more neurite branching	[49]
1% MWCNT pHEMA hydrogels	SHSY5Y neuroblastoma cell	Improved cell communication, optimal elastic moduli, and ideal for decellularised neural tissues	[7]
CNT-collagen scaffolds	Pheochromocytoma rat cells	Improved neuronal growth and differentiation	[17]

TABLE 3.2 *(Continued)*

Neural tissue engineering			
CNT	**Cells**	**Function**	**References**
CNT ropes	Neural stem cells	In vivo nerve regeneration	[31]
Bone tissue engineering			
CNTs reinforcement in alumina composites	Cell culture (osteoblasts, chondrocytes, smooth muscle cells, and fibroblast)	Improved proliferative ability of osteoblasts, chondrocytes, smooth muscle cells, and fibroblasts	–
	Japanese white rabbits	Decreased inflammation	–
CNT-PU composite scaffolds	Osteoblasts	Increased alkaline phosphatase and calcium deposition and osteoblast cell adhesion	[65,69]
10 μA–10 Hz Alternating current to MWCNT-PLLA scaffold	Osteoblasts	Osteoblasts cell proliferation and extracellular calcium deposition	–
Hydroxyapatite crystals on SWCNT and MWCNT	Osteoblasts	Improved proliferation	–
Cardiac and vascular tissue engineering			
CNT-incorporated photo-cross-linkable gelatin methacrylate hydrogels	Neonatal rat cardiomyocytes	Threefold increase in spontaneous synchronous beating rates and 85% lower excitation threshold, compared to those cultured on pristine GelMA hydrogels	[71]
CNT	Cardiomyocytes	Improve cardiomyocyte growth and maturation by gene expression manipulation and no signs of pathological hypertrophy	–

TABLE 3.2 *(Continued)*

Neural tissue engineering			
CNT	**Cells**	**Function**	**References**
SWCNT and MWCNT	Cardiomyocytes	Increase gene expression of focal adhesion kinase	[86–88]
MWCNT	Human umbilical vein endothelial cells (HUVEC)	HUVEC and scaffolds with incorporated MWCNTs showed further ability to inhibit the release of Plasminogen Activator Inhibitor-1, when compared to nanofibrous scaffolds without MWCNTs Also improved endothelial cell arrangement and orientation by stimulating many bioactive factors.	[51]
MWCNTs into alginate vascular conduits		Diffusion of cell media in a radial manner, similar to the flow seen in natural blood vessels	[20]

Limitations

1. One of the major hurdles in using CNT nanotubes is on an ever going debatable topic, that is, toxicity. Hence, it is very essential to establish recognized standard CNT samples to evaluate CNT toxicity.
2. Standardization of dose and administrative route of CNTs which again take us back to the CNT toxicity analysis tissue.[13]
3. CNT-based scaffolds have undiluted properties but have limitation in effective scaffold structure.[21,22]
4. Dispersity of CNT is another major issue which limits their use, though usage of surfactants is conductive to produce stable dispersion of CNTs.[64]

3.6 CONCLUSION

The past few years have witnessed immense potential of CNTs as an innovative nanotool to be utilized for biomedical applications. Due to its broad spectrum, that is, from electronics to biomedical field there is an immense impending application of CNTs. CNTs hold vast potential in dental applications, drug delivery, cancer diagnostics, imaging, tissue engineering, and regenerative medicine. Functionalization of CNTs has brought promising advantages due the superior mechanical and physiochemical properties. Overall, this nanotechnological innovation could revolutionize the present and future therapeutic concepts with a hope for the diagnosis and treatment of many incurable diseases. Despite applauding features of CNTs there are still tremendous opportunities to be explored, significant challenges to be faced and risks to be solved. Therefore, more ideas, innovation and there execution are needed to elaborate functionalized CNTs with high efficacy and safety to be used in future medicinal purposes.

KEYWORDS

- **biocompatibility**
- **biosensor**
- **carbon nanotubes (CNTs)**
- **nano-hydroxyapatite**
- **neural regeneration**
- **tissue engineering**

REFERENCES

1. Abdolahad, M.; Janmaleki, M.; Taghinejad, M.; Taghnejad, H.; Salehi, F.; Mohajerzadeh, S. Single-Cell Resolution Diagnosis of Cancer Cells by Carbon Nanotube Electrical Spectroscopy. *Nanoscale* **2013**, *5* (8), 3421–3427.
2. Abdolahad, M.; Taghinejad, M.; Taghinejad, H.; Janmaleki, M.; Mohajerzadeh, S. A Vertically Aligned Carbon Nanotube-Based Impedance Sensing Biosensor for Rapid and High Sensitive Detection of Cancer Cells. *Lab Chip.* **2012,** *12* (6), 1183–1190.
3. Akasaka, T.; Nakata, K.; Uo, M.; Watari, F. Modification of the Dentin Surface by Using Carbon Nanotubes. *Bio-Med. Mater. Eng.* **2009,** *19* (2–3), 179–185.

4. Anzar, N.; Hasan, R.; Tyagi, M.; Yadav, N.; Narang, J. Carbon Nanotube-a Review on Synthesis, Properties and Plethora of Applications in the Field of Biomedical Science. *Sens. Int.* Jan 1 **2020,** *1*, 100003.
5. Arkan, E.; Saber, R.; Karimi, Z.; Shamsipur, M. A Novel Antibody–Antigen Based Impedimetric Immunosensor for Low Level Detection of HER2 in Serum Samples of Breast Cancer Patients via Modification of a Gold Nanoparticles Decorated Multiwall Carbon Nanotube-Ionic Liquid Electrode. *Analyt. Chim. Acta.* May 18, **2015,** *874*, 66–74.
6. Armentano, I.; Dottori, M.; Fortunati, E.; Mattioli, S.; Kenny, J. M. Biodegradable Polymer Matrix Nanocomposites for Tissue Engineering: A Review. *Polym. Degrad. Stability.* Nov 1, **2010,** *95* (11), 2126–2146.
7. Arslantunali, D. A.; Budak, G.; Hasirci, V. A. Multiwalled CNT-pHEMA Composite Conduit for Peripheral Nerve Repair. *J. Biomed. Mater. Res. Part A* Mar **2014,** *102* (3), 828–841.
8. Balasubramanian, K.; Burghard, M. Biosensors Based on Carbon Nanotubes. *Analyt. Bioanalyt. Chem.* Jun 1, **2006,** *385* (3), 452–468.
9. Barone, P. W.; Baik, S.; Heller, D. A.; Strano, M. S. Near-Infrared Optical Sensors Based on Single-Walled Carbon Nanotubes. *Nat. Mater.* Jan **2005,** *4* (1), 86–92.
10. Bhattacharya, M.; Seong, W. J. Carbon Nanotube-Based Materials—Preparation, Biocompatibility, and Applications in Dentistry. In *Nanobiomaterials in Clinical Dentistry*; Elsevier, 2019; pp 41–76.
11. Boghossian, A. A.; Zhang, J.; Barone, P. W.; Reuel, N. F.; Kim, J. H.; Heller, D. A.; Ahn, J. H.; Hilmer, A. J.; Rwei, A.; Arkalgud, J. R.; Zhang, C. T. Near-Infrared Fluorescent Sensors Based on Single-Walled Carbon Nanotubes for Life Sciences Applications. *ChemSusChem.* Jul 18, **2011,** *4* (7), 848.
12. Bohunicky, B.; Mousa, S. A. Biosensors: The New Wave in Cancer Diagnosis. *Nanotechnol. Sci. App.* **2011,** *4*, 1.
13. Bosi, S.; Fabbro, A.; Ballerini, L.; Prato, M. Carbon Nanotubes: A Promise for Nerve Tissue Engineering? *Nanotechnol. Rev.* Feb 1, **2013,** *2* (1), 47–57.
14. Bottino, M. C.; Thomas, V.; Schmidt, G.; Vohra, Y. K.; Chu, T. M.; Kowolik, M. J.; Janowski, G. M. Recent Advances in the Development of GTR/GBR Membranes for Periodontal Regeneration—a Materials Perspective. *Dental Mater.* July **2012,** *28* (7), 703–721.
15. Chen, J.; Hamon, M. A.; Hu, H.; Chen, Y.; Rao, A. M.; Eklund, P. C.; Haddon, R. C. Solution Properties of Single-Walled Carbon Nanotubes. *Science* Oct 2, **1998,** *282* (5386), 95–98.
16. Chen, Y. S.; Tsou, P. C.; Lo, J. M.; Tsai, H. C.; Wang, Y. Z.; Hsiue, G. H. Poly (N-Isopropylacrylamide) Hydrogels with Interpenetrating Multiwalled Carbon Nanotubes for Cell Sheet Engineering. *Biomaterials* Oct 1, **2013,** *34* (30), 7328–7334.
17. Cho, Y.; Borgens, R. B. The Effect of an Electrically Conductive Carbon Nanotube/ Collagen Composite on Neurite Outgrowth of PC12 Cells. *J. Biomed. Mater. Res. Part A* Nov **2010,** *95* (2), 510–517.
18. Coleman, J. N.; Khan, U.; Blau, W. J.; Gun'ko, Y. K. Small But Strong: A Review of the Mechanical Properties of Carbon Nanotube–Polymer Composites. *Carbon* Aug 1, **2006,** *44* (9), 1624–1652.

19. Dey, R. S.; Bera, R. K.; Raj, C. R. Nanomaterial-Based Functional Scaffolds for Amperometric Sensing of Bioanalytes. *Analyt. Bioanalyt. Chem.* Apr 1 **2013,** *405* (11), 3431–3448.
20. Dolati, F.; Yu, Y.; Zhang, Y.; De Jesus, A. M.; Sander, E. A.; Ozbolat, I. T. In Vitro Evaluation of Carbon-Nanotube-Reinforced Bioprintable Vascular Conduits. *Nanotechnology* Mar 14, **2014,** *25* (14), 145101.
21. Edwards, S. L.; Church, J. S.; Werkmeister, J. A.; Ramshaw, J. A. Tubular Micro-Scale Multiwalled Carbon Nanotube-Based Scaffolds for Tissue Engineering. *Biomaterials* Mar 1 **2009**, *30* (9), 1725–1731.
22. Edwards, S. L.; Werkmeister, J. A.; Ramshaw, J. A. Carbon Nanotubes in Scaffolds for Tissue Engineering. *Expert Rev. Med. Devices* Sep 1, **2009,** *6* (5), 499–505.
23. Eick, S.; Renatus, A.; Heinicke, M.; Pfister, W.; Stratul, S. I.; Jentsch, H. Hyaluronic Acid as an Adjunct after Scaling and Root Planing: A Prospective Randomized Clinical Trial. *J. Periodontol.* Jul **2013**, *84* (7), 941–949.
24. Fan, X.; Jiao, G.; Gao, L.; Jin, P.; Li, X. The Preparation and Drug Delivery of a Graphene–Carbon Nanotube–Fe_3O_4 Nanoparticle Hybrid. *J. Mater. Chem. B* **2013,** *1* (20), 2658–2664.
25. Feng QM, Pan JB, Zhang HR, Xu JJ, Chen HY. Disposable Paper-Based Bipolar Electrode for Sensitive Electrochemiluminescence Detection of a Cancer Biomarker. *Chem. Commun.* Aug 21 **2014,** *50* (75), 10949–10951.
26. Gaunt, N. P.; Patil-Sen, Y.; Baker, M. J.; Kulkarni, C. V. Carbon Nanotubes for Stabilization of Nanostructured Lipid Particles. *Nanoscale* **2015,** *7* (3), 1090–1095.
27. Gheith, M. K.; Pappas, T. C.; Liopo, A. V.; Sinani, V. A.; Shim, B. S.; Motamedi, M.; Wicksted, J. P.; Kotov, N. A. Stimulation of Neural Cells by Lateral Currents in Conductive Layer-By-Layer Films of Single-Walled Carbon Nanotubes. *Adv. Mater.* Nov 17, **2006,** *18* (22), 2975–2979.
28. Holzinger, M.; Le Goff, A.; Cosnier, S. Nanomaterials for Biosensing Applications: A Review. *Front. Chem.* Aug 27, **2014,** *2*, 63.
29. Hopley, E. L.; Salmasi, S.; Kalaskar, D. M.; Seifalian, A. M. Carbon Nanotubes Leading the Way Forward in New Generation 3D Tissue Engineering. *Biotechnol. Adv.* Sep 1, **2014,** *32* (5), 1000–1014.
30. Hsu, Y. T.; Wang, H. L. How to Select Replacement Grafts for Various Periodontal and Implant Indications. *Clin. Adv. Periodontics* Aug **2013**, *3* (3), 167–179.
31. Huang, Y. J.; Wu, H. C.; Tai, N. H.; Wang, T. W. Carbon Nanotube Rope with Electrical Stimulation Promotes the Differentiation and Maturity of Neural Stem Cells. *Small.* Sep 24, **2012,** *8* (18), 2869–2877.
32. Iverson, N. M.; Barone, P. W.; Shandell, M.; Trudel, L. J.; Sen, S.; Sen, F.; Ivanov, V.; Atolia, E.; Farias, E.; McNicholas, T. P.; Reuel, N. In Vivo Biosensing via Tissue-Localizable Near-Infrared-Fluorescent Single-Walled Carbon Nanotubes. *Nat. Nanotechnol.* Nov **2013,** *8* (11), 873–880.
33. Jell, G.; Verdejo, R.; Safinia, L.; Shaffer, M. S.; Stevens, M. M.; Bismarck, A. Carbon Nanotube-Enhanced Polyurethane Scaffolds Fabricated by Thermally Induced Phase Separation. *J. Mater. Chem.* **2008,** *18* (16), 1865–1872.
34. Jin, H.; Heller, D. A.; Kalbacova, M.; Kim, J. H.; Zhang, J.; Boghossian, A. A.; Maheshri, N.; Strano, M. S. Detection of Single-Molecule H_2O_2 Signalling from

Epidermal Growth Factor Receptor Using Fluorescent Single-Walled Carbon Nanotubes *Nat. Nanotechnol.* Apr **2010,** *5* (4), 302–309.
35. Kesharwani, P.; Jain, K.; Jain, N. K. Dendrimer as Nanocarrier for Drug Delivery. *Progr. Polym. Sci.* Feb 1, **2014,** *39* (2), 268–307.
36. Kierny, M. R.; Cunningham, T. D.; Kay, B. K. Detection of Biomarkers Using Recombinant Antibodies Coupled to Nanostructured Platforms. *Nano Rev.* Jan 1, **2012,** *3* (1), 17240.
37. Kumar, S.; Ahlawat, W.; Kumar, R.; Dilbaghi, N. Graphene, Carbon Nanotubes, Zinc Oxide and Gold as Elite Nanomaterials for Fabrication of Biosensors for Healthcare. Biosens. Bioelectron. Aug 15, **2015,** *70*, 498–503.
38. Lerner, M. B.; D'Souza, J.; Pazina, T.; Dailey, J.; Goldsmith, B. R.; Robinson, M. K.; Johnson, A. C. Hybrids of a Genetically Engineered Antibody and a Carbon Nanotube Transistor for Detection of Prostate Cancer Biomarkers. *ACS Nano.* Jun 26 **2012,** *6* (6), 5143–5149.
39. Li, G.; Liao, J. M.; Hu, G. Q.; Ma, N. Z.; Wu, P. J. Study of Carbon Nanotube Modified Biosensor for Monitoring Total Cholesterol in Blood. *Biosens. Bioelectron.* Apr 15, **2005,** *20* (10), 2140–2144.
40. Li, R.; Wu, R. A.; Zhao, L.; Hu, Z.; Guo, S.; Pan, X.; Zou, H. Folate and Iron Difunctionalized Multiwall Carbon Nanotubes as Dual-Targeted Drug Nanocarrier to Cancer Cells. *Carbon* Apr 1. **2011,** *49* (5), 1797–1805.
41. Li, X.; Gao, H.; Uo, M.; Sato, Y.; Akasaka, T.; Feng, Q.; Cui, F.; Liu, X.; Watari, F. Effect of Carbon Nanotubes on Cellular Functions in Vitro. *J. Biomed. Mater. Res. Part A* Oct **2009,** *91* (1), 132–139.
42. Lin, Y.; Lu, F.; Tu, Y.; Ren, Z. Glucose Biosensors Based on Carbon Nanotube Nanoelectrode Ensembles. *Nano Lett.* Feb 11, **2004,** *4* (2), 191–195.
43. Liu, M.; Chen, B.; Xue, Y.; Huang, J.; Zhang, L.; Huang, S.; Li, Q.; Zhang, Z. Polyamidoamine-Grafted Multiwalled Carbon Nanotubes for Gene Delivery: Synthesis, Transfection and Intracellular Trafficking. *Bioconjugate Chem.* Nov 16, **2011,** *22* (11), 2237–2243.
44. Lizundia, E.; Sarasua, J. R.; D'Angelo, F.; Orlacchio, A.; Martino, S.; Kenny, J. M.; Armentano, I. Biocompatible Poly (L-lactide)/MWCNT Nanocomposites: Morphological Characterization, Electrical Properties, and Stem Cell Interaction. *Macromol. Biosci.* Jul **2012,** *12* (7), 870–881.
45. Mackle, J. N.; Blond, D. J.; Mooney, E.; McDonnell, C.; Blau, W. J.; Shaw, G.; Barry, F. P.; Murphy, J. M.; Barron, V. In Vitro Characterization of an Electroactive Carbon-Nanotube-Based Nanofiber Scaffold for Tissue Engineering. *Macromol. Biosci.* Sep 9, **2011,** *11* (9), 1272–1282.
46. Malhotra, R.; Patel, V.; Vaqué, J. P.; Gutkind, J. S.; Rusling, J. F. Ultrasensitive Electrochemical Immunosensor for Oral Cancer Biomarker IL-6 Using Carbon Nanotube Forest Electrodes and Multilabel Amplification. *Analyt. Chem.* Apr 15, **2010,** *82* (8), 3118–3123.
47. Mamalis, A. G.; Vogtländer, L. O.; Markopoulos, A. Nanotechnology and Nanostructured Materials: Trends in Carbon Nanotubes. *Prec. Eng.* Jan 1, **2004,** *28* (1), 16–30.

48. Martins-Júnior, P. A.; Alcântara, C. E.; Resende, R. R.; Ferreira, A. J. Carbon Nanotubes: Directions and Perspectives in Oral Regenerative Medicine. *J. Dental Res.* Jul **2013**, *92* (7), 575–583.
49. Mattson, M. P.; Haddon, R. C.; Rao, A. M. Molecular Functionalization of Carbon Nanotubes and Use as Substrates for Neuronal Growth. *J. Mol. Neurosci.* Jun 1, **2000,** *14* (3), 175–182.
50. Mehra, N. K.; Jain, N. K. Multifunctional Hybrid-Carbon Nanotubes: New Horizon in Drug Delivery and Targeting. *J. Drug Target.* Apr 20, **2016,** *24* (4), 294–308.
51. Mendes, R. M.; Silva, G. A.; Caliari, M. V.; Silva, E. E.; Ladeira, L. O.; Ferreira, A. J. Effects of Single Wall Carbon Nanotubes and Its Functionalization with Sodium Hyaluronate on Bone Repair. *Life Sci.* Aug 14, **2010**, *87* (7–8), 215–222.
52. Mendes, R. M.; Silva, G. A.; Lima, M. F.; Calliari, M. V.; Almeida, A. P.; Alves, J. B.; Ferreira, A. J. Sodium Hyaluronate Accelerates the Healing Process in Tooth Sockets of Rats. *Arch. Oral Biol.* Dec 1, **2008**, *53* (12), 1155–1162.
53. Meng, J.; Han, Z.; Kong, H.; Qi, X.; Wang, C.; Xie, S.; Xu, H. Electrospun Aligned Nanofibrous Composite of MWCNT/Polyurethane to Enhance Vascular Endothelium Cells Proliferation and Function. *J. Biomed. Mater. Res. Part A* Oct **2010,** *95* (1), 312–320.
54. Moghimi, S. M.; Hunter, A. C.; Murray, J. C. Long-Circulating and Target-Specific Nanoparticles: Theory to Practice. *Pharmacol. Rev.* Jun 1, **2001,** *53* (2), 283–318.
55. Moore, V. C.; Strano, M. S.; Haroz, E. H.; Hauge, R. H.; Smalley, R. E.; Schmidt, J.; Talmon Y. Individually Suspended Single-Walled Carbon Nanotubes in Various Surfactants. *Nano Lett.* Oct 8, **2003,** *3* (10), 1379–1382.
56. Murugan, E.; Vimala, G. Effective Functionalization of Multiwalled Carbon Nanotube with Amphiphilic Poly (Propyleneimine) Dendrimer Carrying Silver Nanoparticles for Better Dispersability and Antimicrobial Activity. *Journal Colloid Interf. Sci.* May 15, **2011,** *357* (2), 354–365.
57. Mwenifumbo, S.; Shaffer, M. S.; Stevens, M. M. Exploring Cellular Behaviour with Multi-Walled Carbon Nanotube Constructs. *J. Mater. Chem.* **2007,** *17* (19), 1894–1902.
58. Namgung, S.; Baik, K. Y.; Park, J.; Hong, S. Controlling the Growth and Differentiation of Human Mesenchymal Stem Cells by the Arrangement of Individual Carbon Nanotubes. *ACS Nano* Sep 27, **2011,** *5* (9), 7383–7390.
59. Neelgund, G. M.; Oki, A.; Luo, Z. Antimicrobial Activity of CdS and Ag_2S Quantum Dots Immobilized on Poly (Amidoamine) Grafted Carbon Nanotubes. *Colloids Surf. B: Biointerf.* Dec 1, **2012,** *100*, 215–221.
60. Ovádeková, R.; Jantová, S.; Letašiová, S.; Štepánek, I.; Labuda, J. Nanostructured Electrochemical DNA Biosensors for Detection of the Effect of Berberine on DNA from Cancer Cells. *Analyt. Bioanalyt. Chem.* Dec 1, **2006,** *386* (7–8), 2055–2062.
61. Pan, B. F.; Cui, D. X.; Xu, P.; Huang, T.; Li, Q.; He, R.; Gao, F. Cellular Uptake Enhancement of Polyamidoamine Dendrimer Modified Single Walled Carbon Nanotubes. *J. Biomed. Pharm. Eng.* **2007,** *1* (1), 13–16.
62. Pan, B.; Cui, D.; Xu, P.; Ozkan, C.; Feng, G.; Ozkan, M.; Huang, T.; Chu, B.; Li, Q.; He, R.; Hu, G. Synthesis and Characterization of Polyamidoamine Dendrimer-Coated Multi-Walled Carbon Nanotubes and Their Application in Gene Delivery Systems. *Nanotechnology* Mar 3, **2009,** *20* (12), 125101.

63. Patolsky, F.; Weizmann, Y.; Willner, I. Long-Range Electrical Contacting of Redox Enzymes by SWCNT Connectors. *Angew. Chem.* Apr 13, **2004,** *116* (16), 2165–2169.
64. Pei, B.; Wang, W.; Dunne, N.; Li, X. Applications of Carbon Nanotubes in Bone Tissue Regeneration and Engineering: Superiority, Concerns, Current Advancements, and Prospects. *Nanomaterials* Oct, **2019,** *9* (10), 1501.
65. Price, R. L.; Waid, M. C.; Haberstroh, K. M.; Webster, T. J. Selective Bone Cell Adhesion on Formulations Containing Carbon Nanofibers. *Biomaterials* May 1, **2003,** *24* (11), 1877–1887.
66. Qian, D.; Dickey, E. C.; Andrews, R.; Rantell, T. Load Transfer and Deformation Mechanisms in Carbon Nanotube-Polystyrene Composites. *Appl. Phys. Lett.* May 15, **2000,** *76* (20), 2868–2870.
67. Regev, O.; Barenholz, Y.; Peretz, S.; Zucker, D.; Bavli-Felsen, Y. Can Carbon Nanotube–Liposome Conjugates Address the Issues Associated with Carbon Nanotubes in Drug Delivery? *Fut. Med. Chem.* Apr **2013,** *5* (5), 503–505.
68. Sa, M. A.; Andrade, V. B.; Mendes, R. M.; Caliari, M. V.; Ladeira, L. O.; Silva, E. E.; Silva, G. A.; Corrêa-Júnior, J. D.; Ferreira, A. J. Carbon Nanotubes Functionalized with Sodium Hyaluronate Restore Bone Repair in Diabetic Rat Sockets. *Oral Dis.* Jul **2013,** *19* (5), 484–493.
69. Sato, M.; Webster, T. J. Nanobiotechnology: Implications for the Future of Nanotechnology in Orthopedic Applications. *Expert Rev. Med. Devices* Sep 1, **2004,** *1* (1), 105–114.
70. Shi, X.; Sitharaman, B.; Pham, Q. P.; Liang, F.; Wu, K.; Billups, W. E.; Wilson, L. J.; Mikos, A. G. Fabrication of Porous Ultra-Short Single-Walled Carbon Nanotube Nanocomposite Scaffolds for Bone Tissue Engineering. *Biomaterials* Oct 1, **2007,** *28* (28), 4078–4090.
71. Shin, S. R.; Jung, S. M.; Zalabany, M.; Kim, K.; Zorlutuna, P.; Kim, S. B.; Nikkhah, M.; Khabiry, M.; Azize, M.; Kong, J.; Wan, K. T. Carbon-Nanotube-Embedded Hydrogel Sheets for Engineering Cardiac Constructs and Bioactuators. *ACS Nano* Mar 26, **2013,** *7* (3), 2369–2380.
72. Shobha, B. N.; Muniraj, N. J. Design, Modeling and Performance Analysis of Carbon Nanotube with DNA Strands as Biosensor for Prostate Cancer. *Microsyst. Technolog.* Apr 1, **2015,** *21* (4), 791–800.
73. Smith, A. M.; Mohs, A. M.; Nie, S. Tuning the Optical and Electronic Properties of Colloidal Nanocrystals by Lattice Strain. *Nat. Nanotechnol.* Jan **2009,** *4* (1), 56–63.
74. Tîlmaciu, C. M.; Morris, M. C. Carbon Nanotube Biosensors. *Front. Chem.* **2015,** *3*, 59.
75. Tothill, I. E. Biosensors for Cancer Markers Diagnosis. *Semin. Cell Dev. Biol.* Feb 1, **2009,** *20* (1), 55–62.
76. Tsujita, Y.; Maehashi, K.; Matsumoto, K.; Chikae, M.; Takamura, Y.; Tamiya, E. Microfluidic and Label-Free Multi-Immunosensors Based on Carbon Nanotube Microelectrodes. *Jpn J. Appl. Phys.* Jun 22, **2009,** *48* (6S), 06FJ02.
77. Veetil, J. V.; Ye, K. Development of Immunosensors Using Carbon Nanotubes. *Biotechnol. Progr.* **2007,** *23* (3), 517–531.
78. Wan, Y.; Deng, W.; Su, Y.; Zhu, X.; Peng, C.; Hu, H.; Peng, H.; Song, S.; Fan, C. Carbon Nanotube-Based Ultrasensitive Multiplexing Electrochemical Immunosensor for Cancer Biomarkers. *Biosens. Bioelectron.* Dec 15, **2011,** *30* (1), 93–99.

79. Wang, X.; Wang, C.; Qu, K.; Song, Y.; Ren, J.; Miyoshi, D.; Sugimoto, N.; Qu, X. Ultrasensitive and Selective Detection of a Prognostic Indicator in Early-Stage Cancer Using Graphene Oxide and Carbon Nanotubes. *Adv. Funct. Mater.* Nov 23, **2010,** *20* (22), 3967–3971.
80. Wang, Z.; Dai, Z. Carbon Nanomaterial-Based Electrochemical Biosensors: An Overview. *Nanoscale* **2015,** *7* (15), 6420–6431.
81. Wen, S.; Liu, H.; Cai, H.; Shen, M.; Shi, X. Drug Delivery: Targeted and pH-Responsive Delivery of Doxorubicin to Cancer Cells Using Multifunctional Dendrimer-Modified Multi-Walled Carbon Nanotubes (Adv. Healthcare Mater. 9/2013). *Adv. Healthcare Mater.* Sep **2013,** *2* (9), 1181.
82. You, Y. Z.; Yan, J. J.; Yu, Z. Q.; Cui, M. M.; Hong, C. Y.; Qu, B. J. Multi-Responsive Carbon Nanotube Gel Prepared via Ultrasound-Induced Assembly. *J. Mater. Chem.* **2009,** *19* (41), 7656–7660.
83. Yuan, W.; Jiang, G.; Che, J.; Qi, X.; Xu, R.; Chang, M. W.; Chen, Y.; Lim, S. Y.; Dai, J.; Chan-Park, M. B. Deposition of Silver Nanoparticles on Multiwalled Carbon Nanotubes Grafted with Hyperbranched Poly (Amidoamine) and Their Antimicrobial Effects. *J. Phys. Chem. C* Dec 4, **2008,** *112* (48), 18754–18759.
84. Zelada-Guillén, G. A.; Tweed-Kent, A.; Niemann, M.; Göringer, H. U.; Riu, J.; Rius, F. X. Ultrasensitive and Real-Time Detection of Proteins in Blood Using a Potentiometric Carbon-Nanotube Aptasensor. *Biosens. Bioelectron.* Mar 15, **2013,** *41*, 366–371.
85. Zerda, A. D.; Liu, Z.; Bodapati, S.; Teed, R.; Vaithilingam, S.; Khuri-Yakub, B. T.; Chen, X.; Dai, H.; Gambhir, S. S. Ultrahigh Sensitivity Carbon Nanotube Agents for Photoacoustic Molecular Imaging in Living Mice. *Nano Lett.* Jun 9, **2010,** *10* (6), 2168–2172.
86. Zhang, F.; Xia, Y.; Xu, L.; Gu, N. Surface Modification and Microstructure of Single-Walled Carbon Nanotubes for Dental Resin-Based Composites. *J. Biomed. Mater. Res. B Appl. Biomater.* **2008,** *86* (1), 90–97.
87. Zhang, L.; Gu, F. X.; Chan, J. M.; Wang, A. Z.; Langer, R. S.; Farokhzad, O. C. Nanoparticles in Medicine: Therapeutic Applications and Developments. *Clin. Pharmacol. Therap.* **2008,** *83* (5), 761–769.
88. Zhang, X.; Wang, X.; Lu, Q.; Fu, C. Influence of Carbon Nanotube Scaffolds on Human Cervical Carcinoma HeLa Cell Viability and Focal Adhesion Kinase Expression. *Carbon* Mar 1, **2008,** *46* (3), 453–460.
89. Zhao, B.; Hu, H.; Mandal, S. K.; Haddon, R. C. A Bone Mimic Based on the Self-Assembly of Hydroxyapatite on Chemically Functionalized Single-Walled Carbon Nanotubes. *Chem. Mater.* Jun 14, **2005,** *17* (12), 3235–3241.
90. Zhao, Q.; Gan, Z.; Zhuang, Q. Electrochemical Sensors Based on Carbon Nanotubes. *Electroanalysis: Int. J. Devoted Fundamental Practical Aspects Electroanalysis* Dec **2002,** *14* (23), 1609–1613.

CHAPTER 4

CNT STRUCTURE AND ITS APPLICATION IN DISPLAY TECHNOLOGY

V. N. ARCHANA[1*], N. G. DIVYA[1], V. N. ANJANA[2], and REYHA BENEDICT[3]

[1]*Mar Athanasius College, Kothamangalam, India*

[2]*Sree Sankara Vidyapeetom College, Perumbavoor 683556, Kerala, India*

[3]*St. Teresa's College, Ernakulam 682011, Kerala, India*

**Corresponding author. E-mail: archanavn90@gmail.com*

ABSTRACT

This chapter gives a detailed overview of the structure and some functional characterization of carbon nanotubes (CNTs) in a nutshell. CNTs are a type of carbon-based nanoparticles with an elongated tubular structure of diameter ranging 1–2 nm with the carbon atoms arranged in coaxial cylinders of graphitic sheets. Based on the number of graphite layers, CNTs are classified into single-walled CNTs , double-walled CNTs ,and multiwalled CNTs. The structure of CNTs is unique and it provides unbreakable hexagonal structure which offers unique mechanical, electrical, and thermal properties to the CNTs. Advanced research in the field of CNTs synthesis paved the way for the betterment of synthesis strategies with the high-temperature techniques, such as arc discharge, laser ablation, and chemical vapor deposition have been replaced by low-temperature methods. It is known that the chemical, electrical, and physical properties of CNTs can be altered by inducing change in the structural parameters. This chapter also deals with the fascinating properties of CNTs in the electronic display applications. CNTs are versatile in application since their surface can be chemically modified using various specific functional groups to make it more soluble and biocompatible. CNT offers a platform for the study of intrinsic properties of nanoparticles. The strength and flexibility of CNTs have been widely acclaimed and it makes the CNTs to play a significant role in nanotechnology engineering.

4.1 INTRODUCTION

Carbon nanotube (CNT) is an allotrope of carbon with specialized structures of sp^2 hybridized carbon atoms having a whole range of promising applications due to their incredible features which make them indispensable in many fields. They are macromolecules consisting of carbon atoms arranged in coaxial cylinders of graphitic sheets which can be categorized as a quasi-one-dimensional nanomaterial. Each cylindrical structure of CNTs consists of several hexagons of carbon atoms (honeycomb) arranged in a helical manner about the axis of an elongated tubular structure with both ends usually capped by fullerene-like structures.[1–3]

The CNTs can be broadly categorized into two classes, such as single-walled CNTs (SWCNTs) and multiwalled CNTs (MWCNTs). As the name indicates, the SWCNTs consist of a single layer of sp^2 hybridized carbon atoms with diameters typically in the range of 1–2 nm. And, the MWCNTs consist of several layers of carbon atoms (usually 2–50 layers) rolled up into concentric cylinders that are larger than the SWCNTs.[2] The MWCNTs consist of several single-walled cylindrical structures which are stacked one inside the other with an outer diameter less than 15 nm that distinguish it from carbon fibers, that are not single molecules rather strands of layered graphite sheets.[3]

The MWCNT was discovered accidentally by Iijima in the year 1991 while investigating the surfaces of the graphite electrode used in an electric arc discharge.[2] About 2 years later, he reported the development of SWCNT.[4] Further investigations on the structure and properties of CNTs have given a new dimension to carbon research which also complemented the investigations then prevalent in fullerene research, another allotrope of carbon. During that time itself, the incredible strength and many of the fascinating properties of these tiny CNT structures of carbon appeared to overtake many other nanomaterials in the race to advanced applications.

High quality, uniform size CNTs with low structural and chemical defects are desired for technological applications. Thus, the number of patents and publications on the CNT synthesis methods is increasing rapidly. However, many challenges persist in this regard, such as the mass production of high-quality nanotubes, proper control over the structure and properties of produced CNTs, control over the location and orientation, growth mechanism, etc. Thus, several kinds of researches are still going on in this direction.

The different methods commonly adopted for the synthesis of CNTs include arc discharge, laser ablation, and chemical vapor deposition (CVD), where the applied energy produces fragments of carbon source that can be reorganized to generate CNTs in the presence of a suitable catalyst. The high-temperature techniques, such as arc discharge and laser ablation were used initially to produce CNTs, but recently these synthesis processes have been replaced by low-temperature CVD, because of the alignment, orientation, diameter, nanotube length, purity and density of CNTs that can be precisely controlled in the later growth process. The other CNT growth processes include liquid pyrolysis, bottom-up organic approach, gas-phase methods etc.[5]

Whatever synthesis method is applied, the CNTs always have some impurities whose type and amount depend on the synthesis process. The impurities include other carbonaceous particles, such as graphite, graphene, fullerenes, amorphous carbon, and metal particles introduced as catalysts during synthesis. These impurities usually interfere with the desired properties and cause a serious impediment in applications. Thus, the development of efficient and simple purification methods is one of the important challenges. The most commonly adopted purification process is the acid treatment of the synthesized CNTs.[6,7]

As in the case of any nanoscale materials, the chemical, electrical, and physical properties of CNTs depend heavily on their structure. As an example, the SWCNTs can behave either as a semiconductor or metal depending on the chirality and diameter.[8] Moreover, SWCNTs exhibit ultrahigh intrinsic mobility (approximately 10^5 cm^2/V), high absorption coefficient (10^4–10^5 cm^{-1}), and the excitons diffuse more than 0.1 μm before the recombination, which enables the SWCNT as an ultrathin active material for the next-generation displays and photodetectors.[8,9]

A significant amount of work has been done in the past decades to invoke the unique properties of CNTs.[10–12] They possess exceptional electronic, optical, optoelectronic, thermal, chemical, and mechanical properties, which open up promising platforms for high-performance nanodevices.[8] These unique properties of CNTs arise from its structure and inherent subtlety in the structure associated with the helically arranged hexagonal array of carbon atoms. The metallic or semiconducting properties of CNTs depend on the diameter and arrangement of hexagonal carbon rings along the length of the tubular structure.[3] The helicity and diameter of the CNTs induce significant changes in the electronic density of states which contribute to its unique electronic characteristics.[3] The novel electronic properties provide an eminent platform for the development of electronic device applications.

The unique mechanical properties, such as stiffness, stability, and strength arise from the structure, size, and topology of CNTs. The strength of sp^2 carbon–carbon bonds contributes to the amazing mechanical properties of CNTs. The highest tensile strength or breaking strain of CNT is close to 60 GPa which is almost 50 times higher than steel and even the weakest itself has the strength of several GPa.[13,14] The defect structure of CNTs, filling the nanotubes with foreign materials and functionalization of ends enhance the reactivity of the tube ends, which alters the chemical, thermal, and optical properties.[3] The thermal conductivity of CNTs is very high close to ~3000 W/m/K, which is comparable to that of diamond. Moreover, the small size and large surface area-to-volume ratio of CNTs coupled with the lightness make them a potential candidate for prominent applications in aerospace. The topology or closed nature of individual nanotubes causes to disappear certain aspects of anisotropic properties of graphite that contribute to the physical properties of CNTs.

The interesting physicochemical properties make CNTs promising electron emitters in electronic field emission displays (FEDs), scanning probe microscopy tips, actuators and fillers, reinforcement fillers in nanocomposites as well as molecular-scale components in nano- and microelectronic devices.[3] The CNTs, as a typical one-dimensional (1-D) material, have been widely investigated in various other fields such as field-effect transistors (FET), photodetectors, electronic displays, wastewater treatment, solar cells, gas sensors, etc.[8] The high surface area, antimicrobial activity, active strong adsorption sites, and low toxicity make CNTs widely acceptable for disinfection, adsorption, and microbial control.[1] The CNTs have also been attracting considerable attention due to their potential applications in various display devices, such as field emitters in FEDs, polarizers in LEDs, transparent electrodes, and thin film transistors (TFTs) in OLEDs, incandescence displays, thermochromic displays, and flexible fiber displays.[15] The aligned CNT arrays lead to anisotropic optical absorption, which has been intensively investigated as optical polarizers in LEDs. The aligned CNT films also possess interesting properties suitable for other display applications, such as incandescence displays, thermochromic displays, and flexible fiber displays.[16,17]

The aligned CNTs can be synthesized by CVD, fabric CNTs woven from a CNT forest, and also by using an external shear force or electric/magnetic field to re-orient the CNTs dispersed in a suitable medium.[18] The two-dimensional mesh of SWCNTs have been applied as optically transparent and electrically conducting electrodes for LEDs, liquid crystal displays (LCDs), and touch screens due to their high work function, mechanical flexibility, which is more favorable in comparison with the traditional indium tin oxide (ITO) electrodes in terms of mechanical properties.[19] Widespread development continues for the development of CNT-based transparent conducting films as a promising alternative to ITO which is expensive because of the scarcity of indium, compounded by increasing demand for touch-screen devices, displays, and photovoltaics. Besides cost, the flexibility and transparency of CNTs are major advantages over brittle ITOs for flexible display applications. Moreover, the CNT conductors can be deposited from solution (e.g., ultrasonic spraying, slot-die coating) and patterned by cost-effective nonlithographic methods (e.g., microplotting, screen printing). The recent development for commercial applications has resulted in SWCNT films with a sheet resistivity of 100 ohms per square and 90% transparency. This resistivity is suitable for some applications but higher than that of equally transparent, optimally doped ITOs.[15,20]

To meet the requirements demanded by any specific applications, chemical modification of CNTs are essential. But, it is very difficult to engineer the surface characteristics of CNTs depending on the requirements demanded by certain specific applications. Thus, the controlled surface modifications and interfacial engineering are essential for developing CNT materials for advanced applications. Recent researches on CNTs have been focused on improving the quality and many of these functional properties of constituent nanotubes. Continued researches on

CNTs should overcome some of these major hurdles that these materials are facing in the race to the technological marketplace.

4.2 STRUCTURAL FEATURES OF CNTs

CNTs are a major class of carbon-based NPs with a tubular structure with the diameter ranges in a nanometer scale and length in micrometer scale. Other than CNTs, fullerenes constitute another major class of carbon-based NPs. There were no reports of carbon nanotubes till 1991. CNTs are smaller in size.[2] MWCNTs are the first known carbon tubes synthesized using arc-discharge methods by Bethune et al. in 1991,[21] followed by SWCNTs using the same method by Iijima and Ichihashi[4] independently in 1993. SWCNTs exist in crystalline ropes and consist of tens to hundreds of tubes of similar diameter and chirality, packed in a triangular lattice. CNTs are composed of pure carbon atoms interacting with each other by strong sp^2 chemical bonds and form the unique geometric carbon network which is responsible for the electronic, mechanical, and thermal properties. The strong chemical bonding of carbon network enables the CNTs to have a strong mechanical modulus, high thermal transport, as well as remarkable electrical properties.

CNTs have been classified into single-walled nanotubes, double-walled nanotubes, and multiwalled nanotubes depending on the number of sheets rolled up to make the tubes. If the tube consists of a single graphene sheet, it is called SWCNTs. In DWCNTs, two layers of graphene sheets folded upon each other to form a double layer structure, while in MWCNTs, folding of 2–10 graphene sheets or by rolling a single graphene sheet occurs.

4.2.1 SINGLE-WALLED CARBON NANOTUBES: GEOMETRIC STRUCTURE AND SYMMETRY

SWCNTs are tubular in structure formed by the mapping of the 2-D hexagonal lattice of a single sheet of graphene onto the surface of a cylinder. It was also noted that the boundary conditions of the cylinder will be satisfied only if one of the vectors of the Bravais lattice of the graphene sheet maps to the circumference around the cylinder.[22] Therefore, it is known that lattice vector of the 2D lattice theoretically determines the different ways of rolling up of the graphene sheets to make the tubes and it also has an influence on the properties. It is also known that the properties of nanotubes are due to the diameter and chirality to an extent. Geometric structure and chirality of a nanotube can be determined theoretically from the vector, $c_h = na_1 + ma_2$, where a_1 and a_2 represent the graphene lattice vectors and n and m integers, respectively (Fig. 4.1).

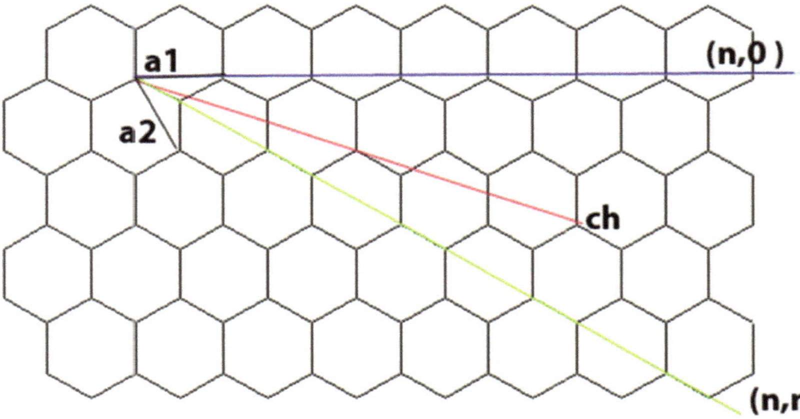

FIGURE 4.1 Schematic diagram of two-dimensional graphene sheet representing lattice vectors a1 and a2, and the roll-up vector ch = na1 + ma2. The vectors representing the cases of (n, 0) zigzag and (n, n) armchair tubes are indicated with blue and green lines, respectively.

It is known that the SWCNTs synthesized with the vector ch have an axial direction perpendicular to the lattice vector and the synthesis of various SWCNTs is based on the values of (n, m) indices in the vector.[23] Various single-walled nanotubes, such as zigzag SWCNTs and armchair SWCNTs are synthesized from graphene sheet rolling up along the directions equivalent to lattice indices of (n, 0), and (n, m) where n = m, respectively. Mapping constructions in directions equivalent to transition indices from (n, 0) to (n, m) by 30 degrees rotation generate chiral (n, m) SWCNTs, where $n \neq m \neq 0$. Typical examples of three types of SWCNTs of (5, 5), (9, 0), and (8, 2), are schematically drawn to illustrate the structures of armchair, zigzag, and chiral nanotubes, respectively.

4.2.2 SINGLE-WALLED CARBON NANOTUBES: ELECTRONIC PROPERTIES

The electronic properties of the CNTs are highly unusual owing to their unique geometric structure and even a small difference can alter the electronic properties of nanotubes. There are reports on the electronic behavior of SWCNTs being correlated with their geometric structure.[24–27] Although graphenes are semiconductors with zero-energy bandgap, CNTs made from these graphenes show an altered energy gap and their properties. CNTs are found to act as metals or semiconductors with tiny energy gaps induced by energy band folding. The metallic or semiconductor properties and the energy gap of carbon nanotubes is also influenced by the diameter and chirality, which represents the (n, m) indices of vector c_h. There are also studies indicating the effect of band folding and electronic properties on the structure of CNTs.[22,28] Graphene possess a unique band structure with a crossing of the states in the Fermi level at two inequivalent points in k-space. The electronic structure of a graphene sheet has been depicted near the Fermi level by an occupied p-band and an empty p*-band which possess a linear dispersion and reported to be meeting at the K-point in the Brillouin zone. An ideal graphene sheet consists of K points at each of its six corners while in the CNTs made from these graphene sheets by rolling up, only few k-states of the planar graphene are found owing to the limitation in the number of periodic boundary conditions imposed in a circumferential direction. Values of the k-vectors of tubes depends upon the diameter and chirality of tubes (5, 5), (7, 1), and (8, 0) allowed k-vectors are shown in solid lines. Various studies have shown that these indices determines the metallic properties if the allowed k contains K-point as shown (5, 5), (7, 1) tubes.

While the K-point is deviated and does not fall in the allowed k-states, for example, a tube (8, 0) behaves as a semiconductor with different energy bandgaps. It is worth noting that for both metallic and semiconducting tubes, the allowed states near the Fermi level are all becoming closest to the K-point, which determines the main electronic and transport properties of tubes. There are reports which claim the calculations to find out the metallicity of CNTs from the parameters, such as diameter and chirality of graphitic rings. For example, consider the case of zigzag SWCNTs with lattice vectors of (n, 0) and when n is divided by 3, zigzag (n, 0) SWCNTs are found to achieve a metallic nature from semiconductor. They are found to achieve armchair (n, n) with a metallic nature if there is a rotation of the vector c_h by 30 degrees relative to (n, 0), where (n =m). Chiral SWCNTs will form as c_h rotates away from (n, 0) to (n, n) and transition forms from zigzag to armchair tubes with various n and m indices are possible. Relative values of n and m determine the behavior of chiral tubes or zigzag tubes whether it is metallic or semiconductor. Typically, the energy of the electronic bandgap of semiconducting tubes is inversely proportional to diameters. In general, as the tube radius, R, increases, the bandgaps of semiconducting tubes will decrease with dependence of 1/R and 1/R,[21] respectively, because of the tube curvature effects.[22]

4.3 DISPLAYS BASED ON CNTs

4.3.1 POLARIZER IN LCDs

The emergence of advanced electronics, such as smartphones, smart televisions, and smart wearable has made a global impact on display devices. LC display is one of the major products that has presently dominated the display device market. It is a sort of light valve allowing the light to pass through when the molecules are made to align in a particular way and while obstructing when aligned in the contrary direction, that is, liquid crystal changes the polarization of the light passing through. These molecules are rod-like and orient themselves by an external manipulation of applying an electric field. An LCD structure predominantly consists of a polarizer, glass substrate, color filter, liquid crystal, pixel electrode, TFT and backlight[29] and works on the property called optical birefringence. A CNT, an elusive 1-D form of carbon having a peculiar geometrical shape with a large rod-like molecule that possess highly desirable and spectacular electrical, mechanical, and optical properties is more akin to a liquid crystal molecule.[30] Incorporating CNTs as guests into an assembly of LC molecules can ameliorate the physical properties of the latter.[31] One-dimensional movement of the electrons makes the CNT to emit polarized light.[32] Its anisotropic property makes it to be utilized as an optical polarizer and thereby finding applications that offer enthralling and remarkable journey in the field of LC displays.

An optical polarizer is always a prerequisite to various industrial applications which has to be flexible enough under different external pressures.

An aligned CNT oriented in a particular direction shows anisotropic optical absorption.[33] A highly ordered CNT will customarily be impotent under various external forces like pressure, bending, etc., and hence it is embedded inside a polymer matrix to necessitate the needed strength. But this can restrict the functionality of the material by decreasing the conductivity of the CNT as well as the anchoring effect of liquid crystals. In the context of this scenario, an active optical polarizer can be fabricated exhibiting multifunctionality, such as high polarization efficiency and high conductance while maintaining the anchoring effect of LC molecules. The polymer embedded CNT sheet (P-ECS) film is flexible, bendable, and transparent indeed to make utilization of in flexible and ultrathin displays.[18]

Theoretical prognostication of CNT as a polarizer suggests that it can switch between the two components, viz., parallel and perpendicular and thereby increasing the possibility of doubling the amount of transmitted information.[34] This can be carried out by utilizing the process called carrier doping where the aligned nanotubes function as a polarization switch.[34,35] It can be rotated virtually by 90° without any spatial rotation by adjusting the number of electrons integrated while doping.[34] Thus, a CNT polarizer can reverse the direction of the polarized light by transmitting the parallel polarized light and absorbing perpendicular polarized light w.r.t the electric vector propagation.[36,37]

Anisotropic absorption of light with the external field happens to be in the visible light, ultra violet and infra-red regions.[38] Considering UV region, CNT polarizer is capable to elongate the absorption of light up to deep UV region in a wider operating wavelength range from 300 to 400 nm.[39] This strongly depends upon the degree of alignment of CNTs in the film. By controlling the light modulation of visible light, it could possibly be applied to dynamic polarizers. CNT-based polarizers can be utilized for the production of profound light in light modulators with a relatively low transmittance.[40]

In order to avoid agglomeration and low volume concentration of CNTs in the film a stretched CNT clusters can be utilized. By regulating the elongation of the CNT clusters and applying an external electric field controlled polarization and tuneable light modulation is possible in the visible region.[40]

A CNT polarizer can also be developed using CNT yarns whereby it absorbs the beam of light parallel to the axis of CNTs. Owing to the small diameter of CNTs, the polarizer could work in the UV regions of wavelength in the range of 10 nm.[37]

A multilayer CNT forms a polarizer which is basically a MWCNT composite film. In the tetra hertz region, its alignment shows a strong dependency of electric field on the orientation of the sample.[41] A multi-layered CNT polarizer shows high stability even under long-time UV exposure. The degree of polarization (DOP) additionally increments with the number of CNT layers. This can result in a better liquid crystal alignment.[39] In a highly aligned absorbing film of SWCNT, absorption shows a high reliance on the tetra hertz polarization w.r.t its direction of alignment axis. A strong absorption can be observed when polarization is parallel to the oriented direction, which elucidates the high degree of alignment of CNTs. Thus, a superior class of tetra hertz polarizer can be composed out of highly aligned SWCNTs.[33]

Most of the LCDs adopt fringe-free switching (FFS) mode with photo-alignment for high transmittance and wide viewing angle.[42] In order to avoid the negative aspects engendered by the rubbing process, such as nonuniformity in alignment, light leakage and broken debris non rubbing photoalignment technique is used.[29,42] A CNT-based UV polarizer can be used as a polarizer is quintessential for photoalignment. By enhancing the efficiency of this, it can further be used to produce high-quality images in LCDs. They produce linearly polarized UV light instead of the conventionally used metallic nanowire-grid polarizer (WGP). It shows strong anisotropic optical property in the areas of UV, visible, and IR regions.[38,39] Thus, a photoaligned FFS cell with CNT-based UV polarizer can reduce the high cost as well as the flexibility toward external pressures.[39]

Other than using CNTs as a polarizer, they can also be incorporated in LCD as transparent electrode, FETs, and TFTs.[43–45]

4.3.2 INCANDESCENCE DISPLAYS

The incandescent display has been the basic process of light generation commonly used as alphanumeric displays in microelectronics. This includes a wide range of display types in which the illumination is provided by an array of incandescent filaments that can be characterized by a bright output, simple technology, and compatibility with ICs, but with very low operating speeds. This can also be used to display rolling characters with very low power consumption. The nanomaterials, especially thin films enhanced the possibility of using films driven by simple addressing circuits in incandescent displays with a very low response time compared with the traditional LCDs. The thin films have opened up the possibility of fabricating incandescent display devices on a large scale using films with uniform properties, which is currently being a topic of wide interest in the field of research in nanoscience.

Incandescent light from carbon fibers was an important source of illumination since the invention of electric light by Thomas Edison. But later, the carbon fibers were upgraded to tungsten as the source of incandescent light due to its long lifetime. The finding of CNTs with the advent of nanotechnology entitles a new opportunity for carbon materials to be used as an incandescent light source due to their unique properties coupled with the benefits arises from their nanosizes. Super aligned CNT arrays are novel materials, and the aligned CNT films can be directly drawn from such arrays because of the strong van der Waals interactions.[46]

The super aligned CNT thin films can be heated up to incandescence in a vacuum with very good response time (about 1 millisecond) even in the incandescence state. The high response speed of the CNT films can be attributed to the ultrasmall heat capacity per unit area (HCPUA) of CNT films, large surface area, and large radiation coefficient of CNTs. The high response speed of such materials renders the classical phenomenon of incandescent light emission which applies to information display. Such displays can be fabricated by developing a CNT film array on a substrate using laser cutting, screen printing, or pipe-free vacuum-sealing technologies.

The highest temperature of CNT films at constant heating current and the ramp-up and cooling-down times for that particular temperature fitted well with those obtained from the thermal conduction equation.[47] The ramp-up time is the time duration from 90% of the maximum of a heating voltage to 90% of the maximum of a photodiode signal. And the cooling-down time is the time duration from turning off of a heating voltage to a value of a photodiode signal slightly larger than its minimum (twice the standard deviation (SD) of the minimum that is larger than the minimum itself). The response of the CNT film is faster in comparison with the traditional tungsten wire since a 15 mm diameter tungsten wire needs about 100 milliseconds to switch into a stable incandescence state when an electric heating signal is applied.[47]

The freestanding aligned CNT film tailored to CNT strips using a laser beam and then shrunk into CNT thin yarns with ~1 μm in diameter are also reported to give incandescence with a fast high-temperature response. These yarns can improve the mechanical strength, electrical conductivity, and thermal conductance compared with the CNT thin films. The electrical heating and natural cooling of the suspended CNT yarns under vacuum causes the emission of an incandescent light from the hot thin yarns. These CNT yarns can be electrically heated in about 0.8 milliseconds up to 2170 K with a cool-down time of about 0.36 millisecond, which can be attributed to the ultra-small mass of the independent single yarn, thermal conductance through the two cool ends and large radiation coefficient.[46]

The CNT-based incandescent electrode structures can be easily developed by adopting the screen printing technology because the thickness of screen-printed electrodes easily reaches several tens of micrometers. The CNT film is to be suspended above the substrate to avoid the possible heat dissipation to the substrate to retain its fast response. A laser beam can be used to cut the as-developed CNT film into separate units in micrometer-scale range with very light mass (nanogram range) to reduce the power consumption.

The CNTs-based incandescent display can overcome several disadvantages of LCDs. The incandescence response time of CNT thin films is shorter than that of the LCDs. At the same time, the CNT film produces almost nonpolarized light, thus free from the viewing-angle problem commonly associated with LCDs. These kinds of CNT thin film displays are more suitable for use in fast-response electronic applications just like an information display.

The CNT thin yarns have advantages over the CNT films in the incandescent display applications, which includes low power consumption and driving voltage. The low driving voltage is due to the reduction of resistance, which is beneficial to reduce the cost of the driving circuits. Moreover, the temperature of the hot yarns is higher than that of the CNT films for the same brightness. Since the radiation power is proportional to T^4, the efficiency of incandescent emission can be improved by increasing the operating temperature. Thus, the reduction of power consumption of CNT yarns can be attributed to the increase in temperature. The CNT yarns can eliminate defects that are more often encountered with CNT films. The localized resistance arises due to the formation of defects which often appear at the edges of the CNT films during the laser-cutting process.

These kinds of defects will be more concealed if the CNT films are densified into yarns, because of the formation of more intertube contacts in yarns that can reduce the ununiformed resistance. Thus, the heat of hot spots can be efficiently dissipated into yarns through the neighboring contacted CNTs. That also contributes to the higher temperature endurance of the CNT yarns. The shrinking process reduces the surface area for carbon evaporation. This makes more stable incandescence emission possible from CNTs.[46]

4.3.3 THERMOCHROMIC DISPLAYS

Thermochromic material is a temperature-sensitive color change material which changes its color reversibly with a change in temperature. This material is widely being used as a temperature indicator in industrial sectors and has also been demonstrated to be useful in information displays. This material is commonly used in industrial sectors as an indicator of temperature variations and reaction heat of a chemical reaction to analyze the temperature distribution of a chemical heating apparatus, and temperature variation indicator in chemical containers, etc.

The CNTs show many fascinating properties that can be used in many electronic display applications, which essentially requires the large-scale production of uniform CNTs. Due to good conductivity, joule-heating effect, CNTs are promising nanoheaters to induce a color change for thermochromic applications. But, the difference in circumstances, temperature ranges, heating dissipation modes, and different load types usually induce a nonuniform thermal behavior to CNTs.[48] The difference in temperature and circumstances induce different modes of heat dissipation from CNTs which will influence its load behavior. The investigations on the influence of a heating load on the thermal response of CNT films as heater under thermal radiation and convection heat dissipation modes revealed that the thermal response slows down with the load. Using a phase change thermochromic pigment load, the thermal response also gets slow down with the phase change.[17] The CNT film microheater array can be developed by screenprinting and laser-cutting technology that can be heated addressable as required with the aid of a driving circuit.

A major problem associated with the use of CNTs for thermochromic display devices is the large-scale production of uniform CNTs. However, the development of super aligned CNT films and yarns overcome this issue which also has got several merits such as fast thermal response compared with films. The fast thermal response of super aligned CNT films and yarns arises due to their ultrasmall HCPUA, large surface area and high thermal emissivity. The flexible and transparent super aligned CNT films possess excellent properties ideal for their use as the low-cost heater materials with medium resistance, high work temperature (air $\approx$ 600 K; vacuum or protective ambient $\approx$ 2000 K), and good tractability. As the substrate and CNT films are flexible and transparent, the CNT-based thermochromic display can work equally well in both flat and curved states.[17]

The freestanding CNT films get destroyed easily, but after being coated on a substrate such as polyethylene terephthalate (PET), the CNT can even pass the scratch test due to the strong adherence to the substrate. However, one side effect of the PET substrate increases the HCPUA of the CNT film heater and causes a decrease in thermal response. In order to obtain a good balance between the mechanical robustness and thermal response, substrate thickness and coverage of CNT films on the substrate should be reduced.[17] It has been found that thinner the substrate, faster is the thermal response. Since heat dissipation increases inversely with the coverage, the decrease in the coverage of CNT film on the PET surface also causes an increase in thermal response.

4.3.4 FLEXIBLE FIBER DISPLAYS

Flexibility is a breakthrough in electronics, and flexible electronics or stretchable electronics is likely to be the future of electronic devices. Many of the leading companies are trying to incorporate this technology into their products, which will enable a paradigm in the design for electronic devices. The potential electronic applications of this technology include wearable electronic devices, compact and portable devices, compact biomedical devices, etc. which require the adaptation of electronic devices for flexibility, from power sources to display screens. Nanomaterials are playing a pivotal role in the development of flexible electronic devices especially displays. The research advancement in this direction has led to the development of CNT-based flexible supercapacitors as an alternative to normal batteries, CNT/polymer composites as an elastic conducting interconnector, CNTs, and copper or silver nanowires as flexible and transparent conductors, etc. for use in flexible devices and displays. In near future, carbon-based materials, such as CNT, graphene, etc. are likely to become a dominant material for the development of flexible electronic devices and displays due to many of their promising properties, such as electrical conductivity, flexibility, and physical strength.

The CNTs are a promising candidate for flexible display applications due to their high conductivity and excellent mechanical deformability which arises from the high aspect ratio of about 333.[49] The availability of direct conductive pathways for the transport of charges, high carrier mobility, chemical stability, and low-defect density associated with the grain boundaries are added benefits of their use as a display device. During the possible deformation, cracks are generated preferentially at the grain boundaries or on other defects. This produces a dramatic increase in their resistance due to the suppression of transport of charges. Thus, it is very significant to prevent the cracks to the maximum possible extent.

According to the percolation theory, the aspect ratio of a conductive material reduces the critical density that is required to form a conductive network. Thus, the 1D nanomaterials such as CNTs can achieve higher conductivity using less material than the materials with other morphology, which is due to their lower percolation threshold and higher aspect ratio. Because of these features, an added benefit of high transparency can be achieved due to the formation of voids in the conducting networks at a submonolayer level. Thus, the conducting network of CNTs intrinsically has good mechanical robustness that can provide both conductivity and transparency for flexible display applications. Several researchers have focused on the mechanical properties of CNTs and tried to use CNTs for electronic display devices. To make such flexible displays, various fabrication methods have been adopted such as spray coating, CVD, etc..[50,51]

Though the CNTs possess properties suitable for flexible and stretchable devices, they also have some obvious disadvantages, such as control of CNT diameter, chirality, and density, which should be resolved for commercial applications. The development of hybrid nanomaterials using CNTs as the organic counterparts is a promising route that complements the inherent disadvantages of these nanomaterials. With this concept, flexible electronic devices based on hybrid nanomaterials show promising performances, but they still require some technological improvements in the large-scale production with uniformity. By solving these issues, they are likely to become vitally important for the production of wearable electronics shortly.[52]

Several CNT/polymer composites have been investigated by dispersing CNTs in polymer matrices which essentially require the uniform dispersion of CNTs. Sekitani et al. reported the development of a printable and flexible elastic conductor by the dispersion of SWCNTs along with an ionic liquid in fluorinated rubber, and they demonstrated a stretchable organic light-emitting diode (OLED) display. However, several drawbacks persist which include the difficulties in synthesizing CNTs with uniform properties.[20] In most of the cases, the synthesized CNTs may be a mixture of metallic and semiconducting nanotubes with different chiralities and lengths. Thus, the selective mass production of CNTs with uniform properties by any of the known growth processes is difficult.[53] Due to the inhomogeneity of CNTs, in several situations, stable performances from the CNT-based devices cannot be expected which restricts the possibility of using CNTs in commercial devices.[6] It is also very difficult to dissolve and separate long tubes, and there is a lack of appropriate techniques for the uniform assembling of CNTs over large areas.[54]

4.4 CONCLUSION

The great piece of technology, carbon nanotube could anticipate the future demands of display technology by using its spectacular properties as various electronic displays have turned out to be ubiquitous in the recent years in different fields. As far as we have described, it is observed that in display devices, CNTs appear as polarizers and new displays like incandescence, thermo chromic, and flexible fibers. In spite of its expected growth during the forecast period, one of the greatest challenges confronted by using the CNTs is that the cost-efficient mass production. Their controlled depositions onto the substrate, as well as dissevering the metallic and semiconducting CNTs are still under challenge. CNTs are under the category of competitive materials in the flexible electronic field, such as organic polymers, inorganic polymers, and metallic nanowires and even with their own family, the 2-D form of carbon, graphene. The cost-effectiveness, the improved power efficiency and lifetime issues are some of the issues needed to be solved for mass production. Efforts are in progress to tackle the issue which could possibly be utilized in mass lighting devices.

Owing to the robustness of CNTs, a CNT-based optical polarizer has potential application to produce profound light even under rigorous chemical or thermal environment. As MWCNTs show higher degree of polarization in UV, visible, and IR regions, it can be used in broadband photonics and electrooptic devices. The use of printed flexible CNTs, that is, a sheet of fully printed high-quality CNT transistors can provide higher mobility with low operating voltage. If the mass production of this can be made possible with printing process, it could drive the demand globally boosting the CNT market in the upcoming period.

KEYWORDS

- **CNT**
- **single-walled**
- **multiwalled**
- **displays**
- **polarizer**
- **incandescent**
- **flexible fiber**

REFERENCES

1. Soni, R.; Pal, A. K.; Tripathi, P.; Lal, J. A.; Kesari, K.; Tripathi, V. An Overview of Nanoscale Materials on the Removal of Wastewater Contaminants. *Appl. Water Sci.* **2020,** *10*, 1–9.
2. Iijima, S. Helical Microtubules of Graphitic Carbon. *Nature* **1991,** *354*, 56–58.
3. Lin, T.; Bajpai, V.; Ji, T.; Dai, L. Chemistry of Carbon Nanotubes. *Aust. J. Chem.* **2003,** *56*, 635–651.
4. Iijima, S.; Ichihashi, T. Single-Shell Carbon Nanotubes of 1-nm Diameter. *Nature* **1993,** *363*, 603–605.
5. Zhang, L.; Wu, Y.; Deng, L.; Zhou, Y.; Liu, C.; Fan, S. Photodetection and Photoswitch Based on Polarized Optical Response of Macroscopically Aligned Carbon Nanotubes. *Nano Lett.* **2016,** *16*, 6378–6382.
6. Prasek, J.; Drbohlavova, J.; Chomoucka, J.; Hubalek, J.; Jasek, O.; Adam, V.; Kizek, R. Methods for Carbon Nanotubes Synthesis. *J. Mater. Chem.* **2011,** *21*, 15872–15884.

7. Yang, L.; Wang, S.; Zeng, Q.; Zhang, Z.; Peng, L. M. Carbon Nanotube Photoelectronic and Photovoltaic Devices and Their Applications in Infrared Detection. *Small* **2013,** *9*, 1225–1236.
8. Liu, P.; Yang, S.-E.; Chen, Y.; Ma, Y.; Liu, S.; Fang, X.; Fan, F.; Han, J. Carbon Nanotube-Based Heterostructures for High-Performance Photodetectors: Recent Progress and Future Prospects. *Ceram. Int.* **2020.**
9. He, X.; Léonard, F.; Kono, J. Uncooled Carbon Nanotube Photodetectors. *Adv. Optic. Mater*. **2015,** *3*, 989–1011.
10. Xiang, L.; Zhang, H.; Hu, Y.; Peng, L.-M. Carbon Nanotube-Based Flexible Electronics. *J. Mater. Chem. C* **2018,** *6*, 7714–7727.
11. Wang, S.; Xu, J.; Wang, W.; Wang, G.-J. N.; Rastak, R.; Molina-Lopez, F.; Chung, J. W.; Niu, S.; Feig, V. R.; Lopez, J. Skin Electronics from Scalable Fabrication of an Intrinsically Stretchable Transistor Array. *Nature* **2018,** *555*, 83–88.
12. Ren, J.; Zhang, Y.; Bai, W.; Chen, X.; Zhang, Z.; Fang, X.; Weng, W.; Wang, Y.; Peng, H. Elastic and Wearable Wire-Shaped Lithium-Ion Battery with High Electrochemical Performance. Angew. *Chem.* **2014,** *126*, 7998–8003.
13. Yu, M.-F.; Files, B. S.; Arepalli, S.; Ruoff, R. S. Tensile Loading of Ropes of Single Wall Carbon Nanotubes and Their Mechanical Properties. *Phys. Rev. Lett.* **2000,** *84,* 5552.
14. Xie, S.; Li, W.; Pan, Z.; Chang, B.; Sun, L. Mechanical and Physical Properties on Carbon Nanotube. *J. Phys. Chem. Solids* **2000,** *61*, 1153–1158.
15. Xiao, L.; Chen, Z.; Feng, C.; Liu, L.; Bai, Z.-Q.; Wang, Y.; Qian, L.; Zhang, Y.; Li, Q.; Jiang, K. Flexible, Stretchable, Transparent Carbon Nanotube Thin Film Loudspeakers. *Nano Lett.* **2008,** *8*, 4539–4545.
16. Hu, L.; Li, J.; Liu, J.; Grüner, G.; Marks, T. Flexible Organic Light-Emitting Diodes with Transparent Carbon Nanotube Electrodes: Problems and Solutions. Nanotechnology **2010,** 21, 155202.
17. Liu, P.; Liu, L.; Jiang, K.; Fan, S. Carbon-Nanotube-Film Microheater on a Polyethylene Terephthalate Substrate and Its Application in Thermochromic Displays. *Small* **2011,** *7*, 732–736.
18. Kim, T. H.; Park, J. G.; Kim, Y. K.; Lim, Y. J.; Kang, J.-W.; Kim, E. S.; Kwon, J. Y.; Lee, Y. H.; Lee, S. H. An Active Carbon-Nanotube Polarizer-Embedded Electrode and Liquid-Crystal Alignment. *Nanoscale* 2020.
19. Hu, L.; Pasta, M.; La Mantia, F.; Cui, L.; Jeong, S.; Deshazer, H. D.; Choi, J. W.; Han, S. M.; Cui, Y. Stretchable, Porous, and Conductive Energy Textiles. *Nano Lett.* **2010,** *10*, 708–714.
20. De Volder, M. F.; Tawfick, S. H.; Baughman, R. H.; Hart, A. J. Carbon Nanotubes: Present and Future Commercial Applications. *Science* **2013,** *339*, 535–539.
21. Bethune, D.; Kiang, C. H.; De Vries, M.; Gorman, G.; Savoy, R.; Vazquez, J.; Beyers, R. Cobalt-Catalysed Growth of Carbon Nanotubes with Single-Atomic-Layer Walls. *Nature* **1993,** *363*, 605–607.
22. Dresselhaus, M.; Dresselhaus, G.; Charlier, J.-C.; Hernandez, E. Electronic, Thermal and Mechanical Properties of Carbon Nanotubes. Phil. Trans. R. Soc. Lond. Ser. A: Math. Phys. *Eng. Sci.* **2004,** *362*, 2065–2098.
23. Rafii-Tabar, H. Computational Modelling of Thermo-Mechanical and Transport Properties of Carbon Nanotubes. *Phys. Rep.* **2004,** *390*, 235–452.

24. Hamada, N.; Sawada, S.-i.; Oshiyama, A. New One-Dimensional Conductors: Graphitic Microtubules. *Phys. Rev. Lett.* **1992,** *68*, 1579.
25. Saito, R.; Fujita, M.; Dresselhaus, G.; Dresselhaus, u. M. Electronic Structure of Chiral Graphene Tubules. *Appl. Phys. Lett.* **1992,** *60*, 2204–2206.
26. Mintmire, J. W.; Dunlap, B. I.; White, C. T. Are Fullerene Tubules Metallic? *Phys. Rev. Lett.* **1992,** *68*, 631.
27. Mintmire, J.; White, C. Electronic and Structural Properties of Carbon Nanotubes. *Carbon* **1995,** *33*, 893–902.
28. Saito, R.; Fujita, M.; Dresselhaus, G.; Dresselhaus, M. S. Electronic Structure of Graphene Tubules Based on C 60. *Phys. Rev. B* **1992,** *46*, 1804.
29. Jiang, K. Carbon Nanotubes for Displaying. In *Industrial Applications of Carbon Nanotubes*; Elsevier, 2017; pp 101–127.
30. Zakri, C. Carbon Nanotubes and Liquid Crystalline Phases. Liquid Cryst. *Today* **2007,** *16,* 1–11.
31. Chen, H.-Y.; Lee, W.; Clark, N. A. Faster Electro-Optical Response Characteristics of a Carbon-Nanotube-Nematic Suspension. *Appl. Phys. Lett.* **2007,** *90,* 033510.
32. Li, P.; Jiang, K.; Liu, M.; Li, Q.; Fan, S.; Sun, J. Polarized Incandescent Light Emission from Carbon Nanotubes. *Appl. Phys. Lett.* **2003,** *82*, 1763–1765.
33. Ren, L.; Pint, C. L.; Booshehri, L. G.; Rice, W. D.; Wang, X.; Hilton, D. J.; Takeya, K.; Kawayama, I.; Tonouchi, M.; Hauge, R. H. Carbon Nanotube Terahertz Polarizer. *Nano Lett.* **2009,** *9*, 2610–2613.
34. Sasaki, K.-i.; Tokura, Y. Theory of a Carbon-Nanotube Polarization Switch. *Phys. Rev. Appl.* **2018,** *9*, 034018.
35. Sasaki, K.-i.; Murakami, S.; Yamamoto, H. Theory of Intraband Plasmons in Doped Carbon Nanotubes: Rolled Surface-Plasmons of Graphene. *Appl. Phys. Lett.* **2016,** *108*, 163109.
36. Chang, C.; Zhao, Y.; Liu, Y.; An, L. Liquid Crystallinity of Carbon Nanotubes. *RSC Adv.* **2018,** *8*, 15780–15795.
37. Jiang, K.; Li, Q.; Fan, S. Spinning Continuous Carbon Nanotube Yarns. *Nature* **2002,** *419*, 801–801.
38. Yang, H.; Fu, B.; Li, D.; Tian, Y.; Chen, Y.; Mattila, M.; Yong, Z.; Li, R.; Hassanien, A.; Yang, C. Broadband Laser Polarization Control with Aligned Carbon Nanotubes. *Nanoscale* **2015,** *7*, 11199–11205.
39. Shin, H. S.; Manda, R.; Kim, T. H.; Park, J. G.; Lim, Y. J.; Kim, B. K.; Lee, Y. H.; Lee, S. H. Multi-Layered Carbon Nanotube UV Polariser for Photo-Alignment of Liquid Crystals. *Liquid Cryst.* **2020,** 1–8.
40. Kang, B. G.; Lim, Y. J.; Jeong, K.-U.; Lee, K.; Lee, Y. H.; Lee, S. H. A Tunable Carbon Nanotube Polarizer. *Nanotechnology* **2010,** *21*, 405202.
41. Dunaevskn, G.; Suslyaev, V.; Zhuravlev, V. Badin, A.; Dorozhkin, K.; Kanygin, M.; Sedelmkova, O.; Bulusheva, L.; Okotrub, A. Electromagnetic Response of Anisotropic Polystyrene Composite Materials Containing Oriented Multiwall Carbon Nanotubes. In *2014 39th International Conference on Infrared, Millimeter, and Terahertz Waves* (IRMMW-THz); IEEE, 2014; pp. 1–2.
42. Lee, S.; Lee, S.; Kim, H.; Kim, J.; Hong, S.; Jeong, Y.; Park, C.; Choi, Y.; Lee, J.; Koh, J. 29.2: 18.1 "Ultra-FFS TFT-LCD with Super Image Quality and Fast Response

Time. In *SID Symposium Digest of Technical Papers*; Wiley Online Library, 2001; pp 484–487.

43. Hecht, D. S.; Hu, L.; Irvin, G. Emerging Transparent Electrodes Based on Thin Films of Carbon Nanotubes, Graphene, and Metallic Nanostructures. *Adv. Mater*. **2011,** *23*, 1482–1513.
44. Kumar, N.; Chen, J.; Kar, M.; Sitaraman, S. K.; Mukhopadhyay, S.; Kumar, S. Multigated Carbon Nanotube Field Effect Transistors-Based Physically Unclonable Functions as Security Keys. *IEEE IoT J.* **2018,** *6*, 325–334.
45. Lee, S. H.; Kim, H. Y.; Lee, S. M.; Hong, S. H.; Kim, J. M.; Koh, J. W.; Lee, J. Y.; Park, H. S. Ultra-FFS TFT-LCD with Super Image Quality, Fast Response Time, and Strong Pressure-Resistant Characteristics. *J. Soc. Info. Display* **2002,** *10*, 117–122.
46. Wei, Y.; Liu, P.; Jiang, K.; Fan, S. A Display Module Implemented by the Fast High-Temperatue Response of Carbon Nanotube Thin Yarns. *Nano Lett.* **2012,** *12*, 2548–2553.
47. Liu, P.; Liu, L.; Wei, Y.; Liu, K.; Chen, Z.; Jiang, K.; Li, Q.; Fan, S. Fast High-Temperature Response of Carbon Nanotube Film and Its Application as an Incandescent Display. *Adv. Mater*. **2009,** *21*, 3563–3566.
48. Liu, P.; Zhou, D.; Wei, Y.; Jiang, K.; Wang, J.; Zhang, L.; Li, Q.; Fan, S. Load Characteristics of a Suspended Carbon Nanotube Film Heater and the Fabrication of a Fast-Response Thermochromic Display Prototype. *ACS Nano* **2015,** *9*, 3753–3759.
49. Hata, K.; Futaba, D. N.; Mizuno, K.; Namai, T.; Yumura, M.; Iijima, S. Water-Assisted Highly Efficient Synthesis of Impurity-Free Single-Walled Carbon Nanotubes. *Science* **2004,** *306*, 1362–1364.
50. Amjadi, M.; Yoon, Y. J.; Park, I. Ultra-Stretchable and Skin-Mountable Strain Sensors Using Carbon Nanotubes–Ecoflex Nanocomposites. *Nanotechnology* **2015,** *26*, 375501.
51. Ma, W.; Song, L.; Yang, R.; Zhang, T.; Zhao, Y.; Sun, L.; Ren, Y.; Liu, D.; Liu, L.; Shen, J. Directly Synthesized Strong, Highly Conducting, Transparent Single-Walled Carbon Nanotube Films. *Nano Lett*. **2007,** *7*, 2307–2311.
52. Park, J.; Hwang, J. C.; Kim, G. G.; Park, J. U. Flexible Electronics Based on One-Dimensional and Two-Dimensional Hybrid Nanomaterials. *InfoMat* **2020,** *2*, 33–56.
53. Wang, C.; Takei, K.; Takahashi, T.; Javey, A. Carbon Nanotube Electronics–Moving Forward. *Chem. Soc. Rev*. **2013,** *42*, 2592–2609.
54. Cai, L.; Wang, C. Carbon Nanotube Flexible and Stretchable Electronics. *Nanoscale Res. Lett.* **2015,** *10*, 1–21.

CHAPTER 5

POTENTIAL OF CARBON NANOTUBES IN HYDROGEN STORAGE

SNEHA MATHEW and BEENA MATHEW*

School of Chemical Sciences, Mahatma Gandhi University, Kottayam, Kerala, India

*Corresponding author. E-mail: beenamscs@gmail.com

ABSTRACT

The ever-increasing need for energy, as well as growing environmental pollution issues have necessitated new and renewable fuels. Hydrogen is the cleanest and most readily usable green energy, which will substitute potentially harmful fossil fuels. Its combustion produces water hence hydrogen is regarded as clean fuel. Considerable research has been ongoing in recent years to the production of new porous materials with high hydrogen storage power, such as carbon nanotubes (CNTs). Carbon nanotubes and their composites are known to be the leading source of hydrogen storage medium due to their special characteristics including high surface area, nanoporous composition, low mass density, chemical stability, hydrogen dissociation, and fast synthesis method. The adsorption properties of CNTs depend on the porous nature of nanotube and the surface area available. This chapter provides the reader the potential of carbon nanotubes in hydrogen storage and modifications of CNTs for better storage of hydrogen.

5.1 INTRODUCTION

World's energy requirements are increasing because of the growing population and changes in living styles, and on the other hand, the natural resources are increasingly depleting at an unprecedented pace amid environmental concerns. The environmental damage caused by the use of fossil fuels has directly contributed to an increase in demand for renewable and sustainable resources. Hydrogen as a fuel falls into the picture in such a situation to satisfy the energy demands.[1] Hydrogen is one of the renewable fuels available since it is lightweight, environmentally sustainable, green and four-fold more effective than petroleum.[2–4] As hydrogen can be synthesized from wind, hydroelectric, solar, thermal, biomass, etc. with no pollution of any kind, it can be regarded as a sustainable energy source.[5] Since the hydrogen's capacity is to bear high gravimetric energy density with zero emissions and regenerative nature, it acts as a renewable energy transporter, making it more environmentally safe. As hydrogen is in gaseous state in ambient condition, its storage is one of the biggest challenges.

Hydrogen can be stored as (1) cryogenic liquid, (2) pressurized gas, (3) solid fuel as chemical (metal hydrate) or physical combination (carbon material).[6] Unlike electric or thermal energy, hydrogen can be processed almost as easily as natural gas on a wide scale, it is environmental friendly and moreover the use of the hydrogen-fuelled internal combustion engine is also an efficient way to minimize emissions dramatically.[7–8] Despite these apparent benefits, however, significant technological challenges concerning the handling or distribution of hydrogen exist as one attempts to use it for fuel cells or combustion engines. Since hydrogen gas needs a considerably greater volume compared with fuel with an equal energy content.[9]

Carbon-based nanostructures have attracted the researchers as one of the auspicious hydrogen storage materials.[10,11] Of these, carbon nanotubes (CNTs) are widely reported as potential hydrogen storage materials.[12–14] Carbon nanotube's high surface-to-volume ratio, chemical stability, and light mass make them potential candidates for hydrogen storage.[15] As materials with high surface area can store large amount of hydrogen, CNT is an ideal material for hydrogen adsorption.[16] Dillon et al. in 1997 were the first to report the high hydrogen storage capacity of single-walled CNTs (SWCNTs). They found that at 273 K and 0.04 MPa achieved 5–10 wt.% meeting the hydrogen storage target set by US DOE (Department of Energy), making CNTs ideal for storing hydrogen. Various researches have been conducted to study the hydrogen storage in CNTs and their results were inconsistent with the standard set by US DOE. The fundamental reason for this inconsistency in results is that the hydrogen storage properties of CNTs is affected by internal and external factors. Internal factors are surface area, purity, etc., while external factors are testing devices, measurement methods, temperature, pressure stability.[17] Cheng et al. noted that the carbon nanotube storage ability increases with CNT diameter optimization and inter-tube spacing. They demonstrated that at 298 K and 10 MPa single-walled CNTs can store hydrogen in excess of 10.0 wt.%.[14] Due to weak van der Waals interaction between hydrogen molecules and CNT, CNT has low hydrogen storage capacity. The hydrogen adsorption capacity of CNT can be improved by co-doping of metal and hetero atom, positively charging, mechanical blending, or activating CNTs.[18]

CNTs play a noteworthy role in energy storage devices, hence, CNTs as candidate for hydrogen storage will not end in the near future. This chapter explores the hydrogen storage potential of CNTs. As well, we have summarized the modifications CNTs for better storage of hydrogen. We have also discussed device performance and its limitations. Finally, we have put forward our perspective to address some of the challenges associated with hydrogen storage by CNTs and their possible remedies.

5.1.1 HYDROGEN STORAGE IN CARBON NANOTUBES

The composition, geometry, structural defects, pretreatment, operating pressure, temperature etc. depend on the ability of CNTs to store hydrogen. The potential storage site for hydrogen can be inside the tube, outside the tube, between tubes (bundles and ropes), and between shells (in the MWCNT case).[1] CNTs may be seen as suitable candidates for hydrogen adsorption with their remarkable chemical stability, large surface area, hollowness, and light mass.[19] These nanometers thickness tube can store hydrogen within them or pores in them (Fig. 5.1).[20] The exact mechanism for hydrogen adsorption by CNT is still unclear, reports showed that hydrogen adsorption occurs through physisorption and chemisorption. Physisorption is due to the van der Waals forces of interaction whereas as chemisorption is due to chemical bond. Chemisorption cannot be beneficial for the practical storage of hydrogen as it can be desorbed only at higher temperature. Chemisorption in CNT occurs when the hydrogen adsorbs exothermically to the top positions of carbon atoms on the tube ends, while physisorption occurs by trapping molecules of hydrogen inside the cylindrical structure of the nanotube. Another parameter that affects hydrogen adsorption by CNT is adsorption energy. A relatively higher adsorption energy favors hydrogen storage. At low adsorption energy, hydrogen will desorb spontaneously at low temperature.[6,21] The adsorption is denoted as a unit of quantity of gas in comparison to a unit of quantity of adsorbent, the most commonly used, weight percent (wt.%).[22]

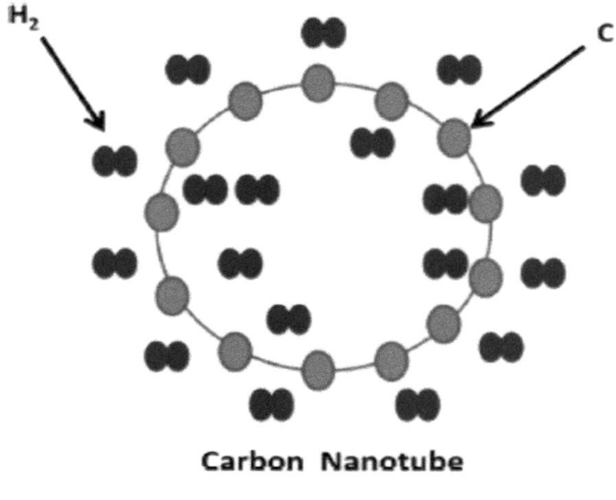

FIGURE 5.1 Schematic depiction of hydrogen storage in carbon nanotubes.
Source: Reprinted with permission from Ref. [20]. © 2015, Elsevier.

Dillon et al. in 1997 asserted 5–10 wt.% hydrogen adsorption on CNT. Rodriguez and his colleagues further recorded unusual qualities of adsorption of hydrogen by carbon nanostructures of up to 10–12 molecules per carbon atom. However, in subsequent correspondence, they reduced this to four molecules per carbon atom.[1] Nikitin et al. stated that by chemisorption, CNTs could store more than 7 wt.% hydrogen.[23] As hydrogen adsorbs to materials from CNTs, three types of geometries are possible. Geometry of the arch type is created as the hydrogen ions adsorb exothermically to the top carbon atom sites on the CNT wall in chemisorption.

In zigzag-type geometry, the hydrogen bonding could be at the exterior and interior of CNT. During chemisorption, the enhancement of the Sp^3 type hybridization in the tube and extension of the nanotube diameter of 6.88–7.78 nm is observed. A more stable geometry is observed when hydrogen is occupied inside the empty space of CNTs.[24] In comparison, multiwalled CNTs (MWCNTs) also appear to have an attractive future for storage of H_2. Chen et al. reported that among alkali-doped MWNT 20 and 14 wt.% hydrogen intake can achieve for Li-doped (653 K) and K-doped (RT) nanotubes.[25] Temperature has a significant effect on hydrogen adsorption by CNT. Lee et al. found that hydrogen storage capacity of activated MWCNTs increases considerably with raise in temperature.[26] Hydrogen stores inside the empty space in nanotubes and this storage capacity increases linearly with tube diameter. It was also found that the diameter of the tube is influenced by the synthesis conditions, and can therefore be controlled by optimizing the conditions of synthesis.[27] Another factor affecting hydrogen adsorption is the presence of defects. The presence of structural defects cause remarkable increase in adsorption binding energy of the order of 50%, and thus, the hydrogen storage potential in CNTs.[28,29] Coiled carbon tubes have built in structural defects and provide larger sites for hydrogen molecules adsorption.[30] Figure 5.2 shows hydrogen storage in nanotube bundles and is clear from the figure as a result of low interaction between CNT and H_2, CNT can store hydrogen only under cryogenic conditions. So storage of large amount of hydrogen at room temperature and at low pressure is very difficult. So modified CNTs are used for storing large quantities of hydrogen at ambient conditions.[31]

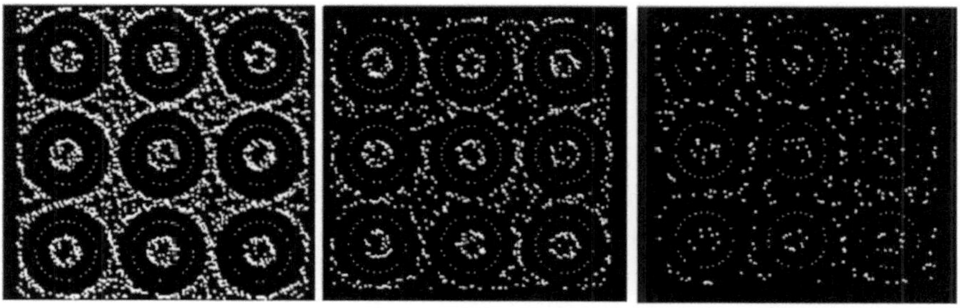

FIGURE 5.2 Hydrogen storage in nanotube bundles. Snapshots from grand canonical Monte Carlo simulations taken under 100 bar pressure at 77 K (left), 175 K (middle), and 293 K (right).

Source: Reprinted with permission from Ref. [31]. © 2011, Elsevier.

5.1.2 APPROACHES TO IMPROVE HYDROGEN ADSORPTION

However, several studies indicate that pure CNTs can adsorb only a very small amount of H_2 molecules with poor binding energy (0.11 eV or less) at room temperature and normal pressure not satisfying US DOE specifications.[32] Numerous modifications are adopted to increase the hydrogen storage properties of CNTs. Chemical activation is one of the adaptations, where samples of CNTs are treated with KOH and NaOH solution. As high surface area and pore volume increases hydrogen adsorption, attempts are made to increase the surface area and pore volume by ultrasonication, ball milling, bromination, and acidification which create defects and other structural disorder and purification to increase the surface area and pore volume. Another attempt has been made to dope metal or metal oxide or hydrides into CNTs.[23] Integration of dopants or functional groups increases binding energy of carbon surfaces and hydrogen and thus increases the availability of adsorption sites. Metal decoration was proposed as a route to improve the reversible hydrogen storage capacity at ambient temperature due to spillover effect. Hydrogen spillover includes the dissociation of chemically extracted H_2 from decorated metal catalysts, preceded by the release of atomic hydrogen onto the carbon supply surface, and eventually, H_2 is deposited reversibly.[33] Doping CNTs with metal elements could improve the electronic structure of the CNTs, which could lead to a stronger impact on hydrogen storage.[34–38] Table 5.1 depicts the hydrogen adsorption capacity of various SWCNTs and MWCNTs under various temperature and pressure ranges.[1]

Both SWNTs and MWNTs were thoroughly investigated for their hydrogen storage ability and it was found that at room temperature, coupled and open MWNTs had a hydrogen storage capacity of 1.97 wt.% at ambient temperature. Decorating with metals like Pt,[46–48] Ti,[49.50] Pd,[51,52] Si[53] increases the hydrogen adsorption by CNT. Scientists have conducted research on MWNTs ability for hydrogen storage with various mean outer diameters in the 13–53 nm range and stated that MWNTs hydrogen storage potential was proportional to their diameter, up to 4.6 wt.% at 293 K and 13.5 MPa.

Hydrogen storage capacity of CNT is also affected by the presence of impurities and the pretreatment process.[54] Coating SWCNTs directly with hydrides and γ irradiation increases structural defects and thus improves the hydrogen adsorption capacity.[24,55] In addition, it has been stated that acid or base treatment, heat treatment, and treatment with fluorine and bromine, hetero atoms, such as N, P, B, Si and hetero atomic mixed-doping increase the hydrogen storage ability of CNTs.[24] Presence of additives alter the electronic structure of the carbon material, increases the inter-layer distance and forms hybrid material and thus improves the hydrogen storage capacity. Use of KOH-modified MWCNTs is one of the most important ways to better establish the textural characteristics and boost the conservation of hydrogen.[56] Metals, especially Ni, Pd, Ag, and Pt, or metal oxides, such as TiO_2 nanocatalysts, decorated on CNTs result in hydrogen storage capacity by a very fascinating process called spillover.[57] Nitrogen doping is currently considered a viable approach to modify the physical and chemical properties of CNTs owing to the presence of nitrogen atoms of extra lone pairs of electrons relative to hydrogen atoms. Numerous studies have indicated the possibility of achieving improved hydrogen storage capacities with nitrogen-doped carbon materials.[59,60] Studies reported that doping with transition metals is unstable as repeated hydrogenation or rehydrogenation causes clustering of transition metals. This problem can be overcome by doping with metals (e.g., Li, Ca) that binds strongly to the substrate than to themselves.[61]

TABLE 5.1 Hydrogen Storage Capacities of Carbon Nanotubes.

Adsorbent	Hydrogen storage capacity (wt.%)	Temperature (K)	Pressure (MPa)	References
SWNT (80–90%)	7	273	0.04	[39]
SWNT (50% purity)	4.2	300	10.1	[40]
SWNT (high purity)	3.5–4.5	298	0.04	[39]
SWNT	10	300	0.04	[22]
MWNT	0.68	A	10	[41]
MWNT	13.8	300	1	[41]
K-doped MWNT	1.8	<313	0.1	[42]
Li-doped MWNT	20	200–400	0.1	[43]
Ni-doped MWCNT	0.298	RT	2	[44]

MWNT, multiwalled nanotube; SWNT, single-walled nanotube; A, ambient temperature.

Ab initio calculations suggest that boron nitride nanotubes are superior to CNTs for hydrogen storage applications because of the heteropolar binding of their atoms.[62] Nanotubes show variability in results, processing uncertainties, release temperatures, low synthetic purity, metal clustering, and material instability problems due to which their use gets limited.[20]

5.2 CONCLUSION

Hydrogen has attracted the attention due to its potential as clean and renewable source of energy. Hydrogen contains much chemical energy per unit mass and contributes significantly to solve energy crisis. Hydrogen is believed to gradually replace fossil fuels in the future as the latter cause serious environmental issues. The search for an ideal material for hydrogen storage by the scientific community hit at CNTs. Since Dillon et al. have documented the storage of hydrogen in CNTs, the CNTs have drawn considerable interest as a candidate for hydrogen storage. Thanks to its interesting structural features and properties. CNTs and their modified forms are used as promising candidate for energy storage in the recent years. Large surface area, light weight, and chemical stability of CNTs favor them as a potential candidate for hydrogen storage. CNTs have proved to be capable of promoting effective heterogeneous catalyst manufacturing processes in hydrogen processing. In addition, CNTs are versatile adsorbent material which may form the basis for technologically feasible hydrogen storage systems. Nonetheless, the high potential of absorption of hydrogen in CNTs could not be replicated, and recent studies have shown that it is difficult for CNTs to store a significant volume of H_2 simply through poor molecular adsorption. So various modifications are added to CNTs to promote higher hydrogen storage. Decorating with metals, hydrides, ion irradiation, acid or base treatments, heat treatments are some of the modifications added. Both single-walled and multiwalled CNTs are found effective in hydrogen adsorption. Still more research has to be done in this field to explore the hydrogen storage properties of carbon nanotubes.

KEYWORDS

- **carbon nanotubes**
- **hydrogen storage**
- **physisorption**
- **chemisorption**
- **SWCNT**
- **MWCNT**
- **doping**

REFERENCES

1. Mohan, M.; Sharma, V. K.; Kumar, E. A.; Gayathri, V. Hydrogen Storage in Carbon Materials: A review. *Energy Storage* **2019,** *1*, 1–26.
2. Rajaura, R. S.; Srivastava, S.; Sharma, P. K.; Mathur, S.; Shrivastava, R.; Sharma, S. S.; Vijay, Y. K. Structural and Surface Modification of Carbon Nanotubes for Enhanced Hydrogen Storage Density. *Energy Storage* **2018,** *14*, 57–65.
3. Yürüm, Y., Taralp, A.; Veziroglu, T. N. Storage of Hydrogen in Nanostructured Carbon Materials. *Int. J. Hydrog. Energy* **2009,** *34*, 3784–3798.
4. Esswein, A. J.; Nocera, D. G. Hydrogen Production by Molecular Photocatalysis. *Chem. Rev* **2007,** *107*, 4022–4047.
5. Cheng, H.-M.; Yang, Q.-H.; Liu, C. Hydrogen Storage in Carbon Nanotubes. *Carbon* **2001,** *39*, 1447–1454.
6. Yao, Y. Hydrogen Storage Using Carbon Nanotubes. *Carbon Nanotubes*. IntechOpen. 2010.
7. Zhang, K., Liu, S.; Liu, M.; Zhang, H.; Lu, J.; Cao, Z.; Zhang, H. Research Developments of Hydrogen Energy (in Chinese). *Mater. Rev*. **2011,** *25*, 116–119.
8. Liu, M.; Li, D.; Qiao, J.; Li, W.; Wu, Z. The Use of Hydrogen Energy and Hydrogen Adsorption Storage Technology of Carbonaceous Materials (in Chinese). *Chem. Ind. Times* **2013,** *27*, 35–38.
9. Shin, W. H.; Jung, H. M.; Choi, Y. J.; Miyasaka, K.; Kang, J. K. Bimetallic Catalysts Selectively Grown via N-Doped Carbon Nanotubes for Hydrogen Generation. *J. Mater. Chem.* **2010,** *20*, 6544.
10. Li, J.; Terumi, F.; Hajime, G.; Toshiyuki, O.; Yoshiya, F.; Sidney, Y. Theoretical Evaluation of Hydrogen Storage Capacity in Pure Carbon Nanostructures. *J. Chem. Phys.* **2003,** *119*, 2376–2385.
11. Zuttel, A.; Sudan, P.; Mauron, P.; Kyobayashi, T.; Emenenrgger, C.; Schlapbach, L. Hydrogen Storage in Carbon Nanostructures. *Int. J. Hydrog. Energy* **2002,** *27*, 203–212.

12. Muniz, A. R.; Singh, T.; Maroudas, D. Effects of Hydrogen Chemisorption on the Structure and Deformation of Single-Walled Carbon Nanotubes. *Appl. Phys. Lett.* **2009,** *94*, 103108.
13. Ming-Wen, Z.; Yue-Yuan, X.; Yu-Chen, M.; Min-Ju, Y.; Xiang-Dong, L.; Liang-Mo, M. Tunable Adsorption and Desorption of Hydrogen Atoms on Single-Walled Carbon Nanotubes. *Chinese Phys. Lett.* **2002,** *19*, 1498–1500.
14. Cheng, J.; Yuan, X.; Zhao, L.; Huang, D.; Zhao, M.; Dai, L.; Ding, R. GCMC Simulation of Hydrogen Physisorption on Carbon Nanotubes and Nanotube Arrays. *Carbon* **2004,** *42*, 2019–2024.
15. Surya, V. J.; Iyakutti, K.; Venkataramanan, N.; Mizuseki, H.; Kawazoe, Y. The Role of Li and Ni Metals in the Adsorbate Complex and Their Effect on the Hydrogen Storage Capacity of Single Walled Carbon Nanotubes Coated with Metal Hydrides, LiH and NiH_2. *Int. J. Hydrog. Energy* **2010,** *35*, 2368–2376.
16. Kim, B.; Lee. Y.; Park, S. Novel Porous Carbons Synthesized from Polymeric Precursors for Hydrogen Storage. *Int. J. Hydrog. Energy* **2008,** *33*, 2254–2259.
17. Lyu, J.; Kudiiarov, V.; Lider, A. An Overview of the Recent Progress in Modifications of Carbon Nanotubes for Hydrogen Adsorption. *Nanomaterials* **2020,** *10*, 255.
18. Silambarasan, D.; Surya, V. J.; Vasu, V.; Iyakutti, K. Reversible and Reproducible Hydrogen Storage in Single- Walled Carbon Nanotubes Functionalized with Borane. *Carbon Nanotubes—Recent Progress*. IntechOpen. 2018.
19. Rakhi, R.; Sethupathi K.; Ramaprabhu, S. Synthesis and Hydrogen Storage Properties of Carbon Nanotubes. *Int. J. Hydrog. Energy*. **2008,** *33*, 381–386.
20. Niaz, S.; Manzoor, T.; Pandith, A. H. Hydrogen Storage: Materials, Methods and Perspectives. *Renew. Sust. Energ. Rev*. **2015,** *50*, 457–469.
21. Volpe, M.; Cleri, F. Chemisorption of Atomic Hydrogen in Graphite and Carbon Nanotubes. *Surf. Sci.* **2003,** *544*, 24–34.
22. Darkrim Lamari, F.; Levesque, D. Hydrogen Adsorption on Functionalized Graphene. *Carbon* **2011,** *49*, 5196–5200.
23. Nikitin, A.; Li, X.; Zhang, Z.; Ogasawara, H.; Dai, H.; Nilsson, A. Hydrogen Storage in Carbon Nanotubes through the Formation of Stable C−H Bonds. *Nano Lett.* **2008,** *8*, 162–167.
24. Rather, S. u. Preparation, Characterization and Hydrogen Storage Studies of Carbon Nanotubes And Their Composites: A Review. *Int. J. Hydrog. Energy* **2020,** *45*, 4653–4672.
25. Chen, P. High H_2 Uptake by Alkali-Doped Carbon Nanotubes under Ambient Pressure and Moderate Temperatures. *Science* **1999,** *285*, 91–93.
26. Lee, S.-Y.; Park, S.-J. Effect of Temperature on Activated Carbon Nanotubes for Hydrogen Storage Behaviors. *Int. J. Hydrog. Energy* **2010,** *35*, 6757–6762.
27. Lee, S. M.; Lee, Y. H. Hydrogen Storage in Single-Walled Carbon Nanotubes. *Appl. Phys. Lett.* **2000,** *76*, 2877–2879.
28. Gayathri, V.; Geetha, R. Hydrogen Adsorption in Defected Carbon Nanotubes. *Adsorption* **2007,** *13*, 53–59.
29. Geetha, R.; Gayathri, V. Comparative Study on Gas Adsorption in Defected Carbon and Boron Nitride Nanotube. *Curr. Nanosci.* **2010,** *6*, 131–136.
30. Gayathri, V.; Devi, N. R.; Geetha, R. Hydrogen Storage in Coiled Carbon Nanotubes. *Int. J. Hydrog. Energy* **2010,** *35*, 1313–1320.

31. Froudakis, G. E. Hydrogen Storage in Nanotubes & Nanostructures. *Mater. Today* **2011,** *14*, 324–328.
32. Barghi, S. H.; Tsotsis, T. T.; Sahimi, M. Chemisorption, Physisorption and Hysteresis during Hydrogen Storage in Carbon Nanotubes. *Int. J. Hydrog. Energy* **2014,** *39*, 1390–1397.
33. Yu, X.; Tang, Z.; Sun, D.; Ouyang, L.; Zhu, M. Recent Advances and Remaining Challenges of Nanostructured Materials for Hydrogen Storage Applications. *Prog. Mater. Sci.* **2017,** *88*, 1–48.
34. Hosseini, S. V.; Arabi, H.; Kompany, A. Comparison of Hydrogen Absorption in Metallic and Semiconductor Single-Walled Ge- and GeC_2-Doped Carbon Nanotubes. *Int. J. Hydrog. Energy* **2017,** *42*, 969–977.
35. Mei-Yan, N.; Xian-Long, W.; Zhi, Z. Interaction of Hydrogen Molecules on Ni-Doped Single-Walled Carbon Nanotube. *Chinese Phys. B.* **2009,** *18*, 357–362.
36. Luna, C. R.; Bechthold, P.; Brizuela, G.; Juan, A.; Pistonesi, C. The Adsorption of CO, O_2 and H_2 on Li–Doped Defective (8,0) SWCNT: A DFT Study. *Appl. Surf. Sci.* **2018,** *459*, 201–207.
37. Li, C. Hydrogen Storage in Single-Walled Carbon Nanotubes- Revisited. *Carbon* **2000,** *38*, 623–626.
38. Li, M.; Li, Y.; Zhou, Z.; Shen, P.; Chen, Z. Ca-Coated Boron Fullerenes and Nanotubes as Superior Hydrogen Storage Materials. *Nano Lett.* **2009,** *9*, 1944–1948.
39. Liu, C.; Chen, Y.; Wu, C.-Z.; Xu, S.-T.; Cheng, H.-M. Hydrogen Storage in Carbon Nanotubes Revisited. *Carbon* **2010,** *48*, 452–455.
40. Liu, C. Hydrogen Storage in Single-Walled Carbon Nanotubes at Room Temperature. *Science* **1999,** *286*, 1127–1129.
41. Lim, K. L.; Kazemian, H.; Yaakob, Z.; Daud, W. R. W. Solid-State Materials and Methods for Hydrogen Storage: A Critical Review. *Chem. Eng. Technol.* **2010,** *33*, 213–226.
42. Chambers, A.; Park, C.; Baker, R. T. K.; Rodriguez, N. M. Hydrogen Storage in Graphite Nanofibers. *J. Phys. Chem. B* **1998,** *102*, 4253–4256.
43. Dillon, A. C.; Jones, K. M.; Bekkedahl, T. A.; Kiang, C. H.; Bethune, D. S.; Heben, M. J. Storage of Hydrogen in Single-Walled Carbon Nanotubes. *Nature* **1997,** *386*, 377–379.
44. Kaskun, S.; Kayfeci, M. The Synthesized Nickel-Doped Multi-Walled Carbon Nanotubes for Hydrogen Storage under Moderate Pressures. *Int. J. Hydrog. Energy* **2018,** *43*, 10773–10778.
45. Lee H, Kang YS, Kim SH, Lee JY. Hydrogen Desorption Properties of Multiwall Carbon Nanotubes with Closed and Open Structures. *Appl. Phys. Lett.* **2002,** *80*, 577–579.
46. Sharma, A. Investigation on Platinum Loaded Multi-Walled Carbon Nanotubes for Hydrogen Storage Applications. *Int. J. Hydrog. Energy* **2020,** *45*, 2967–2974.
47. Park, S.-J.; Lee, S.-Y. Hydrogen Storage Behaviors of Platinum-Supported Multi-Walled Carbon Nanotubes. *Int. J. Hydrog. Energy* **2010,** *35*, 13048–13054.
48. Yang, S. J.; Cho, J. H.; Nahm, K. S.; Park, C. R. Enhanced Hydrogen Storage Capacity of Pt-Loaded CNT@MOF-5 Hybrid Composites. *Int. J. Hydrog. Energy* **2010,** *35*, 13062–13067.

49. Yang, L.; Yu, L. L.; Wei, H. W.; Li, W. Q.; Zhou, X.; Tian, W. Q. Hydrogen Storage of Dual-Ti-Doped Single-Walled Carbon Nanotubes. *Int. J. Hydrog. Energy* **2019,** *44*, 2960–2975.
50. Yildirim, T.; Ciraci. Titanium-Decorated Carbon Nanotubes as a Potential High-Capacity Hydrogen Storage Medium. *Phys. Rev. Lett.* **2005,** *94*.
51. Reyhani, A.; Mortazavi, S. Z.; Mirershadi, S.; Moshfegh, A. Z.; Parvin, P.; Golikand, A. N. Hydrogen Storage in Decorated Multiwalled Carbon Nanotubes by Ca, Co, Fe, Ni, and Pd Nanoparticles under Ambient Conditions. *J. Phys. Chem. C.* **2011,** *115*, 6994–7001.
52. Erünal, E.; Ulusal, F.; Aslan, M. Y.; Güzel, B.; Üner, D. Enhancement of Hydrogen Storage Capacity of Multi-Walled Carbon Nanotubes with Palladium Doping Prepared through Supercritical CO_2 Deposition Method. *Int. J. Hydrog. Energy* **2018,** *43*, 10755–10764.
53. Cho, J. H.; Yang, S. J.; Lee, K.; Park, C. R. Si-Doping Effect on the Enhanced Hydrogen Storage of Single Walled Carbon Nanotubes and Graphene. *Int. J. Hydrog. Energy* **2020,** *36*, 12286–12295.
54. Rashidi, A. M.; Nouralishahi, A.; Khodadadi, A. A.; Mortazavi, Y.; Karimi, A.; Kashefi, K. Modification of Single Wall Carbon Nanotubes (SWNT) for Hydrogen Storage. *Int. J. Hydrog. Energy* **2010,** *35*, 9489–9495.
55. Silambarasan, D.; Surya, V. J.; Iyakutti, K.; Asokan, K.; Vasu, V.; Kawazoe, Y. Gamma (γ)-Ray Irradiated Multi-Walled Carbon Nanotubes (MWCNTs) for Hydrogen Storage. *Appl. Surf. Sci.* **2017,** *418*, 49–55.
56. Elyassi, M.; Rashidi, A.; Hantehzadeh, M. R.; Elahi, S. M. Hydrogen Storage Behaviors by Adsorption on Multi-Walled Carbon Nanotubes. *J. Inorg. Organomet. Polym.* **2016,** *27*, 285–295.
57. Akbarzadeh, R.; Ghaedi, M.; Nasiri Kokhdan, S.; Vashaee, D. Remarkably Improved Electrochemical Hydrogen Storage by Multi-Walled Carbon Nanotubes Decorated with Nanoporous Bimetallic Fe–Ag/TiO_2 Nanoparticles. *Dalton Trans.* **2019,** *48*, 898–907.
58. Chen, L.; Xia, K.; Huang, L.; Li, L.; Pei, L.; Fei, S. Facile Synthesis and Hydrogen Storage Application of Nitrogen-Doped Carbon Nanotubes with Bamboo-Like Structure. *Int. J. Hydrog. Energy* **2013,** *38*, 3297–3303.
59. Park, S. S.; Chu, S.-W.; Xue, C.; Zhao, D.; Ha, C.-S. Facile Synthesis of Mesoporous Carbon Nitrides Using the Incipient Wetness Method and the Application as Hydrogen Adsorbent. *J. Mater. Chem.* **2011,** *21*, 10801.
60. Zheng, Z.; Gao, Q.; Jiang, J. High Hydrogen Uptake Capacity of Mesoporous Nitrogen-Doped Carbons Activated Using Potassium Hydroxide. *Carbon* **2010,** *48*, 2968–2973.
61. Jena, P. Materials for Hydrogen Storage: Past, Present, and Future. *J. Phys. Chem. Lett.* **2011,** *2*, 206–211.
62. Mpourmpakis, G.; Froudakis, G. E. Why Boron Nitride Nanotubes Are Preferable to Carbon Nanotubes for Hydrogen Storage? An Ab Initio Theoretical Study. *Catal. Today* **2007,** *120*, 341–345.

CHAPTER 6

SYNTHESIS AND HETEROGENOUS CATALYTIC APPLICATIONS OF NOBLE METAL NANOPARTICLES/CARBON NANOTUBES NANOCOMPOSITES

MAMATHA SUSAN PUNNOOSE and BEENA MATHEW*

School of Chemical Sciences, Mahatma Gandhi University, Kottayam, India

Corresponding author. E-mail: beenamscs@gmail.com

ABSTRACT

With the advancement of nanotechnology, nanoparticles in particular, noble metal nanoparticles, such as Au, Ag, Pd, Pt, and Ru show better catalytic performance than their bulk counterparts particularly because of their smaller particle size and nano effects. The unique physiochemical properties and their potential applications of noble metal nanoparticles in the fields of biomedicine, sensing, surface-enhanced Raman scattering (SERS), and especially catalysis can be elaborated to carbon nanotubes (CNTs) by the incorporation of noble metal nanoparticles with CNTs surface by various methods of synthesis. This chapter focuses on the recent progress in the synthesis of CNTs functionalized by noble metal nanoparticles and its advancement towards heterogenous catalysis.

6.1 INTRODUCTION

Research in the field of porous nanocarbon materials and their applications are gaining much attention in the recent years. Both single-walled carbon nanotubes (SWCNTs) and multiwalled carbon nanotubes (MWCNTs) show exceptional surface area, electrical conductance, and high chemical and electrochemical stabilities.[1–4] All these properties along with good electrical and mechanical properties contribute toward the special interest of CNTs in the modern science and nanotechnology. CNTs have been used in rechargeable lithium battery applications,[5] plasmonic sensor,[6] water treatment,[7] drug delivery,[8] and environmental pollutant degradations.[9]

Though CNTs are promising materials with vivid applications, their solubility in aqueous and organic solvents is much less due to the strong vander waals interactions. Low dispersibility of CNTs can be rectified by their surface functionalizations using organic moieties, such as polymers, dendrimers, or surfactants. Noncovalent functionalization of CNTs with different surfactant molecules occur via hydrophobic or π-π interactions, which result in a net positive or negative charge on the carbon nanotube surface, and thus upgrades the solubility of CNTs in aqueous solution.[10] Also, the high dispersive efficiency of surfactants reduces the aggregation of CNTs in aqueous medium, and thus improves the processability of CNTs toward diverse applications.

Incorporation of heteroatoms, different metal oxides, and mixed metal oxides, such as MgO, ZnO, TiO_2, and Mg-Al or Ce-Zr is also preferred for the modification of the CNTs surface. TiO_2- and SiO_2-functionalized CNTs composites are efficient candidates for the removal of hazardous dye effluents. Bismuth oxide iodide nanosheet-coated MWCNTs are potent photocatalysts for the degradation of methyl orange (MO), rhodamine B (RhB), and para-chlorophenol.[11] BiOCl–Au–CdS photocatalyst shows excellent photocatalytic activity toward the degradation of methyl orange, rhodamine B, phenol, and antibiotic sulfadiazine.[12] Alkaline metal (Li, Na, K, Cs)-doped MWCNTs act as efficient catalyst for the Knoevenagel condensation,[13] transesterification of triglyc-erides[14] and aldol condensations reactions.[15]

The immobilization of noble metal nanoparticles (NPs) onto the surface functionalized CNTs are of greater attention from the past decades. The surfactant plays the role of stabilizers for the growth of metal NPs. Also, they successfully lower the interface tension and thus amplifies the dispersibility of NPs.[16,17] The nanostructure hydrids of noble metal NPs with CNTs have emerged as a new class of composite materials that are widely known for their catalytic and biomedical applications. The unique electronic, optical, magnetic, and catalytic properties of metal NPs in addition to their large surface-to-volume proportions and particle's size-dependent surface morphologies are incorporated into the CNTs which lead to the generation of new unique functions due to interactive effects among CNTs and noble metal NPs.[18–20] These nanohybrids have much importance in various fields, such as fuel cells, hydrogen storage, heterogeneous catalysis, drug delivery, electronic nanodevices, and chemo/biosensors.[21,22] The ever first report on the synthesis and application of noble metal–CNTs hybrid was presented by Ajayan and coworkers in 1994.[23] SWCNTs decorated with ruthenium NPs acted as efficient catalysts in heterogenous catalysis reactions like hydrogenation of cinnamaldehyde. The following section discusses on the various approaches toward the noble metal NPS/CNTs nanocomposite synthesis and their applications in the field of heterogenous catalysis.

6.2 SYNTHESIS OF NOBLE METAL NPS/CNTs NANOHYBRIDS

The hybrid nanocomposites of CNTs with NPs can be achieved mainly by two approaches. Firstly, the NPs are grown directly or deposited on the CNTs surface. Second approach involves the connection of the preformed NPs on to the CNTs via covalent linkages.

6.2.1 PREPARATION OF METAL NPS DIRECTLY ON CNTS

6.2.1.1 REDUCTION

Various salts of noble metals act as precursors for NPs generation, which are usually attained by simple reductions. If the reduction process is carried out in the presence of CNTs, the NPs can be effectively deposited onto the surface walls of CNTs via van der Waals forces. The metal cation reductions are mostly achieved by heat, light, reducing agents, or evaporation techniques. The incorporation of metals onto the CNTs substrate enhances their extensive properties.

Chen and coworkers described the growth of Pt, Ag, Au, Pd, and Cu NPs on the surface of MWCNTs.[24] The originally dispersed metals on the wall surface are reduced by thermal treatment. And the number of NPs deposited on CNTs was found to be much greater than that for amorphous or graphitic carbon by the same method. Xin and coworkers reported two different reduction approaches for the preparation of Pt NPs/MWCNTs nanocomposites.[25] The first one involves the reduction of hexachloroplatinic acid with aqueous formaldehyde solution. The second approach includes the reduction of the Pt precursor with an EG solution. Improvement in catalytic efficiency of DMFC was attributed toward the better homogeneous dispersion of spherical PtNPs with a narrower particle size distribution of nanocomposites obtained from the EG method. Immobilization of Au, Pt, and Rh NPs on SWNT surfaces by mild reduction of the metal salt precursors using polyethylene glycol-200 (PEG200) as reducing agent has been reported.[26] The noble metal nanoclusters dispersed on carbon nanotube (CNT) surface walls are coated with oleylamine, which enhances the colloidal dispersion of SWCNTs in organic solvents such as chloroform and to increase their advantages in the fields of homogenous and heterogenous catalysis. The AuNPs with less than 9 nm diameter were spatially well dispersed on the CNTs by the method of metal salt reduction in the presence of citric acid. The addition of ethanol reduces the interfacial tension between CNTs and AuNPs.

6.2.1.2 *ELECTROCHEMICAL DEPOSITION*

Electrochemical deposition technique involves the interaction of CNTs with precursor salt suspensions, such as $HAuCl_4$, K_2PtCl_4, or $(NH_4)_2PdCl_4$ for the deposition of NPs of Au, Pt, and Pd respectively on the surface of CNTs.[27–29] The size and distribution of metal NPs on the sidewalls of CNTs are controlled by the metal salt concentration and various deposition variables like nucleation potential, pulse rate, and deposition time. High purity and good adherence to CNTs supporting material are the highlights on this approach.

Electrochemical deposition of Pt or bimetallic Pt–Ru NPs onto CNTs with the high diameters of 60–80 nm was reported.[28,31] Tsai et al. successfully synthesized Pt–Ru NPs/CNTs nanohybrids of uniformly dispersed small NPs size on the surface of CNTs by means of potentiostatic electrodeposition in 0.5 M ethylene glycol/0.25 M H_2SO_4 aqueous dispersion. The metal precursors $H_2PtCl_6 \cdot 6H_2O$ and $RuCl_3 \cdot xH_2O$ of Pt and Ru were respectively optimized in the 2:1 ratio has shown the high efficiency of methanol oxidation. The well-dispersed Pt–Ru NPs show a diameter size of 3.1–5.6 nm. Addition of ethylene glycol (EG) prevented the agglomeration of NPs during the whole process of electrochemical deposition and maintained their fine particle size.[32,33] The pulsed electrodeposition was adapted for the production of 2–3 nm-sized Pt NPs with high dispersion on CNTs, by restricting the viscosity of H_2PtCl_6 solution by means of glycerol addition. And the glycerol is ingeniously removed from the NPs surface by rinsing with hot water.[34] The deposition of noble metals NPs like Au, Pt, and Pd on SWCNTs under direct potential control was done by Quinn et al. The surface coverage and particle size of NPs are controlled by the potential, deposition time, and metal salt concentration.[35]

The method of ultrasonic-electrodeposition of AuPt NPs on the surface of MWCNT–ionic liquid composite was reported by Xiao et al. The well-dispersed nanocluster shown particles of 70 nm size in SEM. It showed high electrochemical oxidation of amino acid, cysteine.[36] Huang et al. reported the preparation of well-dispersed Pt NPs on MWCNTs via the co-electrodeposition/stripping (CS) protocol, for high performance electrocatalytic oxidation of methanol in both acidic and alkaline media.[37]

6.2.1.3 ELECTROLESS DEPOSITION

Many reports on the noble metal NPs/CNTs formation by the electroless deposition method are available. Dai et al. reported the spontaneous deposition of Au and Pt NPs on SWCNTs.[4] The procedure includes the immersion of SWCNTs into the corresponding metal precursor solution without any reducing agent. The proposed mechanism involved a direct redox reaction between the metal ions and CNTs.

During the deposition process, metal ions can be transformed to metal NPs on the CNT substrate only when the redox potential of metal ions is higher than that of CNTs. Qu et al. reported a novel process of substrate-enhanced electroless deposition (SEED) for spontaneous deposition of metal NPs on CNTs in the absence of any additional reducing agent. Some metals like Ag having lower redox potential than that of CNTs are included in this method. CNTs act as cathodes and templates for metal deposition from the corresponding metal salts.[38] Reports on the gold deposition onto CNTs using HAuCl4 as precursor salt and its application toward pollutant degradation was noted.[39] One-pot synthesis method for noble metal NPs/CNTs nanohybrids via a redox reaction between metal ions and reduced CNTs was reported.[40] The process involved two steps. In the first step, CNTs are reduced by metallic Na in an aprotic organic solvent, which resulted in a solution of exfoliated negatively charged CNTs. Second step involved the immediate reduction of metals into NPs just upon the addition of metal cations into the CNTs solution. The fast reduction was acquired due to the availability of electrons in excess on the nanocarbon surface. This methodology does not require any metal ion with a redox potential higher than that of CNTs. Au and Pd NPs are successfully anchored to the surface of MWCNTs and SWCNTs by this method without the presence of any surfactants or pretreatment surface functionalization of CNTs.

6.2.1.4 PHYSICAL METHODS

The different physical methods used for the noble metal NPs/CNTs nanohybrid preparation includes ion/electron beam irradiation, sputtering, and evaporation deposition techniques.

Efficient sputtering deposition method for the extensive preparation of ultrafine platinum NPs on CNTs external surface were highly reported.[41–43] Uniform-size noble metal NPs are deposited on the external surface wall of CNTs by controlling the current and appropriate sample exposure time while using selective metal cathodes. Platinum NPs of 3–5 nm size range were uniformly deposited on a vertically aligned CNT (VACNT) arrays via a DC sputtering system.[42] The electrochemical properties and structure of the well-separated Pt NPs of an average diameter of 2 nm on nitrogen-containing CNTs composites directly grown on Si substrates were investigated. Microwave–plasma-enhanced chemical vapor deposition (CVD) method was adopted for growing nanotube arrays and it then acted as template support for Pt dispersion in the sputtering process. And the nanocomposite possesses a high potential for micro direct methanol fuel cell (DMFC) applications.[43]

AuNPs/MWCNTs composite electrodes were prepared by sputter deposition of gold in argon atmosphere on MWCNTs followed by heat-treatment of the catalyst at different temperatures. Au NPs with 20 nm diameter were dispersed on side walls and at the tips of nanotubes. Surface characterization and electrochemical studies of the composites demonstrate their electrocatalytic properties toward O_2 reduction in acid media.[44] Vapor phase growth of gold metal NPs on MWCNT buckypaper by sputtering was investigated. Approximately, 5–30 nm diameter range of gold NPs can be produced depending on the intrinsic sputter process parameters like metal ion flux and kinetic energy. The defect density of MWCNT surfaces can be controlled by annealing prior to sputtering.[45]

The ion and electron beam irradiation deposition technique introduces noble metal NPs on the surface of CNTs. Wang et al. reported the decoration of 2.5–4.0 nm PtNPs on MWCNTs were obtained by γ-irradiation-induced synthesis. These Pt NPs/CNTs nanohybrids showed good electrocatalytic properties as the cathodic material and have promising applications in proton exchange membrane fuel cells (PEMFCs).[46] Uniform dispersion of Pd NPs with an average diameter of 2 nm occurs on the CNT surface by electron beam irradiation method. Homogenous reduction and nucleation which lead to the production of monodispersed small noble metal NPs was the advantage of irradiation deposition over chemical methods.[47] The Pd-coated nanotubes show excellent electrocatalytic performances toward ethanol oxidation in 1 M KOH and were evaluated using cyclic voltammetric studies.[48]

Uniform and highly dense Au NPs are intercalated and encapsulated onto the SWCNTs and double-walled CNTs (DWCNTs) are synthesized using a plasma technique combined by electron irradiation in imidazolium type ionic liquid, under low gas pressures. Instead of the Au NPs, SWNTs can also be impregnated with monodispersed palladium NPs which are generated by the electron irradiation by the static plasma/ionic liquid interface using palladium chloride ($PdCl_2$). These hybrid materials are used for the substantial advancement in nanoelectronic devices.[49]

Thermal or evaporation deposition methods are commonly used techniques for the nanometal/CNTs hybrid production. MWCNTs grown by chemical vapor deposition method modified with Au NPs by thermal evaporation method result in metal nanotubes hybrids having intense applications in catalysis, fuel cells, and electrochemical sensors. The TEM analysis reveals the dependence of evaporation rate, nominal film thickness, and substrate temperature on the nanoparticle size and distribution of AuNPs decorated on the CNTs. The NPs were spatially well dispersed on the CNTs and structures were observed within a range from small 4 nm diameter spherical particles to 150 nm long wire-like structures.[50] The functionalized MWCNTs were covered with silver or gold nanoclusters by means of thermal evaporation. Metal decorated nanotubes act as highly selective sensors for NO_2.[51] A thermal CVD method was adopted for decorating Au NPs on the surface of ethylene diamine functionalized SWCNTs.[52] Thermal and electron beam evaporation technique was adopted for the selective electroplating of Pd, Pt, Rh, and Au metals on isolated networks of SWNTs which were integrated into Si chips by CVD and photolithography processes resulted in an electronic sensor array. The resultant metal nanohybrids act as efficient gas sensors for the detection and identification of toxic/combustible gases like H_2, CH_4, CO, and H_2S for personal safety and air pollution monitoring.[53]

6.2.1.5 MICROEMULSION

Direct deposition of palladium, rhodium, and bimetallic Pd/Rh NPs on the surfaces of functionalized multiwalled CNTs is possible through the water-in-hexane microemulsion technique on alumina. The CNT-supported Pd NPs show unusual catalytic activity for hydrogenation of olefins, carbon–carbon bond formation, and for carbon–oxygen bond cleavage reactions. Pd–CNT composites successfully catalyzed the Heck coupling reaction between iodobenzene and styrene. The CNT-supported Rh NPs are highly active catalysts for hydrogenation of arenes. And Rh–Pd NPs deposited on MWNTs are more active than the single metal NPs on CNTs in the catalytic hydrogenation of anthracene.[54]

6.2.1.6 GRAFTING

The novel rhodium-supported carbon material (Rh/MWNT-COONa) was used as a selective catalyst for the hydrogenation of *trans*-cinnamaldehyde and the hydroformylation of hex-1-ene in the liquid phase. The dimeric complex $[Rh_2Cl_2(CO)_4]$ was grafted to MWNTs surface previously oxidized with nitric acid and was then treated with sodium carbonate to produce carboxylate groups on their outer surface. The grafting mechanism involves the bridge-splitting and carboxylation of MWCNTs surface, followed by the reduction step attained the highly dispersed rhodium NPs of size 1.5–2.5 nm in the presence dihydrogen at 573 K.[55] Composites of Pt-supported MWCNTs are extensively used as cathode electrocatalysts in PEMFC for oxygen reduction reactions. These Pt/MWCNTs were prepared by chemical reduction method involving hexachloroplatinic acid solution and functionalized MWCNTs. The homogeneous distribution of small Pt NPs around 5–8 nm in diameter on the pretreated MWCNTs surface resulted in the better adhesion between the nanocomposite, leading to enhanced performance.[56] Pt-based alloy catalysts show high poison tolerance and greater methanol oxidation activity. A higher electrochemical activity performance was observed for platinum/ruthenium NPs decorated on CNTs in supercritical carbon dioxide using Pt(II)-acetylacetonate as the metal precursor. Experimental results prove that the bimetallic nanocomposite based on CNT act as promising catalyst for methanol oxidation in DMFCs due to the high surface area of CNTs and the lower overpotential for methanol oxidation.[57]

6.2.1.7 SONOCHEMICAL

A sonochemical process was developed for the uniform deposition of PtNPs with a diameter less than 5 nm on the CNT surface. It is clear from the cyclic voltammetry measurements that the PtNPs–CNT composites are more active in the electrochemical adsorption and desorption of hydrogen than PtNPs on carbon black. This enhancement of electrochemical activity is attributed to the uniqueness of CNTs and its interactions between the PtNPs. The synthesized nanocomposites can be used as high loading catalysts for the cathode of polymer electrolyte membrane (PEM) fuel cells.[58]

6.2.1.8 MICROWAVE

Microwave irradiation technique is a well-known noninvasive and clean processing tool for the introduction of desired functional groups like carboxyl, hydroxyl, carbonyl, and allyl termini on CNTs surface, in the absence of strong oxidants or ultrasonication.

Microwave-assisted uniform dispersion of gold NPs on the MWCNTs using $HAuCl_4$ and EG as reducing agent has been reported. The residual negative charge on the functional groups eases the nucleation of AuNPs at the functionalized sites on the CNTs by the $HAuCl_4$ reduction by EG.[59] A microwave-assisted polyol reduction method was adopted for the decoration of Fe oxides on PtNPs/CNTs. The Fe promoted PtNPs catalyst show high turnover frequency and selectivity in the hydrogenation of cinnamaldehyde due to the availability of highly active sites formed at the interface of two metals.[60] N,N-dimethylformamide is used as a dispersant for CNTs and reductant for the Pt ions for the preparation of CNTs-supported PtNPs in microwave heating conditions. The face-centered cubic crystal PtNPs show a particle size of 2–4 nm.[61]

6.2.2 FABRICATION OF NOBLE METAL NPs ON SURFACE FUNCTIONALIZED CNTs

The drawback of insufficient binding sites for anchoring the metal ion precursors or metal NPs on the CNTs can be easily overcomed by its surface modifications. CNTs surface functionalizations can be attained by covalent bonding of functional groups to the Л-conjugated carbon skeleton of the CNTs or by means of noncovalent attachment like Л- Л stacking, hydrophobic interaction, or electrostatic attractions.

The common covalent functionalizations of CNTs include the carboxyl or carbonyl group addition onto the nanotube surface by means of oxidation treatment via HNO_3 or HNO_3/H_2SO_4 mixture.[62,63] And these functionalized CNTs are grafted with noble metal NPs with high dispersion. Amine terminated ionic liquids/carboxyl-functionalized MWCNTs were grafted by the amide linkage between the reactants. AuNPs are uniformly dispersed on the nanohybrid through the electrostatic and ionic exchange interactions of HAuCl4 with the ionic liquid-functionalized CNTs.[64] MWCNTs covalently modified with fourth generation NH_2-terminated polyamidoamine dendrimers (CNT/DEN) are an efficient support for AuNPs. The possible interactions involve the encapsulation of NPs in dendrimers and coordination of NPs onto the surface of dendrimers. In situ synthesis of Ag, Pt, and Ag/Au NPs-coated CNT/DEN is also possible.[65]

The binding of AuNPs on the amine and thioether-functionalized MWCNTs has also been reported.[66,67] PdNPs were dispersed on thiol group, benzenesulfonic, and ionic liquid-functionalized MWCNTs via covalent linkages.[68,69] Hydrophobic and hydrogen bond interactions are used to anchor monolayer dodecanethiol, octanethiols, and mixed-monolayer decanethiol/mercaptoundecanoic acid-protected AuNPs clusters to functionalized nanosurfaces of CNTs for device applications.[70–72] Nitrogen-doped CNTs (NCNTs) effectively anchor and disperse noble metal Au or Pt NPs on their surface.[73] The Pt/NCNTs nanohybrids showed high catalytic activity toward the hydrogenation of cinammaldehyde to cinammyl alcohol and hydrocinnamaldehyde.[74]

The pyrene, thionine, or triphenylphosphine and its derivatives noncovalently bind to the sidewalls of CNTs via Л- Л stacking interactions. These molecules upon modification with thiol, amine, or carboxyl groups can act as linkers for anchoring noble metal NPs on CNT surface.[75–77] Recently reported Pt-CNTs nanocomposites act as anode catalyst for methanol oxidation fuel cells. The PtNPs are absorbed on the CNTs by the aid of triphenylphosphine. 1-aminopyrene is used as a linker for PtRu/MWCNTs nanohybrids.[78] These bimetallic nanohybrids shown higher electrocatalytic activity and better stability for the electrochemical oxidation of methanol.

Also imidazolium-salt-based ionic liquids (IS-ILs) and aminothiols or bifunctional thiols act as linkers for the preparation of CNTs/IS-ILs/Pt NPs and AuNPs/SWCNTs, respectively.[79–81] Au–CNT nanocomposites with AuNPs of diameter 2–4 nm is prepared using 1-pyrenemethylamine as the linker.[75]

The polymers are used to anchor noble metal NPs to the surface of CNTs. Pt and PtPd NPs are immobilized on the surface of polypyrrole-modified CNTs (CNTs-PPY) and Au NPs were uniformly dispersed on the surface of polyaniline (PANI) or polythiophene-modified CNTs.[82–84] The polyvinylpyrrolidone (PVP)-modified CNTs provide a better dispersion of Pt and PtRu NPs with smaller average sizes on the nanosurface.[85] Uniformly dispersed PtRu and Pt NPs on the ionic liquid polymer-functionalized CNTs (CNTs-PIL) showed high performance in direct electrooxidation of methanol.[86] The polymer film introduces a large number of surface functional groups on the CNTs in order to anchor Pt and Ru precursors via coordination and electrostatic interactions.

Cationic polyelectrolyte, poly(diallyldimethylammonium)chloride (PDADMAC), were adsorbed on the nanotube surface by electrostatic mode of interaction between carboxyl groups on the oxidized CNTs surface and polyelectrolyte chains. Negatively charged AuNPs (10 nm) from a gold colloid suspension were anchored to the CNTs surface by the electrostatic interaction between the polyelectrolyte and AuNPs.[87] An amino-rich cationic polyelectrolyte, polyethyleneimine (PEI) are used as a reducing and protecting agent for the preparation of AuNPs–CNTs composite.[88] In order to accept the positively charged NPs, the CNT surface is modified to negative charge by varying the kind of polyelectrolytes used.

Many groups reported the electrostatic layer-by-layer (LBL) self-assembly-based synthesis of noble metal NPs/CNTs nanohybrids.[89–91] This method involves the polyelectrolytes for wrapping CNTs and providing them with adsorption sites for electrostatically driven NP deposition. Acid-functionlized MWCNTs were covered with a positively charged poly(diallyldimethylammonium chloride) polymer [PDDA] followed by a negatively charged polymer layer of poly(sodium 4-styrenesulfonate)] [PSS]. The AuNPs of positive charge are then subsequently anchored to PSS layer by means of electrostatic interactions.[92] Marzán's group reported the wrapping of CNT with a negatively charged PSS polyelectrolyte, then by a self-assembly of positively charged PDDA by electrostatic interactions serve as a real template for the adsorption of AuNPs.[89] Negatively charged tungstophosphoric acid ($H_3PW_{12}O_{40}$ HPW) was electrostatically self-assembled on the positively charged PDDA-functionalized MWCNTs. This is followed by the homogenous dispersion of PdNPs with an average size of 2 nm and the obtained nanohybrid is denoted as Pd/HPW/PDDA/ MWCNTs. The incorporation of HPW into the PDDA-MWCNTs significantly reduced the particle size and enhanced the uniform deposition of Pd NPs on MWCNTs surface.[91]

Homogeneous coating of Pd NPs on CNTs surface by means of supramolecular auto-organization of amphiphilic molecules as template was reported. These nanohybrids show superior electrocatalysis activity in ethanol oxidation.[48] Supramolecular self-assembly of sodium dodecyl sulfate (SDS) was used as a template for the selective deposition of Pd and Pt NPs on CNTs.[92,93] Hexadecyltrimethylammonium bromide (CTAB) capped seed-mediated method was adopted for the alignment of uniform Au nanorods over MWCNTs, where nanorods are connected by end- to -end contacts.[94]

6.3 HETEROGENOUS CATALYSIS OF NOBLE METAL NPS/CNTs

Recently, immobilization of metal NPs on CNTs and its applications are studied extensively. The size and shape-dependent nature of metal NPs showed broad range of applications from catalysis to biomedical field. The heterogenous catalytic activity of CNT nanocomposites in the various reactions is briefly discussed here.

High mechanical strength, large surface-to-volume ratio and good durability even in harsh conditions make CNT an efficient support material for metal NPs. The interaction of noble metal catalyst with CNTs surface results in the modification of the electron density of the noble metal nanoclusters and hence enhances the catalytic activities.

PdNP/SWCNTs exhibit higher catalytic activity than palladium-activated carbon (Pd/C) for the Heck reaction of styrene and iodobenzene and also for the Suzuki coupling of phenylboronic and iodobenzene. The activity of the palladium catalyst for C-C bond forming reactions including hydrogenation and oxidation are deeply influenced by their particle size (Fig. 6.1).[95,96] Efficient Suzuki-Miyaura coupling of aryl-bromides are efficiently catalyzed by Pd/CNT under ligandless and additive-free conditions in aqueous media.[97] The effective hydrodehalogenation of aryl halides was promoted by Pd/CNT nanocomposites at a very low Pd content (~2.3%) and in the absence of any ligands. Thiol groups were used as linkers to anchor the PdNPs on the CNTs surface without agglomeration.[98]

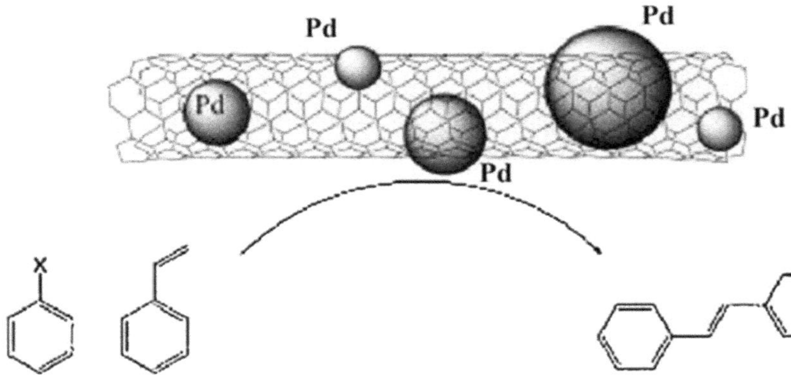

FIGURE 6.1 Scheme of Pd/SWCNTs nanocomposite for Heck coupling reaction.
Source: Reproduced with permission from [96], ©2011 Elsevier.

The heterogeneous catalysis of RuNPs/CNTs nanohybrids involved in the hydrogenation of cinnamaldehyde is the very first report on the application of metal NPs/CNTs.[23] About 4–6 nm-sized homogenous PdNPs-deposited MWCNTs catalyst was tested for the selective hydrogenation of cinnamaldehyde which contains both a C=C and a C=O bond. The PdNPs/MWCNTs exhibited highly selective catalytic activity toward the C=C bond hydrogenation when compared with the commercial Pd/C catalyst due to the faster desorption of the C=C hydrogenated products from their active sites.[99]

Chun et al. reported the deposition of PdNPs onto the imidazolium bromide-functionalized ionic MWCNTs through hydrogen reduction of Na_2PdCl_4 in water.[69] These highly active heterogenous catalysts show efficient hydrogenation of trans-stilbene in ionic solvents. And the catalysts were stable upto 50 times of recycling. RuNPs dispersed on the exterior nanotube surface shows higher activity in the ammonia synthesis than the CNT-confined Ru, although both have a similar metal particle size.[100] N_2 physisorption, H_2 chemisorption, temperature-programmed reduction, and CO adsorption microcalorimetry characterizations suggested that the exterior of Ru exhibits a higher electron density than the interior of Ru. The dissociative adsorption of N_2 is the rate-determining step of ammonia synthesis and is more facile over the exterior of RuNPs than that over the interior of Ru. Deposition of utrafine PtNPs onto MWNTs with the aid of tip sonication, shows excellent activity in the hydrogenation of nitrobenzene to aniline under solvent-free conditions. The nanocomposites catalyzed reaction resulted in a high turnover frequency.[101] The catalytic efficiency of AgNP/CNTs for trypan blue reduction reaction resulted in a rate constant value of 14.05×10^{-2} min^{-1}.[102]

6.4 CONCLUSIONS

Noble metal NPs/CNTs nanohybrids are a new class of composite materials with extraordinary properties which are inherited from both the noble metal NPs and CNTs. A brief description on the various synthetic approaches toward noble metal NP/CNTs including electrochemical deposition, electroless deposition, physical methods, microwave, sonochemical, and dispersion of noble metal NPs on the functionalized CNTs have been presented. The main advantages of CNTs as a supporting material involve the high surface-to-volume area, thermal properties, chemical inertness, and electrical conductivity. The results so far reported show that noble metal NP/CNTs are very promising due to the enhanced activity and selectivity in heterogeneous catalysis. The use of noble metal NP/CNTs as catalysts is of growing interest toward efficient industrial applications.

KEYWORDS

- **single-walled carbon nanotubes**
- **multi-walled carbon nanotubes**
- **functionalization**
- **heterogenous catalysis**
- **noble metal nanoparticles**
- **nanocomposites**

REFERENCES

1. Chu, H.; Wei, L.; Cui, R.; Wang, J.; Li, Y. Carbon Nanotubes Combined with Inorganic Nanomaterials: Preparations and Applications. *Coordination Chem. Rev.* **2010,** *254* (9–10), 1117–1134.
2. Gong, K.; Du, F.; Xia, Z.; Durstock, M.; Dai, L. Nitrogen-Doped Carbon Nanotube Arrays with High Electrocatalytic Activity for Oxygen Reduction. *Science* **2009,** *323* (5915), 760–764.
3. Qu, L.; Dai, L.; Osawa, E. *J. Am. Chem. Soc.* **2006,** *128* (16), 5523–5532.
4. Choi, H. C.; Shim, M.; Bangsaruntip, S.; Dai, H. Spontaneous Reduction of Metal Ions on the Sidewalls of Carbon Nanotubes. *J. Am. Chem. Soc.* **2002,** *124* (31), 9058–9059.
5. Yuan, W.; Zhang, Y.; Cheng, L.; Wu, H.; Zheng, L. Zhao, D. The Applications of Carbon Nanotubes and Graphene in Advanced Rechargeable Lithium Batteries. *J. Mater. Chem. A* **2016,** *4* (23), 8932–8951.
6. Gupta, B. D.; Pathak, A.; Semwal, V. Carbon-Based Nanomaterials for Plasmonic Sensors: A Review. *Sensors* **2019,** *19* (16), 3536.
7. Ahsan, M. A.; Jabbari, V.; Islam, M. T.; Turley, R. S.; Dominguez, N.; Kim, H.; Castro, E.; Hernandez-Viezcas, J. A.; Curry, M. L.; Lopez, J.; Gardea-Torresdey, J. L.; Noveron, J. C. Sustainable Synthesis and Remarkable Adsorption Capacity of MOF/Graphene Oxide and MOF/CNT Based Hybrid Nanocomposites for the Removal of Bisphenol A from Water. *Sci. Total Environ.* **2019,** *673*, 306–317.
8. Daniyal, M.; Liu, B.; Wang, W. Comprehensive Review on Graphene Oxide for Use in Drug Delivery System. *Curr. Med. Chem.* **2020,** *27* (22), 3665–3685.
9. Zhang, F.; Li, Y.-H.; Li, J.-Y.; Tang, Z.-R.; Xu, Y.-J. 3D graphene-Based Gel Photocatalysts for Environmental Pollutants Degradation. *Environ. Pollut.* **2019,** *253*, 365–376.
10. Strano, M. S.; Moore, V. C.; Miller, M. K.; Allen, M. J. , Haroz, E. H.; Kittrell, C.; Hauge, R. H.; Smalley, R. E. The Role of Surfactant Adsorption during Ultrasonication in the Dispersion of Single-Walled Carbon Nanotubes. *J. Nanosci. Nanotechnol.* **2003,** *3* (1), 81–86.

11. Li, S.; Hu, S.; Xu, K.; Jiang, W.; Liu, J.; Wang, Z. A Novel Heterostructure of BiOI Nanosheets Anchored onto MWCNTs with Excellent Visible-Light Photocatalytic Activity. *Nanomaterials* **2017,** *7* (1), 22.
12. Li, Q.; Guan, Z.; Wu, D.; Zhao, X.; Bao, S.; Tian, B.; Zhang, J. Z-Scheme BiOCl-Au-CdS Heterostructure with Enhanced Sunlight-Driven Photocatalytic Activity in Degrading Water Dyes and Antibiotics. *ACS Sustain. Chem. Eng.* (2017), *5* (8), 6958–6968.
13. Fujita, S.; Katagiri, A.; Watanabe, H.; Asano, S. Yoshida, H.; Arai, M. Preparation of Nitrogen-Doped Carbon from Polyacrylonitrile and its Application as a Solid-Base Catalyst. *ChemCatChem* **2015,** *7* (18), 2965–2970.
14. Tessonnier, J.-P.; Villa, A.; Majoulet, O.; Su, D. S.; Schlögl, R. Defect-Mediated Functionalization of Carbon Nanotubes as a Route to Design Single-Site Basic Heterogeneous Catalysts for Biomass Conversion. *Angew. Chem. Int. Ed.* **2009,** *48* (35), 6543–6546.
15. Ramirez-Barria, C.; Guerrero-Ruiz, A.; Castillejos-López, E.; Rodríguez-Ramos, I.; Durand, J.; Volkman, J.; Serp, P. Surface Properties of Amphiphilic Carbon Nanotubes and Study of Their Applicability as Basic Catalysts. *RSC Adv.* **2016,** *6* (59), 54293–54298.
16. Aiad, I.; Shaban, S. M.; Tawfik, S. M.; Khalil, M. M. H.; El-Wakeel, N. Effect of Some Prepared Surfactants on Silver Nanoparticles Formation and Surface Solution Behavior and Their Biological Activity. *J. Mol. Liquids* **2018,** *266*, 381–392.
17. Morsy, S. M. Role of Surfactants in Nanotechnology and Their Applications. *Int. J. Curr. Microbiol. Appl. Sci.* **2014,** *3* (5), 237–260.
18. Cozzoli, P. D.; Pellegrino, T.; Manna, L. *Chem. Soc. Rev.* **2006,** *35* (11), 1195.
19. Yin, Y.; Alivisatos, A. P. Colloidal Nanocrystal Synthesis and the Organic–Inorganic Interface. *Nature* **2004,** *437* (7059), 664–670.
20. Schlögl, R.; Abd Hamid, S. B. Nanocatalysis: Mature Science Revisited or Something Really New. *Angew. Chem. Int. Ed.* **2004,** *43* (13), 1628–1637.
21. Guldi, D. M.; Rahman, G. M. A.; Zerbetto, F.; Prato, M. Carbon Nanotubes in Electron Donor−Acceptor Nanocomposites. *Acc. Chem. Res.* **2005,** *38* (11), 871–878.
22. Baughman, R. H. Carbon Nanotubes the Route toward Applications. *Science* **2002,** *297* (5582), 787–792.
23. Planeix, J. M.; Coustel, N.; Coq, B.; Brotons, V.; Kumbhar, P. S.; Dutartre, R.; Geneste, P.; Bernier, P.; Ajayan, P. M. Application of Carbon Nanotubes as Supports in Heterogeneous Catalysis. *J. Am. Chem. Soc.* **1994,** *116* (17), 7935–7936.
24. Xue, B.; Chen, P.; Hong, Q.; Lin, J.; Tan, K. L. Growth of Pd, Pt, Ag and Au nanoparticles on carbon nanotubes. *J. Mater. Chem.* **2001,** *11* (9), 2378–2381.
25. Li, W.; Liang, C.; Zhou, W.; Qiu, J.; Zhou, Sun, G.; Xin, Q. Preparation and Characterization of Multiwalled Carbon Nanotube-Supported Platinum for Cathode Catalysts of Direct Methanol Fuel Cells. *J. Phys. Chem B.* **2003,** *107* (26), 6292–6299.
26. Tzitzios, V.; Georgakilas, V.; Oikonomou, E.; Karakassides, M.; Petridis, D. Synthesis and Characterization of Carbon Nanotube/Metal Nanoparticle Composites Well Dispersed in Organic Media. *Carbon* **2006,** *44* (5), 848–853.
27. Dobrzanski, L. A.; Pawlyta, M.; Krzton, A.; Liszka, B.; Tai, C. W.; Kwasny, W. Synthesis and Characterization of Carbon Nanotubes Decorated with Gold Nanoparticles. *Acta Phys. Polonica A.* **2010,** *118* (3), 483–486.

28. He, Z.; Chen, J.; Liu, D.; Zhou, H.; Kuang, Y. Electrodeposition of Pt–Ru Nanoparticles on Carbon Nanotubes and Their Electrocatalytic Properties for Methanol Electrooxidation. *Diamond Relat. Mater*. **2004,** *13* (10), 1764–1770.
29. Zhao, Y.; Fan, L.; Zhong, H.; Li, Y.; Yang, S. Platinum Nanoparticle Clusters Immobilized on Multiwalled Carbon Nanotubes: Electrodeposition and Enhanced Electrocatalytic Activity for Methanol Oxidation. *Adv. Funct. Mater*. **2007,** *17* (9), 1537–1541.
30. Zhao, Y.; Y. E, L. Fan, Qiu, Y.; Yang, S. A New Route for the Electrodeposition of Platinum–Nickel Alloy Nanoparticles on Multi-Walled Carbon Nanotubes. *Electrochim. Acta*. **2007,** *52* (19), 5873–5878.
31. He, Z.; Chen, J.; Liu, D.; Tang, H.; Deng, W.; Kuang, Y. Deposition and Electrocatalytic Properties of Platinum Nanoparticals on Carbon Nanotubes for Methanol Electrooxidation. *Mater. Chem. Phys*. **2004,** *85* (2–3), 396–401.
32. Tsai, M.-C.; Yeh, T.-K.; Tsai, C.-H. An Improved Electrodeposition Technique for Preparing Platinum and Platinum–Ruthenium Nanoparticles on Carbon Nanotubes Directly Grown on Carbon Cloth for Methanol Oxidation. *Electrochem. Commun*. **2006,** *8* (9), 1445–1452.
33. Tsai, M.-C.; Yeh, T.-K.; Tsai, C.-H. Electrodeposition of Platinum-Ruthenium Nanoparticles on Carbon Nanotubes Directly Grown on Carbon Cloths for Methanol Oxidation. *Mater. Chem. Phys*. **2008,** *109* (2–3), 422–428.
34. Chen, X.; Li, N.; Eckhard, K.; Stoica, L.; Xia, W.; Assmann, J.; Muhler, M.; Schuhmann, W. Pulsed Electrodeposition of Pt Nanoclusters on Carbon Nanotubes Modified Carbon Materials Using Diffusion Restricting Viscous Electrolytes. *Electrochem. Commun*. **2007,** *9* (6), 1348–1354.
35. Quinn, B. M.; Dekker, C.;Lemay, S. G. Electrodeposition of Noble Metal Nanoparticles on Carbon Nanotubes. *J. Am. Chem. Soc*. **2005,** *127* (17), 6146–6147.
36. Xiao, F.; Mo, Z.; Zhao, F.; Zeng, B. Ultrasonic-Electrodeposition of Gold–Platinum Alloy Nanoparticles on Multi-Walled Carbon Nanotubes – Ionic Liquid Composite Film and Their Electrocatalysis towards the Oxidation of Nitrite. *Electrochem. Commun*. **2008,** *10* (11), 1740–1743.
37. Huang, J.; Xie, Q.; Tan, Y.; Fu, Y.; Su, Z.; Huang, Y.; Yao, S. Preparation of Pt/Multiwalled Carbon Nanotubes Modified Au Electrodes via Pt–Cu co-Electrodeposition/Cu Stripping Protocol for High-Performance Electrocatalytic Oxidation of Methanol. *Mater. Chem. Phys*. **2009,** *118* (2–3), 371–378.
38. Qu, L.; Dai, L. Substrate-Enhanced Electroless Deposition of Metal Nanoparticles on Carbon Nanotubes. *J. Am. Chem. Soc*. **2005,** *127* (31), 10806–10807.
39. Ghose, S.; Watson, K. A.; Elliott, H. A.; Working, D. C., Criss, J. M.; Dudley, K. L.; Siochi, E. J.; Connell, J. W. Fabrication and Characterization of High Temperature Resin/Carbon Nanofiller Composites. *Multifunct. Nanocompos*. **2006**.
40. Lorençon, E.; Ferlauto, A. S.; de Oliveira, S.; Miquita, D. R.; Resende, R. R.; Lacerda, R. G.; Ladeira, L. O. Direct Production of Carbon Nanotubes/Metal Nanoparticles Hybrids from a Redox Reaction between Metal Ions and Reduced Carbon Nanotubes. *ACS Appl. Mater. Interf.* **2009,** *1* (10), 2104–2106.
41. Soin, N.; Roy, S. S.; Karlsson, L.; McLaughlin, J. A. Sputter Deposition of Highly Dispersed Platinum Nanoparticles on Carbon Nanotube Arrays for Fuel Cell Electrode Material. *Diamond Relat. Mater*. **2010,** *19* (5–6), 595–598.

42. Zacharia, R.; Rather, S.; Hwang, S. W.; Nahm, K. S. Spillover of Physisorbed Hydrogen from Sputter-Deposited Arrays of Platinum Nanoparticles to Multi-Walled Carbon Nanotubes. *Chem. Phys. Lett.* **2007,** *434* (4–6), 286–291.
43. Sun, C.-L.; Chen, L.-C.; Su, M.-C.; Hong, L.-S.; Chyan, O.; Hsu, C.-Y.; Chen, K.-H.; Chang, T.-F.; Chang, L. Ultrafine Platinum Nanoparticles Uniformly Dispersed on Arrayed CNxNanotubes with High Electrochemical Activity. *Chem. Mater.* **2005,** *17* (14), 3749–3753.
44. Alexeyeva, N.; Kozlova, J.; Sammelselg, V.; Ritslaid, P.; Mändar, H.; Tammeveski, K. Electrochemical and Surface Characterisation of Gold Nanoparticle Decorated Multi-Walled Carbon Nanotubes. *Appl. Surf. Sci.* **2010,** *256* (10), 3040–3046.
45. Muratore, C.; Reed, A. N.; Bultman, J. E.; Ganguli, S.; Cola, B. A.; Voevodin, A. A. Nanoparticle Decoration of Carbon Nanotubes by Sputtering. *Carbon* **2013,** *57*, 274–281.
46. Wang, H.; Sun, X.; Ye, Y.; Qiu, S. Radiation Induced Synthesis of Pt Nanoparticles Supported on Carbon Nanotubes. *J. Power Sourc.* **2006,** *161* (2), 839–842.
47. Belloni, J.; Mostafavi, M.; Remita, H.; Marignier, J.-L.; Delcourt, M.-O. Radiation-Induced Synthesis of Mono- and Multi-Metallic Clusters and Nanocolloids. *N. J. Chem.* **1998,** *22* (11), 1239–1255.
48. Mackiewicz, N.; Surendran, G.; Remita, H.; Keita, B.; Zhang, G.; Nadjo, L.; Hagège, A.; Doris, E.; Mioskowski, C. Supramolecular Self-Assembly of Amphiphiles on Carbon Nanotubes: A Versatile Strategy for the Construction of CNT/Metal Nanohybrids, Application to Electrocatalysis. *J. Am. Chem. Soc.* **2008,** *130* (26), 8110–8111.
49. Baba, K.; Kaneko, T.; Hatakeyama, R.; Motomiya, K.; Tohji, K. Synthesis of Monodispersed Nanoparticles Functionalized Carbon Nanotubes in Plasma-Ionic Liquid Interfacial Fields. *Chem. Commun.* **2010,** *46* (2), 255–257.
50. Gingery, D.; Bühlmann, P. Formation of Gold Nanoparticles on Multiwalled Carbon Nanotubes by Thermal Evaporation. *Carbon* **2008,** *46* (14), 1966–1972.
51. Espinosa, E. H.; Ionescu, R.; Bittencourt, C.; Felten, A.; Erni, R.; Van Tendeloo, G.; Pireaux, J.-J.; Llobet, E. Metal-Decorated Multi-Wall Carbon Nanotubes for Low Temperature Gas Sensing. *Thin Solid Films* **2007,** *515* (23), 8322–8327.
52. Jeong, G.-H. Surface Functionalization of Single-Walled Carbon Nanotubes Using Metal Nanoparticles. *Trans. Nonferrous Metals Soc. China* **2009,** *19* (4), 1009–1012.
53. Star, A.; Joshi, V.; Skarupo, S.; Thomas, D.; Gabriel, J.-C. P. Gas Sensor Array Based on Metal-Decorated Carbon Nanotubes. *J. Phys. Chem. B* **2006,** *110* (42), 21014–21020.
54. Yoon, B.; Wai, C. M. Microemulsion-Templated Synthesis of Carbon Nanotube-Supported Pd and Rh Nanoparticles for Catalytic Applications. *J. Am. Chem. Soc.* **2005,** *127* (49), 17174–17175.
55. Giordano, R.; Serp, P.; Kalck, P.; Kihn, Y.; Schreiber, J.; Marhic, C.; Duvail, J.-L. Preparation of Rhodium Catalysts Supported on Carbon Nanotubes by a Surface Mediated Organometallic Reaction. *Eur. J. Inorg. Chem.* **2003,** *2003* (4), 610–617.
56. Yu, R.; Chen, L.; Liu, Q.; Lin, J.; Tan, K.-L.; Ng, S. C.; Chan, H. S. O.; Xu, G.-Q.; Hor, T. S. A. Platinum Deposition on Carbon Nanotubes via Chemical Modification. *Chem. Mater.* **1998,** *10* (3), 718–722.

57. Lin, Y.; Cui, X.; Yen, C.; Wai, C. M. Platinum/Carbon Nanotube Nanocomposite Synthesized in Supercritical Fluid as Electrocatalysts for Low-Temperature Fuel Cells. *J. Phys. Chem. B*. **2005,** *109* (30), 14410–14415.
58. Xing, Y. Synthesis and Electrochemical Characterization of Uniformly-Dispersed High Loading Pt Nanoparticles on Sonochemically-Treated Carbon Nanotubes. *J. Phys. Chem. B* **2004,** *108* (50), 19255–19259.
59. Raghuveer, M. S.; Agrawal, S.; Bishop, N.; Ramanath, G. Microwave-Assisted Single-Step Functionalization and in Situ Derivatization of Carbon Nanotubes with Gold Nanoparticles. *Chem. Mater*. **2006,** *18* (6), 1390–1393.
60. Guo, Z.; Zhou, C.; Shi, D.; Wang, Y.; Jia, X.; Chang, J.; Borgna, A.; Wang, C.; Yang, Y. Toward the Decoration of Pt Nanoparticles Supported on Carbon Nanotubes with Fe Oxides and Its Effect on the Catalytic Reaction. *Appl. Catalys. A Gen*. **2012,** *435*–436,131–140.
61. Xu, J. B.; Zhao, T. S. Synthesis of Well-Dispersed Pt/Carbon Nanotubes Catalyst Using Dimethylformamide as a Cross-Link. *J. Power Sourc*. **2010,** *195* (4), 1071–1075.
62. Hou, X.; Wang, L.; Zhou, F.; Li, L.; Li, Z. High-Density Assembly of Gold Nanoparticles to Multiwalled Carbon Nanotubes Using Ionic Liquid as Interlinker. *Mater. Lett*. **2009,** *63* (8), 697–699.
63. Xu, H.; Zeng, L.; Xing, S.; Shi, G.; Chen, J.; Xian, Y.; Jin, L. Highly Ordered Platinum-Nanotube Arrays for Oxidative Determination of Trace Arsenic (III). *Electrochem. Commun*. **2008,** *10* (12), 1893–1896.
64. Wang, Z.; Zhang, Q.; Kuehner, D.; Xu, X.; Ivaska, A.; Niu, L. The Synthesis of Ionic-Liquid-Functionalized Multiwalled Carbon Nanotubes Decorated with Highly Dispersed Au Nanoparticles and Their Use in Oxygen Reduction by Electrocatalysis. *Carbon* **2008,** *46* (13), 1687–1692.
65. Lu, X.; Imae, T. Size-Controlled in situ Synthesis of Metal Nanoparticles on Dendrimer-Modified Carbon Nanotubes. *J. Phys. Chem. C* **2007,** *111* (6), 2416–2420.
66. Hou, X.; Wang, L.; Zhou, F.; Wang, F. High-Density Attachment of Gold Nanoparticles on Functionalized Multiwalled Carbon Nanotubes Using Ion Exchange. *Carbon* **2009,** *47* (5), 1209–1213.
67. Yang, S.; Zhang, X.; Mi, H.; Ye, X. Pd Nanoparticles Supported on Functionalized Multi-Walled Carbon Nanotubes (MWCNTs) and Electrooxidation for Formic Acid. *J. Power Sourc*. **2008,** *175* (1), 26–32.
68. Kim, J. Y.; Jo, Y.; Kook, S.-K.; Lee, S.; Choi, H. C. Synthesis of Carbon Nanotube Supported Pd Catalysts and Evaluation of Their Catalytic Properties for CC Bond Forming Reactions. *J. Mol. Catalys. A Chem*. **2010,** *323* (1–2), 28–32.
69. Chun, Y. S ; Shin, J. Y.; Song, C. E.; Lee, S. Palladium Nanoparticles Supported onto Ionic Carbon Nanotubes as Robust Recyclable Catalysts in an Ionic Liquid. *Chem. Commun*. **2008,** *8* (8), 942–944.
70. Rahman, G. M.; Guldi, D.; Zambon, E.; Pasquato, L.; Tagmatarchis, N.; Prato, M. Dispersable Carbon Nanotube/Gold Nanohybrids: Evidence for Strong Electronic Interactions. *Small* **2005,** *1* (5), 527–530.
71. Ellis, A. V.; Vijayamohanan, K.; Goswami, R.; Chakrapani, N.; Ramanathan, L. S.; Ajayan, P. M.; Ramanath, G. Hydrophobic Anchoring of Monolayer-Protected Gold Nanoclusters to Carbon Nanotubes. *Nano Lett*. **2003,** *3* (3), 279–282.

72. Han, L.; Wu, W.; Kirk, F. L.; Luo, J.; Maye, M. M.; Kariuki, N. N.; Lin, Y.; Wang, C.; Zhong, C.-J. A Direct Route toward Assembly of Nanoparticle–Carbon Nanotube Composite Materials. *Langmuir* **2004,** *20* (14), 6019–6025.
73. Li, X.; Liu, Y.; Fu, L.; Cao, L.; Wei, D.; Yu, G.; Zhu, D. Direct Route to High-Density and Uniform Assembly of Au Nanoparticles on Carbon Nanotubes. *Carbon* **2006,** *44* (14), 3139–3142.
74. Lepró, X.; Terrés, E.; Vega-Cantú, Y.; Rodríguez-Macías, F. J.; Muramatsu, H.; Kim, Y. A.; Hayahsi, T.; Endo, M.; Torres, R. M.; Terrones, M. Efficient anchorage of Pt Clusters on N-Doped Carbon Nanotubes and Their Catalytic Activity. *Chem. Phys. Lett.* **2008,** *463* (1–3), 124–129.
75. Liu, L.; Wang, T.; Li, J.; Guo, Z.-X.; Dai, L.; Zhang, D.; Zhu, D. Self-Assembly of Gold Nanoparticles to Carbon Nanotubes Using a Thiol-Terminated Pyrene as Interlinker. *Chem. Phys. Lett.* **2003,** *367* (5–6), 747–752.
76. Ou, Y.-Y.; Huang, M. H. High-Density Assembly of Gold Nanoparticles on Multiwalled Carbon Nanotubes Using 1-Pyrenemethylamine as Interlinker. *J. Phys. Chem. B* **2006,** *110* (5), 2031–2036.
77. Mu, Y.; Liang, H.; Hu, J.; Jiang, L.; Wan, L. Controllable Pt Nanoparticle Deposition on Carbon Nanotubes as an Anode Catalyst for Direct Methanol Fuel Cells. *J. Phys. Chem. B* **2005,** *109* (47), 22212–22216.
78. Wang, S.; Wang, X.; Jiang, S. P. PtRu Nanoparticles Supported on 1-Aminopyrene-Functionalized Multiwalled Carbon Nanotubes and Their Electrocatalytic Activity for Methanol Oxidation. *Langmuir* **2008,** *24* (18), 10505–10512.
79. Azamian, B. R.; Coleman, K. S.; Davis, J. J.; Hanson, N.; Green, M. L. H. Directly Observed Covalent Coupling of Quantum Dots to Single-Wall Carbon Nanotubes. *Chem. Commun.* **2002,** *4*, 366–367.
80. Zanella, R.; Basiuk, E. V.; Santiago, P.; Basiuk, V. A.; Mireles, E.; Puente-Lee, I.; Saniger, J. M. Deposition of Gold Nanoparticles onto Thiol-Functionalized Multiwalled Carbon Nanotubes. *J. Phys. Chem. B* **2005,** *109* (34), 16290–16295.
81. Guo, S.; Dong, S.; Wang, E. Constructing Carbon Nanotube/Pt Nanoparticle Hybrids Using an Imidazolium-Salt-Based Ionic Liquid as a Linker. *Adv. Mater.* **2010,** *22* (11), 1269–1272.
82. Selvaraj, V.; Alagar, M.; Kumar, K. Synthesis and Characterization of Metal Nanoparticles-Decorated PPY–CNT Composite and Their Electrocatalytic Oxidation of Formic Acid and Formaldehyde for Fuel Cell Applications. *Appl. Cataly. B Environ.* **2007,** *75* (1–2), 129–138.
83. Oliveira, M. M.; Zarbin, A. J. G. Carbon Nanotubes Decorated with both Gold Nanoparticles and Polythiophene. *J. Phys. Chem. C* **2008,** *112* (48), 18783–18786.
84. P. Santhosh, A. Gopalan, K. Lee. Gold Nanoparticles Dispersed Polyaniline Grafted Multiwall Carbon Nanotubes as Newer Electrocatalysts: Preparation and Performances for Methanol Oxidation. *J. Cataly.* **2006,** *238* (1), 177–185.
85. Hsin, Y. L.; Hwang, K. C.; Yeh, C.-T. Poly (vinylpyrrolidone)-Modified Graphite Carbon Nanofibers as Promising Supports for PtRu Catalysts in Direct Methanol Fuel Cells. *J. Am. Chem. Soc.* **2007,** *129* (32), 9999–10010.
86. Wu, B.; Hu, D.; Kuang, Y.; Liu, B.; Zhang, X.; Chen, J. Functionalization of Carbon Nanotubes by an Ionic-Liquid Polymer: Dispersion of Pt and PtRu Nanoparticles on

Carbon Nanotubes and Their Electrocatalytic Oxidation of Methanol. *Angew. Chem. Int. Ed.* **2009,** *48* (26), 4751–4754.

87. Jiang, K.; Eitan, A.; Schadler, L. S.; Ajayan, P. M.; Siegel, R. W.; Grobert, N.; Mayne, M.; Reyes-Reyes, M.; Terrones, H.; Terrones, M. Selective Attachment of Gold Nanoparticles to Nitrogen-Doped Carbon Nanotubes. *Nano Lett.* **2003,** *3* (3), 275–277.
88. Wang, T.; Hu, X.; Qu, X.; Dong, S. Noncovalent Functionalization of Multiwalled Carbon Nanotubes: Application in Hybrid Nanostructures. *J. Phys. Chem. B* **2006,** *110* (13), 6631–6636.
89. Wang, L.; Guo, S.; Huang, L.; Dong, S. Alternate Assemblies of Polyelectrolyte Functionalized Carbon Nanotubes and Platinum Nanoparticles as Tunable Electrocatalysts for Dioxygen Reduction. *Electrochem. Commun.* **2007,** *9* (4), 827–832.
90. Correa-Duarte, M. A.; Liz-Marzán, L. M. Carbon Nanotubes as Templates for One-Dimensional Nanoparticle Assemblies. *J. Mater. Chem.* **2006,** *16* (1), 22–25.
91. Wang, D.; Lu, S.; Jiang, S. P. Pd/HPW-PDDA-MWCNTs as Effective Non-Pt Electrocatalysts for Oxygen Reduction Reaction of Fuel Cells. *Chem. Commun.* **2010,** *46* (12), 2058.
92. Kim, B.; Sigmund, W. M. Functionalized Multiwall Carbon Nanotube/Gold Nanoparticle Composites. *Langmuir* **2004,** *20* (19), 8239–8242.
93. Tan, Z.; Abe, H.; Naito, M.; Ohara, S. Arrangement of Palladium Nanoparticles Templated by Supramolecular Self-Assembly of SDS Wrapped on Single-Walled Carbon Nanotubes. *Chem. Commun.* **2010,** *46* (24), 4363.
94. Lin, J. F.; Mason, C. W.; Adame, A.; Liu, X.; Peng, X. H.; Kannan, A. M. Synthesis of Pt Nanocatalyst with Micelle-Encapsulated Multi-Walled Carbon Nanotubes as Support for Proton Exchange Membrane Fuel Cells. *Electrochim. Acta.* **2010,** *55* (22), 6496–6500.
95. Corma, A.; Garcia, H.; Leyva, A. Catalytic Activity of Palladium Supported on Single Wall Carbon Nanotubes Compared to Palladium Supported on Activated Carbon. *J. Mol. Catal. A Chem.* **2005,** *230* (1–2), 97–105.
96. Wu, B.; Kuang, Y.; Zhang, X.; Chen, J. Noble Metal Nanoparticles/Carbon Nanotubes Nanohybrids: Synthesis and Applications. *Nano Today* **2011,** *6* (1), 75–90.
97. Zhang, P.-P. , Zhang, X.-X.; Sun, H.-X.; Liu, R.-H.; Wang, B.; Lin, Y.-H. Pd–CNT-Catalyzed Ligandless and Additive-Free Heterogeneous Suzuki–Miyaura Cross-Coupling of Arylbromides. *Tetrahedron Lett.* **2009,** *50* (31), 4455–4458.
98. Kim, J. Y.; Jo, Y.; Lee, S.; Choi, H. C. Synthesis of Pd–CNT Nanocomposites and Investigation of Their Catalytic Behavior in the Hydrodehalogenation of Aryl Halides. *Tetrahedron Lett.* **2009,** *50* (46), 6290–6292.
99. Tessonnier, J.-P.; Pesant, L.; Ehret, G.; Ledoux, M. J.; Pham-Huu, C. Pd Nanoparticles Introduced Inside Multi-Walled Carbon Nanotubes for Selective Hydrogenation of Cinnamaldehyde into Hydrocinnamaldehyde. *Appl. Catalys. A Gen.* **2005,** *288* (1–2), 203–210.
100. Guo, S.; Pan, X.; Gao, H.; Yang, Z.; Zhao, J.; Bao, X. Probing the Electronic Effect of Carbon Nanotubes in Catalysis: NH_3 Synthesis with Ru Nanoparticles. *Chem.—A Eur. J.* **2010,** *16* (18), 5379–5384.

101. Sun, Z.; Zhao, Y.; Xie, Y.; Tao, R. , Zhang, H.; Huang, C.; Liu, Z. The Solvent-Free Selective Hydrogenation of Nitrobenzene to Aniline: An Unexpected Catalytic Activity of Ultrafine Pt Nanoparticles Deposited on Carbon Nanotubes. *Green Chem.* **2010,** *12* (6), 1007.
102. Arumugam, S.; Chakkarapani, L. D. Metal Nanoparticles Functionalized Carbon Nanotubes for Efficient Catalytic Application. *Mater. Res. Express* **2019,** *6* (10), 1050e3.

CHAPTER 7

CNTS AS NEW EMERGING LUBRICANT ADDITIVES FOR ENHANCING ENERGY EFFICIENCY

AVINASH V. BORGAONKAR[1,*] and SHITAL B. POTDAR[2]

[1]*Department of Mechanical Engineering, National Institute of Technology Warangal, Warangal 506004, Telangana, India*

[2]*Department of Chemical Engineering, National Institute of Technology Warangal, Warangal 506004, Telangana, India*

[*]*Corresponding author. E-mail: avi.borgaonkar@gmail.com*

ABSTRACT

Almost 30% of energy is lost in machinery and engine parts due to wear and frictional loss. The semi-solid and liquid lubricants used in machinery and engine parts to minimize friction and wear are unable to sustain at high load and high operating temperatures. To improve the tribological performance and load wear index, antiwear (AW), and extreme pressure (EP), additives are dispersed into lubricating oil. These additives provide cushioning effect in between the contacting surfaces by forming thin films in turns preventing severe wear and seizure. In the past two decades, researchers tried different additives, such as graphite, fullerenes, nanoonions, CNT, and graphene due to their extraordinary properties. The sp2 C-C bond of CNTs made them a strongest candidate among solid materials. Because of stress-free surface modification and larger surface area compared with other existing inorganic nanomaterials, the use of CNTs aggrandized the attention of researchers for tribological applications.

7.1 INTRODUCTION

Recently, the published literature reported that friction and wear are the major reasons toward failure of machineries and their components as well as systems used in military, aerospace, and industrial applications. In automobile engines, almost 5% of the total energy produced is lost to overcome frictional resistance.[1] With regard to this issue, there is necessity to formulate novel lubricants with doping of various nanoadditives which help to improve the tribological performance. Undoubtedly, this will have immediate impact on energy efficiency, sustainability, and increased mechanical performance.

The research in the field of energy efficient solutions is attributed toward the development of novel nanomaterials, instead of the use of conventional nanoadditives. Presently, different types of nanomaterials are under examination as base oil additives. The nanomaterials have been evolved as a relatively new class of lubricant additives in the current development of lubricant additives. Nanolubricants mainly contain base oil or fully formulated lubricant with suspended colloidal solid particles within them. The tribological tests can be performed employing different test rigs like pin-on disc, four ball tester, ball-on-disc, ball-on-plate, block-on-ring, twin-disc test rig, etc. This section represents an overview on the tribological properties of nanomaterials type, compatibility of base oil/nanomaterial additive, optimum wt.% of nanoparticles and environmental tests conditions.

The different test rigs used to carry out the tribological analysis are represented in Figure 7.1.

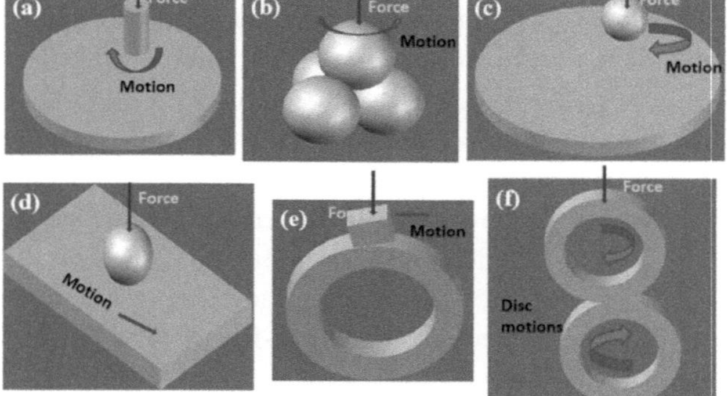

FIGURE 7.1 Different test rigs used for tribological analysis: (a) pin-on-disc test rig, (b) four-ball test rig, (c) ball-on-disc test rig, (d) ball-on-flat tester, (e) block-on-ring test rig, and (f) twin-disc test rig.

TABLE 7.1 Tribological Test Carried Out Using Pin-on-Disc Test Rig.

Particles	Base oil	Findings	References
Surface-modified MWCNTs	Pure liquid paraffin	Effective membrane of MWCNTs dispersed lubricant developed on the surface leads to reduction in friction and wear	[2]
MWCNTs	Lithium grease	At lower speed (150 RPM) with 0.5 wt.% addition of MWCNTs reduces COF by 42.6%	[3]
Cu nanoparticles	Commercial lubricant (500SN)	Heat generated during friction brings about the deposition of the Cu nanoparticles, which results in mending effect	[4]
Ni-containing SWCT and carbon nanoonions	PAO oil	The presence of a Ni metallic core inside CNTs shown significant improvement in tribological performance under boundary lubrication conditions	[5]
Polyphenylene sulfide (PPS) composites filled with CuO particles	Water	The agglomeration of CuO nanoparticles takes place on the contacting substrate surface provides protection from wear	[6]
Nanodiamonds	Ethylene glycol	The decrease in friction and wear was observed with both evenly distributed and agglomerated nanodiamonds	[7]
Ag-MoS_2 nanoparticles	25 cst 75W140 Gear oil and PAO with 8, 25, and 40 cst viscosity	Ag–MoS_2 nanolubricant has significantly reduces COF and wear. Besides, Ag nanoparticles prolonged the life of pure MoS_2 nanoparticles at higher temperatures	[8]
Cu nanoparticles	Mineral and synthetic ester base oils	The Cu dispersed mineral oil exhibits reduced COF and enhanced antiwear characteristics at 0.3 wt.%	[9]
Titanium oxide (TiO_2) nanoparticles	Mineral-based multigrade engine oil	The TiO_2-dispersed engine oil shows significant reduction in COF and wear rate	[10]

TABLE 7.2 Tribological Testing Using Four Ball Test Rig.

Particles	Base oil	Findings	References
TiO_2, Al_2O_3, CuO, and MWCNTs	Metkool, Metkut, Montgomery, Ecodraw	CuO-dispersed Metkut lubricant shows reduction in wear scar diameter and friction coefficient by 86% and 7%, respectively	[11]
Long and short MWCNTs	EP 140 gear oil	The stability of the MWCNTs dispersed lubricant improved significantly. This leads to enhanced tribological and extreme pressure properties	[12,13,15]
CNTs	Calcium grease	With 3 wt.% CNT addition friction is reduced by about 50%, the wear scar diameter (WSD) decreased to 32% and the EP properties enhance about 38%	[14]
Surface coated by LaF_3 nanoparticles	Liquid paraffin	Enhanced load-carrying capacity and antiwear property have been observed with LaF_3 dispersed oil. Besides significant reduction in friction was observed as compared with ZDDP	[16]
MoS_2 and WS_2 nanoparticles	EP 140 grade transmission oil	The 0.5 wt.% nanoparticles-dispersed lubricant enhances tribological properties and EP properties lubricant	[17]
Tetrafluorobenzoic acid-modified TiO_2 nanoparticles	Liquid paraffin	The nanoparticle-dispersed lubricant enhances tribological properties and EP properties lubricant	[18]
WS_2, Chloroparaffins (T302) and MoS_2	Petrol engine oil	The WS_2-dispersed petrol engine oil exhibits excellent tribological properties	[19]
CuO, ZnO, and ZrO_2 nanoparticles	PAO oil	CuO and ZnO nanoparticles exhibited best EP behavior whereas ZrO_2 shown worst behavior	[20]

TABLE 7.2 *(Continued)*

Particles	Base oil	Findings	References
Surface-modified Cu nanoparticles	50 CC oil	A thin Cu film having low elastic modulus and hardness was observed on the contacting surface resulting in the enhanced frictional and wear properties; especially at higher oil temperature	[21]
Modified graphene platelets (MGP)	350SN base oil	The load-carrying capacity and wear resistance significantly improved with 0.075 wt.%. MGP dispersed oil. The friction has been significantly reduced with MGP-dispersed oil compared with pure base oil	[22]

TABLE 7.3 Tribological Test Carried Out Using Ball-on-Disc Test Rig.

Particles	Base oil	Findings	References
MWCNTs, MWCNTs with surfactant (Triton X-100)	Mineral oil (0.2228 cPs)	The nanolubricant containing MWCNT dispersed with surfactant exhibits minimum wear and COF	[24]
Rhenium-doped inorganic fullerene MoS_2 nanoparticles	Polyalpha olefin PAO oil	In case of both boundary and mixed lubrication regimes, great improvement in the tribological properties have been observed with formulated oil under severe operating conditions	[25]
Surface-modified palladium (Pd) nanoparticles	Liquid paraffin+ tetrabutylammonium acetate (TBA) as surfactant	The Pd-nanoparticles protected by a surfactant layer exhibits better lubrication of metallic contacts possessing lower values of electrical resistivity. The formed protective layer on the ball surface heals the wear tracks developed on the disc surface. This resulted into enhanced load-carrying capacity and reduction in COF	[26]

TABLE 7.3 *(Continued)*

Particles	Base oil	Findings	References
MWNTs	PAO oil	The MWNTs-dispersed oil reduces the wear rate by 5–9 times and friction coefficient by 2 times	[27]
TiO_2 nanoadditive	Water	The TiO_2-dispersed water lubricant exhibits enhanced antifriction and antiwear properties	[28]
Silver nanoparticles	polyethylene-glycol (PEG) lubricant	The addition of silver nanoparticles shows significant reduction in wear	[29]
MoS_2 nanoparticles	Castor oil	The MoS_2 nanoparticle-dispersed lubricant reduces the possibility of asperities direct contact, resulting in the reduction of COF and adhesive wear	[30]
Composite Si_3N_4/SiC nanoparticles	Mineral oil mixed with nanoparticles of diamond	The lubricant tends to decrease the COF by more than 40% and also significantly decreases the wear coefficient	[31]
TiO_2 nanoparticles	Sodium dodecyl benzene sulfonate (SDBS) + glycerol + water	The formulated nanolubricant with 4 wt.% of TiO_2 and 0.4 wt.% of SDBS exhibited enhanced frictional and wear properties by reducing friction coefficient and wear rate up to 70.5 and 84.3%, respectively	[32]
h-BN nanoparticles	Ester base oil	The lubricant reduces COF by 25%, scar width by 9%, scar depth by 14% and transversal area by 22% all with respect to pure oil	[33]

TABLE 7.4 Tribological Test Carried Out Using Ball-on-Flat Tester.

Particles	Base oil	Findings	References
Inorganic fullerene-like (IF) nanoparticles of WS_2	Mineral oil SN 90, SN 150, and bright stock	The COF has been reduced by 50% in the mixed lubrication regime with the use of formulated oil	[34]

TABLE 7.4 *(Continued)*

Particles	Base oil	Findings	References
WS_2 (MoS_2) inorganic nanotubes and fullerene nanoparticles	Paraffin oil	WS_2-dispersed lubricant results into reduced COF	[35]
		This lubricant shows significant reduction in the run-in period. The friction and wear properties also have been significantly improved	[36]
CuO, TiO_2, and diamond nanoparticles	PI-SF engine oil and base oil	The CuO nanoparticles-dispersed API-SF engine oil and the Base oil decreases the COF by 18.4 and 5.8%, respectively, and reduces the worn scar depth by 16.7 and 78.8%, respectively in comparison with pure base oil	[37]
MoS_2 and SiO_2 nanoparticles	EOT5 engine oil	The MoS_2-dispersed lubricant exhibits excellent stability compared with SiO_2 nanolubricants. Also MoS_2 dispersed lubricant exhibited significant improvement in tribological properties and load-carrying capacity	[38]
Fluorographene (FG)	GTL-8-based lubricant oil	The FG-dispersed oil with 0.3 mg/ml reduces COF and the wear rate by 35 and 90% respectively	[39]
Perfluoropolyether (PFPE) nanolubricant		The COF of the counter-disc bonded with PEPE film decreases linearly with increase in the surface coverage	[40]
α-pinene tribo-polymers		The vapor of α-pinene and pinnae produces a polymer-like film during sliding between the contacting surfaces. This developed film play a role of an effective lubricant in the absence of continuous supply of the organic vapor	[41]

TABLE 7.4 *(Continued)*

Particles	Base oil	Findings	References
α-FeOOH nanorods	PAO lubricating oils	Closely packed nanorods have significantly reduces wear and friction. COF decreases by 22%	[42]
MoS_2, SiO_2, ND, CNTs, α-Al_2O_3, and ZrO_2).	Palm oil	The Al_2O_3 nanofluid exhibits significant antifriction property. The COF reduced by 20%	[43]

TABLE 7.5 Tribological Test Carried Out Using Block-on-Ring Test Rig.

Particles	Base oil	Findings	References
Nanodiamond, onion-like carbon (OLC), (SWCNTs/ MWCNTs) or nanographene platelets	Polyalphaolefin (PAO) oil and PAO + molybdenum dialkyldithiophosphate (MoDDP)	The polishing capability of nanodiamonds along with the protective action of MoDDP greatly reduces the COF and wear rate in comparison with pure PAO	[44]
Modified nanodiamond (ND)	SAE10W engine oil	ND dispersed lubricant helps to improve the frictional and wear properties and reduces the oil temperature. The antiwear mechanism attributed due to developed hard and porous layer between the contacting surfaces	[45]
Ni nanoparticles	PAO6 oil	Ni nanoparticles-dispersed oil helps to improve the tribological properties and also improved the load-carrying capacity	[46]
Carbon-coated Cu nanoparticles	PAO6 oil	The carbon-coated Cu nanoparticles-dispersed oil leads to enhance the load-carrying capacity and reduces the wear	[47]

TABLE 7.5 *(Continued)*

Particles	Base oil	Findings	References
Multilayer graphene impregnated with copper (MLG-Cu), and polyaniline (MLG-PANI)	Commercial motor oil	The MLG-Cu with 0.5 and 2 wt.% exhibits reductions in COF and wear up to 43% and 63%, respectively	[48]
Nanodiamond-PTFE-MoDDP	Polyalphaolefins and engine oils	Significant reductions in the COF and wear have been observed. With the DND dispersed oil 35% reduction in surface roughness has been observed compared with pure oil	[49]
CuO nanoparticles	Polyalphaolefin (PAO6)	Due to CuO-dispersed oil reduction in COF and wear have been observed. This is attributed because of third body and tribo-sinterization mechanism	[50]
Nanodiamond	Commercial oil (CPC circulation oil R68, Taiwan)	3 wt.% addition of the ND exhibits significant reduction in the friction compared to base oil	[51]

TABLE 7.6 Tribological Test Carried Out Using Twin-Disc Test Rig.

Particles	Base oil	Findings	References
Graphite Carbon nanofibres, CNTs	Mineral oil (Super gear EP220, SK, Korea)	Nanooils with the fibrous particle additives (e.g., CNFs/CNTs) showed higher COF than the nanooils with spherical nanoparticle additives, since the fibrous nanoparticles have higher aspect ratios	[52]
Functionalized SWCNTs/MWCNTs modified with carboxylic acid	Oil and water	Reduced in both COF and wear caused due to the formation of an amorphous carbon film transferred from the CNT's	[53]

TABLE 7.6 *(Continued)*

Particles	Base oil	Findings	References
Graphite nanoparticles	Industrial gear oil (viscosity 220 cSt)	The graphite-dispersed lubricant exhibits enhanced tribological properties	[54]
CuO nanoparticles	Mineral oil + sodium oleate (SOA) surfactant	The CuO/SOA nanoparticle additives reduce the friction. The effect of nanoparticles is more influential at higher loads and concentrations. The steady-state temperature of the lubricant is lower in the presence of nanoparticles	[55]
Carbon nanopearls (CNPs) and MWCNTs	Mixture of a synthetic ester resin and a rosin ester derived from pine tar, and linseed oil	Biolubricant additivation with raw unpurified MWCNTs and CNPs for enhanced tribological performance	[56]
Fullerene nanoparticles	Mineral oil-based lubricant	In the friction test, the friction coefficient of the disc specimen immersed in the nanooil was significantly lower than that of the disc specimen immersed in the mineral oil	[57]
Sulfur and phosphorus	EAL-based oil and commercial heavy truck gearbox lubricant	The EAL additives reacted with the surface to a greater extent than the commercial lubricant additives. An increase in the boundary layer, using the EAL, which itself needs to be sheared off, can explain the higher friction	[58]
Antioxidants, dispersants, detergent, EP, AW and FM compounds + 5% of MoS_2	Two commercial greases (LiCaM220 and LiM460)	The lithium thickened grease (LiM50) showed lower friction values than the Calcium-thickened grease (CaM50). LiM50 provides significant wear reduction while CaM50 generates high wear	[59]

7.2 ELEMENTS OF NANOLUBRICANTS

7.2.1 LUBRICANT CLASSIFICATION

7.2.1.1 VEGETABLE AND ANIMAL OIL

These oils are extracted from the fats of vegetables and animals. These oils have a tendency to stick on the contacting surfaces effectively at higher loads and high temperatures due to good oiliness. These oils are used for the purpose of lubrication. But with humid air or water, they easily undergo oxidation and petroleum-based lubricants are blended with these agents to get enhanced oiliness.

7.2.1.2 PETROLEUM OILS OR MINERAL

These oils have about 12–50 atoms of carbon. They are available in abundance and are economical. Due to their stability under working conditions, they are extensively used. It has less oiliness compared with animal and vegetable oils.

7.2.1.3 BLENDED OILS

The oils do not have all the required properties that are required for a lubricant, hence, they are added with sufficient additives to enhance their properties and for better performance. The additives that are added to lubricating to enhance the properties of oils are named as blended oils.

Functions of lubricant

- To absorb heat generated between two mating surfaces.
- To reduce friction and wear of the contacting surfaces.
- To restrict the expansion of the contacting surfaces due to heat generated.
- To smoothen the relative motion.
- To reduce the maintenance cost.
- To reduce power losses in turn minimizes the energy losses.

Desirable features of lubricating oil

- It should possess high boiling point and low-freezing point.
- It should possess optimum viscosity for proper functioning.
- It should possess high oxidation and high heat resistance.
- It should possess anticorrosive properties.
- It should possess higher stability.

7.2.1.4 GREASE OR SEMI-SOLID LUBRICANTS

Greases are the semi-solid lubricants that are obtained by mixing lubricating oil with thickening agents. The principal component in the greases is lubricating oil and it can be from synthetic hydrocarbon or petroleum oil which is of high to low viscosity. The thickeners may consist of soaps and nonsoap types. At lower speed, grease can support a heavier load. Grease has high internal resistance when compared with the lubricating oils. Lubricating oils can efficiently disperse heat from the bearings when compared with grease. So greases work at a relatively lower temperature.

7.2.1.5 NANOMATERIALS

At present, research on nanomaterials is of great scientific interest because of huge potential in mechanical, biomedical, optical, and electronic field applications. Material with a size between 1 and 100 nm is classified as nanomaterial, while the behavior of materials depend on their size. Hence, as the size is reduced to nanoscale and as the quantity of the atoms on the surface of the material becomes significant, change in properties of materials is witnessed.

There are a number of nanoparticles that can be utilized as grease-added substances. These are listed as follows.

Molybdenum disulfide, carbon nanotubes, tungsten disulfide, polymers, zinc oxide, gold or silver nanoparticles, dendrimers, liposomes, graphene nanoplatelets, quantum dabs, iron oxide, perflurocarbon, micelles, ferritin, lanthanide particles.

Nanoparticles have surprising optical properties so that they are used in electrons to produce quantum effects. Due to differences in densities, the nanoparticles can be suspended in solvents. The surface interaction has a high impact on the suspension, which results in either sinking in or floating on a liquid. At elevated temperatures, the volume of the nanoparticles plays a major role in diffusion. There is a vast development in the last 10 years and has many applications in the industrial, medical, and electronics field.

7.2.2 SELECTION OF BASE LUBRICANT

A lot of research is being done by many researchers to understand the effect of MWCNTs as additives in various lubricating oils. Lubricating oils provide the best wear reduction and friction coefficient and so is the best choice for an engine application. However, normal lubricants are not sufficient as in automotive systems due to their poor wear, friction, and load-bearing capacity, heat transfer coefficient, and corrosion properties. So, in addition to the lubricating oil, a small number of other ingredients, including corrosion inhibitors, antifriction, antiwear, and other additives are added to the lubricating oil.

Traditional lubricant typically contains 73–80 wt.% of aliphatic hydrocarbons, 11–15% of monoaromatic hydrocarbons, 4–8% polyaromatic hydrocarbons, and 2–5% diaromatic hydrocarbons. Approximately, 20% of the lubricants contain additives like zinc diaryl, zinc dithiophosphate, molybdenum disulfide and other organic metallic compounds. These lubricants gained international popularity for having an unsurpassed property for extended time intervals. The formulation process used for preparing a nanolubricant is depicted in Figure 7.2.

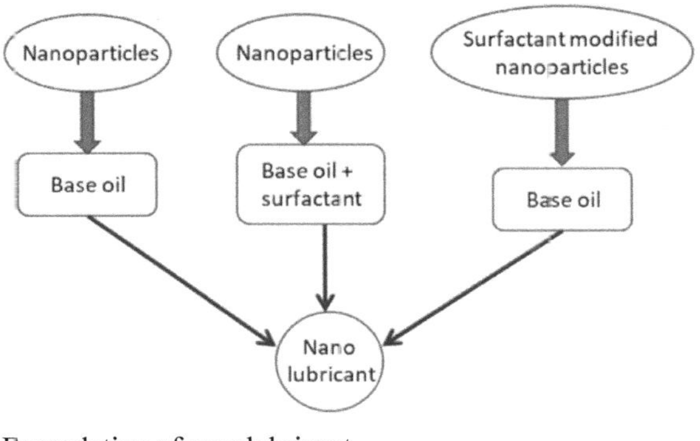

FIGURE 7.2 Formulation of nanolubricant.

7.3 LUBRICATION MECHANISMS

The lubrication mechanism analysis is an important step to understand the tribological behavior of nanoparticles. Various types of mechanisms have been suggested by researchers employing the surface analysis techniques to elaborate the lubrication enhancement by the nanoparticle-suspended lubricating oil. These mechanisms reported in the literature listed as ball-bearing effect,[37,60,61] development of protective film,[62–64] mending effect,[4] and polishing effect.[65] These mechanisms have been briefly elaborated as follows.

7.3.1 BALL-BEARING EFFECT

The quasi-spherical and spherical nanoparticles are assumed to function like minute ball-bearings which roll in between the contacting surfaces. Such shapes of nanoparticles lead to change the sliding friction toward a combination of sliding and rolling friction. Such type of mechanism occurs in the case of tribological pair having steady low-load conditions between the shearing surfaces to manage the shape and rigidity of the nanoparticles. The ball-bearing mechanism is illustrated in Figure 7.3.

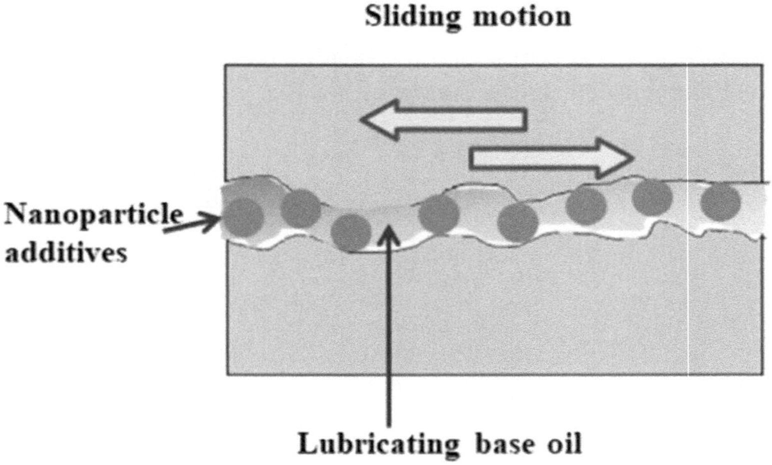

FIGURE 7.3 Schematic representation of ball-bearing mechanism with nanolubricant.

7.3.2 DEVELOPMENT OF PROTECTIVE FILM

The developed tribofilms on the contacting surfaces govern tribological behaviors of the contacting surfaces. This development of the film is initialized due to chemical reaction between the substrate material, base oil, and the additives under the set operating conditions.[46] The formation of a durable tribofilm with addition of Mo, S, and P elements reduces the friction, wear, and seizure for MoS_2 nanoparticles dispersed lubricant.[35,58] Figure 7.4 illustrates the tribofilm development, which protects the surfaces from crack propagation by reducing friction between the contacting asperities.

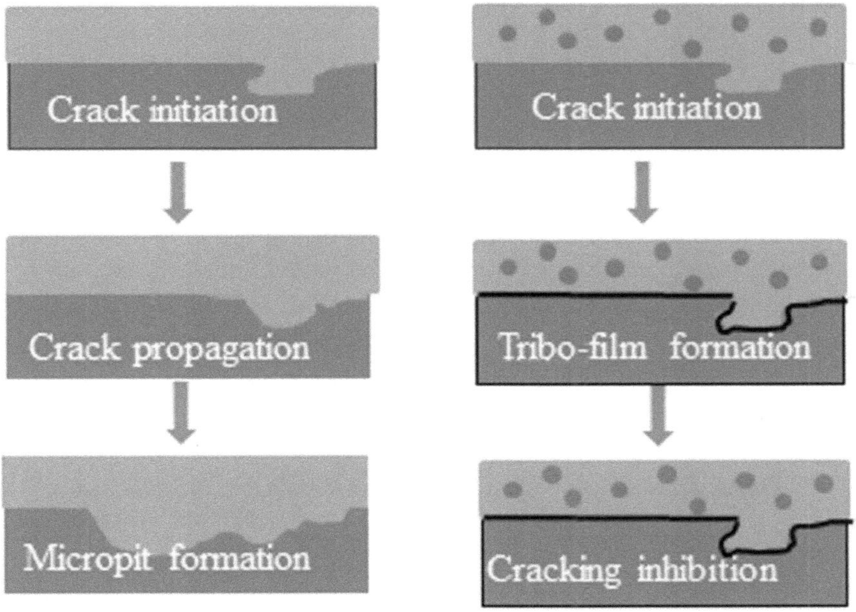

FIGURE 7.4 Schematic representation of development of tribofilm with nanolubricant.

7.3.3 MENDING EFFECT

The self-repairing effect or mending effect is analyzed by the deposition of the nanoparticles on the contacting surfaces and wear compensation. In this phenomenon, the nanoparticles get deposited on the comparatively soft surface to reduce abrasion.[66] Nanoparticles dispersed in the base oil have the tendency to fill grooves and asperities of the frictional surface. Figure 7.5 represents schematically the filled up asperities and grooves with nanoparticles on the frictional surface.

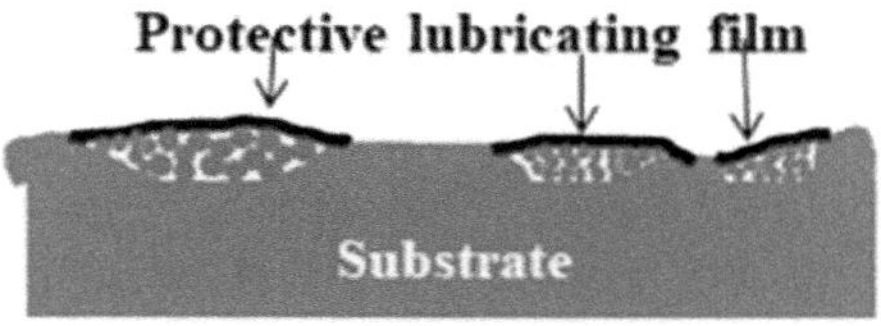

FIGURE 7.5 Schematics of nanoparticles filled up asperities and grooves on frictional surface.

7.3.4 POLISHING EFFECT

This is also defined as smoothing effect. In this phenomenon, it is hypothesized that the roughness of the mating surfaces is reduced by nanoparticle-assisted abrasion. The nanoparticles fill the rough asperities, which may act as reservoirs for nanolubricants within the contact, as depicted schematically in Figure 7.5. The process of filling up the rough asperities termed as the smoothing process. This polishing process or artificial smoothing resulted into improved tribological performance mainly because of the decreased surface roughness.[67]

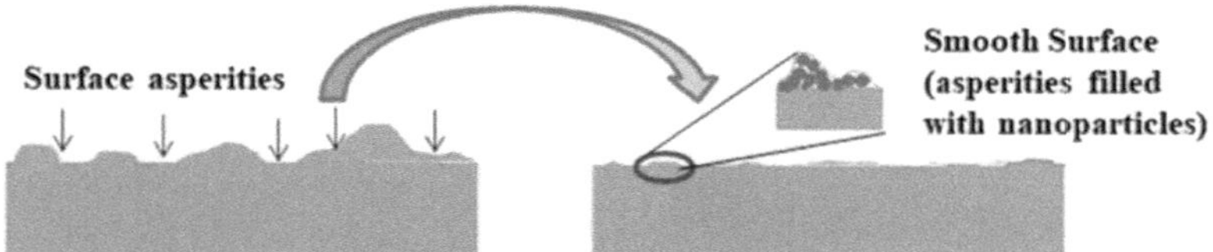

FIGURE 7.6 Schematics of polishing/smoothing effect due to filling up of rough valleys of contacting surfaces with nanoparticles.

7.4 SUMMARY

The referred literature reported that CNTs have been popularly used as a nanomaterial in order to improve the tribological performance of the contacting surfaces. Figure 7.7 shows that the carbon nanotubes (CNTs) are the potential candidate for various tribological applications.

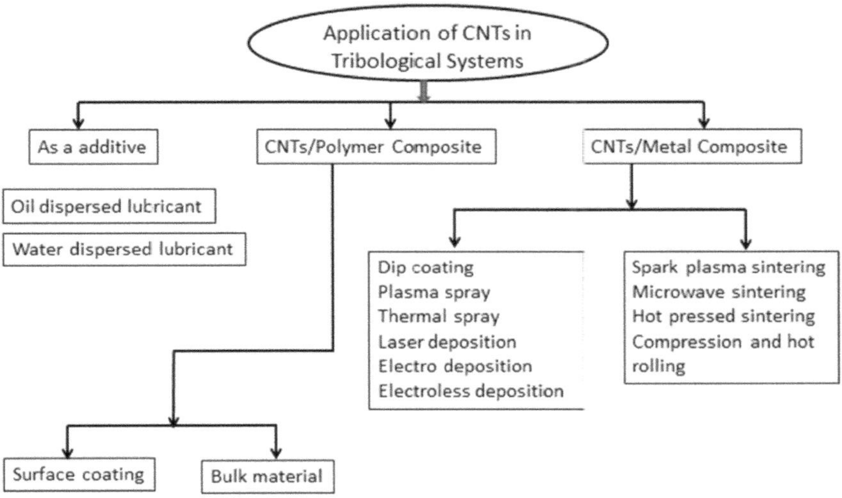

FIGURE 7.7 Application of CNTs in tribological systems.

The usage of different nanoparticles dispersed lubricants help to reduce friction and wear of the contacting surfaces having different relative motions. The reduction in friction in wear resulted into energy saving and enhanced efficiency of the tribological applications. The literature review represented in Table 7.7 shows that CNTs possessing the interesting antifriction and antiwear behavior make them suitable candidates as additives to improve the tribological performance. The Percentage (%) contribution is determined on the basis of nanoparticles repetitively used in cited articles in the current section.

TABLE 7.7 Different Nanomaterial Additives Used for the Tribological Studies.

Sr. No.	Nanomaterial additives	Percentage (%) contribution	References (from year 2000 to 2020)
1	LaF_3	1.2	[16]
2	Ni	2.4	[23,46]
3	TiO_2	7.1	[10,11,18,28,32,37]

TABLE 7.7 *(Continued)*

Sr. No.	Nanomaterial additives	Percentage (%) contribution	References (from year 2000 to 2020)
4	Perfluoropolyether PFPE	1.2	[40]
5	Cu	5.9	[4,9,21,47,48]
6	WS_2	3.6	[17,19,34]
7	MWNTs	19.05	[2,3,11–15,24,27,43,44,52,53,56]
8	WS_2/MoS_2	2.4	[35,36]
9	Chloroparaffins	1.2	[19]
10	MoS_2	7.1	[17,19,30,38,43,59]
11	PPS-CuO	1.2	[6]
12	CuO	5.9	[11,20,37,50,55]
13	ZnO	1.2	[20]
14	ZrO_2	2.4	[20,43]
15	Palladium (Pd)	1.2	[26]
16	Nanodiamond	7.1	[7,37,43,44,49,51]
17	Ni-SWCNT	1.2	[5]
18	Carbon nanoonions	3.6	[5,44,56]
19	Graphite	4.8	[13,39,52,54]
20	Fullerene	2.4	[25,57]
21	Modified graphene platelets (MGP)	2.4	[22,44]
22	Ag–MoS_2	2.4	[8,44]
23	Al_2O_3	2.4	[11,43]
24	Nanodiamond-PTFE	1.2	[49]
25	Composite Si_3N_4/SiC	1.2	[31]
26	SiO_2	2.4	[38,43]
27	α-pinene tribo-polymers	1.2	[41]
28	Polyaniline MLG-PANI	1.2	[48]
29	α-FeOOH nanorods	1.2	[42]
30	h-BN	1.2	[33]
31	Sulfur and phosphorus	1.2	[58]

Due to sulfur and phosphorus content traditional lubricants leads to health and environmental hazards. These sensitive issues can be tackled by replacing conventional base oils by bio-based base stocks dispersed with different nanomaterials suits for the specific applications. In this regard, the dispersion stability and tribological compatibility of various nanomaterials as additives to bio-based base stocks, is required to be investigated for different applications.

KEYWORDS

- **energy**
- **friction**
- **MWCNTs**
- **SWCNTs**
- **tribological properties**
- **wear**

REFERENCES

1. Holmberg, K.; Erdemir, A. Influence of Tribology on Global Energy Consumption, Costs and Emissions. *Friction* **2017,** *5* (3), 263–284.
2. Chen, C. S.; Chen, X. H.; Xu, L. S.; Yang, Z.; Li, W. H. Modification of Multi-Walled Carbon Nanotubes with Fatty Acid and Their Tribological Properties as Lubricant Additive. *Carbon* **2005,** *43* (8), 1660–1666.
3. El-Adly, R. A.; Hussein, M .F.; Mohamed, A. Preparation and Evaluation of Lubricating Greases Based on Carbon Nanotube. *Int. J. Adv. Pharm. Biol. Chem.* **2015,** *4* (3), 675–684.
4. Liu, G.; Li, X.; Qin, B.; Xing, D.; Guo, Y.; Fan, R. Investigation of the mending effect and mechanism of copper nano-particles on a tribologically stressed surface. Tribology Letters **2004,** *17* (4), 961–966.
5. Joly-Pottuz, L.; Vacher, B.; Le Mogne, T.; Martin, J. M.; Mieno, T.; He, C. N.; Zhao, N. Q. The Role of Nickel in Ni-Containing Nanotubes and Onions as Lubricant Additives. *Tribol. Lett.* **2008,** *29* (3), 213–219.
6. Cho, M. H.; Bahadur, S. Friction and Wear of Polyphenylene Sulfide Composites Filled with Micro and Nano CuO Particles in Water-Lubricated Sliding. *Tribol. Lett.* **2007,** *27* (1), 45–52.
7. Elomaa, O ; Oksanen, J.; Hakala, T. J.; Shenderova, O.; Koskinen, J. A Comparison of Tribological Properties of Evenly Distributed and Agglomerated Diamond

Nanoparticles in Lubricated High-Load Steel–Steel Contact. *Tribol. Int.* **2014,** *71*, 62–68.
8. Zhang, W.; Demydov, D.; Jahan, M. P.; Mistry, K.; Erdemir, A.; Malshe, A. P. Fundamental Understanding of the Tribological and Thermal Behavior of Ag–MoS_2 Nanoparticle-Based Multi-Component Lubricating System. *Wear* **2012,** *288*, 9–16.
9. Borda, F. L. G.; de Oliveira, S. J. R.; Lazaro, L. M. S. M.; Leiroz, A. J. K. Experimental Investigation of the Tribological Behavior of Lubricants with Additive Containing Copper Nanoparticles. *Tribol. Int.* **2018,** *117*, 52–58.
10. Laad, M.; Jatti, V. K. S. Titanium Oxide Nanoparticles as Additives in Engine Oil. *J. King Saud Univ. –Eng. Sci.* **2018,** *30* (2), 116–122.
11. Pena-Paras, L.; Taha-Tijerina, J.; Garcia, A.; Maldonado, D.; Gonzalez, J. A.; Molina, D.; Palacios, E.; Cantu, P. Antiwear and Extreme Pressure Properties of Nanofluids for Industrial Applications. *Tribol. Trans.* **2014,** *57* (6), 1072–1076.
12. Chebattina, K. R. R.; Srinivas, V.; Rao, N. M. Effect of Size of Multiwalled Carbon Nanotubes Dispersed in Gear Oils for Improvement of Tribological Properties. *Adv. Tribol.* **2018,** *V*2018, 1–13.
13. Bhaumik, S.; Prabhu, S.; Singh, K. J. Analysis of Tribological Behavior of Carbon Nanotube Based Industrial Mineral Gear Oil 250 cSt Viscosity. *Adv. Tribol.* **2014,** 1–8.
14. Kamel, B. M., Mohamed, A., El Sherbiny, M. and Abed, K. A. Tribological Behaviour of Calcium Grease Containing Carbon Nanotubes Additives. *Indust. Lubricat. Tribol.* **2016,** *68* (6), 723–728.
15. Khalil, W.; Mohamed, A.; Bayoumi, M.; Osman, T. A. Tribological Properties of Dispersed Carbon Nanotubes in Lubricant. *Fullerenes, Nanotubes Carbon Nanostruct.* **2016,** *24* (7), 479–485.
16. Zhou, J.; Wu, Z.; Zhang, Z.; Liu, W.; Dang, H. Study on an Antiwear and Extreme Pressure Additive of Surface Coated LaF3 Nanoparticles in Liquid Paraffin. *Wear* **2001,** *249* (6), 333–337.
17. Srinivas, V.; Thakur, R. N.; Jain, A. K.; Saratchandra Babu, M. Tribological Studies of Transmission Oil Dispersed with Molybdenum Disulfide and Tungsten Disulfide Nanoparticles. *J. Tribol.* **2017,** *139* (4), 1–6.
18. Ye, W.; Cheng, T.; Ye, Q.; Guo, X.; Zhang, Z.; Dang, H. Preparation and Tribological Properties of Tetrafluorobenzoic Acid-Modified TiO_2 Nanoparticles as Lubricant Additives. *Mater. Sci. Eng. A* **2003,** *359* (2), 82–85.
19. Chen, S. H. I.; Da-Heng, M. A. O.; Hao, F. E. N. G. Study on Tribological Characteristics of Ultrafine WS_2 Powders in Petrol Engine Oil [J]. *China Tungsten Indust.* **2006,** 1–3.
20. Battez, A. H.; González, R.; Felgueroso, D.; Fernández, J. E.; del Rocío Fernandez, M.; García, M. A.; Penuelas, I. Wear Prevention Behaviour of Nanoparticle Suspension under Extreme Pressure Conditions. *Wear* **2007,** *263* (12), 1568–1574.
21. Yu, H.; Xu, Y.; Shi, P. Tribological Properties and Lubricating Mechanisms of Cu Nanoparticles in Lubricant. *Trans. Nonferrous Met. Soc. China* **2008,** *18* (1), 636–641.
22. Lin, J.; Wang, L.; Chen, G. Modification of Graphene Platelets and Their Tribological Properties as a Lubricant Additive. *Tribol. Lett.* **2011,** *41* (1), 209–215.

23. Qiu, S.; Zhou, Z.; Dong, J.; Chen, G. Preparation of Ni Nanoparticles and Evaluation of Their Tribological Performance as Potential Additives in Oils. *J. Trib.* **2001,** *123* (3), 441–443.
24. Lijesh, K. P.; Muzakkir, S. M.; Hirani, H. Experimental Tribological Performance Evaluation of Nano Lubricant Using Multi-Walled Carbon Nano-Tubes (MWCNT). *Int. J. Appl. Eng. Res.* **2015,** *10* (6), 14543–14550.
25. Yadgarov, L.; Petrone, V.; Rosentsveig, R.; Feldman, Y.; Tenne, R.; Senatore, A. Tribological Studies of Rhenium Doped Fullerene-Like MoS_2 Nanoparticles in Boundary, Mixed and Elasto-Hydrodynamic Lubrication Conditions. *Wear* **2013,** *297* (1–2), 1103–1110.
26. Kolodziejczyk, L.; Martinez-Martinez, D.; Rojas, T. C.; Fernandez, A.; Sanchez-Lopez, J. C. Surface-Modified Pd Nanoparticles as a Superior Additive for Lubrication. *J. Nanoparticle Res.* **2007,** *9* (4), 639–645.
27. Kalin, M.; Kogovsek, J.; Remskar, M. Mechanisms and Improvements in the Friction and Wear Behavior Using MoS_2 Nanotubes as Potential Oil Additives. *Wear* **2012,** *280*, 36–45.
28. Wu, H.; Zhao, J.; Cheng, X.; Xia, W.; He, A.; Yun, J. H.; Huang, S.; Wang, L.; Huang, H.; Jiao, S.; Jiang, Z. Friction and Wear Characteristics of TiO_2 Nano-Additive Water-Based Lubricant on Ferritic Stainless Steel. *Tribol. Int.* **2018,** *117*, 24–38.
29. Neuffer, H.; Ghaednia, H.; Jackson, R. Wear Volume Analysis Using a Nano-Lubricant for Ball-on-Disk Testing. *Tribol. Lubricat. Technol.* **2014,** *2*, 1–3.
30. Yu, R.; Liu, J.; Zhou, Y. Experimental Study on Tribological Property of MoS_2 Nanoparticle in Castor Oil. *J. Tribol.* **2019,** *141* (10), 1–5.
31. Charoo, M. S.; Wani, M. F. Friction and Wear Properties of Nano-Si_3N_4/Nano-SiC Composite under Nanolubricated Conditions. *J. Adv. Ceram.* **2016,** *5* (2), 145–152.
32. Wu, H.; Jia, F.; Li, Z.; Lin, F.; Huo, M.; Huang, S.; Sayyar, S.; Jiao, S.; Huang, H.; Jiang, Z. Novel Water-Based Nanolubricant with Superior Tribological Performance in Hot Steel Rolling. *Int. J. Extreme Manuf.* **2020,** *2* (2), 1–11.
33. Lineira del Rio, J. M.; Guimarey, M. J.; Comunas, M. J.; Lopez, E. R.; Prado, J. I.; Lugo, L.; Fernandez, J. Tribological and Thermophysical Properties of Environmentally-Friendly Lubricants Based on Trimethylolpropane Trioleate with Hexagonal Boron Nitride Nanoparticles as an Additive. *Coatings* **2019,** *9* (8), 1–17.
34. Greenberg, R.; Halperin, G.; Etsion, I.; Tenne, R. The Effect of WS_2 Nanoparticles on Friction Reduction in Various Lubrication Regimes. *Tribol. Lett.* **2004,** *17* (2), 179–186.
35. Rapoport, L.; Fleischer, N.; Tenne, R. Applications of WS_2 (MoS_2) Inorganic Nanotubes and Fullerene-Like Nanoparticles for Solid Lubrication and for Structural Nanocomposites. *J. Mater. Chem.* **2005,** *15* (18), 1782–1788.
36. Perfiliev, V.; Moshkovith, A.; Verdyan, A.; Tenne, R.; Rapoport, L. A New Way to Feed Nanoparticles to Friction Interfaces. *Tribol. Lett.* **2006,** *21* (2), 89–93.
37. Wu, Y. Y.; Tsui, W. C.; Liu, T. C. Experimental Analysis of Tribological Properties of Lubricating Oils with Nanoparticle Additives. *Wear* **2007,** *262* (8), 819–825.
38. Xie, H.; Jiang, B.; He, J.; Xia, X.; Pan, F. Lubrication Performance of MoS_2 and SiO_2 Nanoparticles as Lubricant Additives in Magnesium Alloy-Steel Contacts. *Tribol. Int.* **2016,** *93*, 63–70.

39. Ci, X.; Zhao, W.; Luo, J.; Wu, Y.; Ge, T.; Shen, L.; Gao, X.; Fang, Z. Revealing the Lubrication Mechanism of Fluorographene Nanosheets Enhanced GTL-8 Based Nanolubricant Oil. *Tribol. Int.* **2019,** *138*, 174–183.
40. Choi, J.; Kawaguchi, M.; Kato, T. The Surface Coverage Effect on the Frictional Properties of Patterned PFPE Nanolubricant Films in HDI. *IEEE Trans. Magnet.* **2003,** *39* (5), 2492–2494.
41. He, X.; Barthel, A. J.; Kim, S. H. Tribochemical Synthesis of Nano-Lubricant Films from Adsorbed Molecules at Sliding Solid Interface: Tribo-Polymers from α-Pinene, Pinane, and n-Decane. *Surf. Sci.* **2016,** *648*, 352–359.
42. Zin, V.; Agresti, F.; Barison, S.; Litti, L.; Fedele, L.; Meneghetti, M.; Fabrizio, M. Effect of External Magnetic Field on Tribological Properties of Goethite (a-FeOOH) Based Nanofluids. *Tribol. Int.* **2018,** *127*, 341–350.
43. Wang, Y.; Li, C.; Zhang, Y.; Li, B.; Yang, M.; Zhang, X.; Guo, S.; Liu, G.; Zhai, M. Comparative Evaluation of the Lubricating Properties of Vegetable-Oil-Based Nanofluids between Frictional Test and Grinding Experiment. *J. Manuf. Process.* **2017,** *26*, 94–104.
44. Nunn, N.; Mahbooba, Z.; Ivanov, M. G.; Ivanov, D. M.; Brenner, D. W.; Shenderova, O. Tribological Properties of Polyalphaolefin Oil Modified with Nanocarbon Additives. *Diamond Relat. Mater.* **2015,** *54*, 97–102.
45. Puzyr, A. P.; Burov, A. E.; Selyutin, G. E.; Voroshilov, V. A.; Bondar, V. S. Modified Nanodiamonds as Antiwear Additives to Commercial Oils. *Tribol. Trans.* **2012,** *55* (1), 149–154.
46. Chou, R.; Battez, A. H.; Cabello, J. J.; Viesca, J. L.; Osorio, A.; Sagastume, A. Tribological Behavior of Polyalphaolefin with the Addition of Nickel Nanoparticles. *Tribol. Int.* **2010,** *43* (12), 2327–2332.
47. Viesca, J. L.; Battez, A. H.; González, R.; Chou, R.; Cabello, J. J. Antiwear Properties of Carbon-Coated Copper Nanoparticles Used as an Additive to a Polyalphaolefin. *Tribol. Int.* **2011,** *44* (8), 829–833.
48. Ramon-Raygoza, E. D.; Rivera-Solorio, C. I.; Gimenez-Torres, E.; Maldonado-Cortes, D.; Cardenas-Aleman, E.; Cue-Sampedro, R. Development of Nanolubricant Based on Impregnated Multilayer Graphene for Automotive Applications: Analysis of Tribological Properties. *Powder Technol.* **2016,** *302*, 363–371.
49. Shenderova, O.; Vargas, A.; Turner, S.; Ivanov, D. M.; Ivanov, M. G. Nanodiamond-Based Nanolubricants: Investigation of Friction Surfaces. *Tribol. Trans.* **2014,** *57* (6), 1051–1057.
50. ChuBattez, A. H.; Viesca, J. L.; Gonzalez, R.; Blanco, D.; Asedegbega, E.; Osorio, A. Friction Reduction Properties of a CuO Nanolubricant Used as Lubricant for a NiCrBSi Coating. *Wear* **2010,** *268* (2), 325–328.
51. Chu, H. Y.; Hsu, W. C.; Lin, J. F. The Anti-Scuffing Performance of Diamond Nano-Particles as an Oil Additive. *Wear* **2010,** *268* (8), 960–967.
52. Y. Hwang, C. Lee, Y. Choi, S. Cheong, D. Kim, K. Lee, J. Lee, S. H. Kim, Effect of the Size and Morphology of Particles Dispersed in Nano-Oil on Friction Performance between Rotating Discs. *J. Mech. Sci. Technol.* **2011,** *25*, 2853–2857.
53. Cornelio, J. A. C.; Cuervo, P. A.; Hoyos-Palacio, L. M.; Lara-Romero, J.; Toro, A. Tribological Properties of Carbon Nanotubes as Lubricant Additive in Oil and Water for a Wheel–Rail System. *J. Mater. Res. Technol.* **2016,** *5* (1), 68–76.

54. Lee, C. G.; Hwang, Y. J.; Choi, Y. M.; Lee, J. K.; Choi, C.; Oh, J. M. A Study on the Tribological Characteristics of Graphite Nano Lubricants. *Int. J. Prec. Eng. Manuf.* **2009,** *10* (1), 85–90.
55. Ghaednia, H.; Jackson, R. L.; Khodadadi, J. M. Experimental Analysis of Stable CuO Nanoparticle Enhanced Lubricants. *J. Exp. Nanosci.* **2015,** *10* (1), 1–18.
56. Perez-de Brito, A. F.; Ponce, S.; Perez-Robles, J. F.; Higuera-Ciapara, I.; Toro, A.; Esparza, R.; Medina, D. I.; Villaseñor-Ortega, F.; Luna-Barcenas, G. Linseed and Complex Rosin Ester Oils Additivated with MWCNTs and Nanopearls for Gears/Wheel-Rail Systems. *Eur. J. Lipid Sci. Technol.* **2020,** *122* (2), 1–15.
57. Lee, K.; Hwang, Y.; Cheong, S.; Choi, Y.; Kwon, L.; Lee, J.; Kim, S. H. Understanding the Role of Nanoparticles in Nano-Oil Lubrication. *Tribol. Lett.* **2009,** *35* (2), 127–131.
58. Ellen, B.; Zhu, Y.; Söderberg, A. Study of Surface Roughness on Friction in Rolling/Sliding Contacts: Ball-on-Disc versus Twin-Disc. *Tribol. Lett.* **2020,** *68* (2), 1–15.
59. Biazon, L.; Ferrer, B. P.; Toro, A.; Cousseau, T. Correlations between Rail Grease Formulation and Friction, Wear and RCF of a Wheel/Rail Tribological Pair. *Tribol. Int.* **2020,** 1–10.
60. Rapoport, L.; Leshchinsky, V.; Lvovsky, M.; Nepomnyashchy, O.; Volovik, Y.; Tenne, R. Mechanism of Friction of Fullerenes. *Indust Lubricat. Tribol.* **2002,** *54* (4), 171–176.
61. Chinas-Castillo, F.; Spikes, H. Mechanism of Action of Colloidal Solid Dispersions. *J. Tribol.* **2003,** *125* (3), 552–557.
62. Ginzburg, B.; Shibaev, L.; Kireenko, O.; Shepelevskii, A.; Baidakova, M.; Sitnikova, A. Antiwear Effect of Fullerene C_{60} Additives to Lubricating Oils. *Russian J. Appl. Chem.* **2002,** *75* (8), 1330–1335.
63. Hu, Z. S.; Lai, R.; Lou, F.; Wang, L. G.; Chen, Z. L.; Chen, G. X.; et al. Preparation and Tribological Properties of Nanometer Magnesium Borate as Lubricating Oil Additive. *Wear* **2002,** *252* (6), 370–374.
64. Xiaodong, Z.; Xun, F.; Huaqiang, S.; Zhengshui, H. Lubricating Properties of Cyanex 302-Modified MoS_2 Microspheres in Base Oil 500SN. *Lubricati. Sci.* **2007,** *19* (1), 71–79.
65. Sui, T.; Song, B.; Zhang, F.; Yang, Q. Effect of Particle Size and Ligand on the Tribological Properties of Amino Functionalized Hairy Silica Nanoparticles as an Additive to Polyalphaolefin. *J. Nanomater.* **2015,** 2015, 1–9.
66. Song, X.; Zheng, S.; Zhang, J.; Li, W.; Chen, Q.; Cao, B. Synthesis of Monodispersed $ZnAl_2O_4$ Nanoparticles and Their Tribology Properties as Lubricant Additives. *Mater. Res. Bull.* **2012,** *47* (12), 4305–4310.
67. Kheireddin, B. A. Tribological Properties of Nanoparticle-Based Lubrication Systems; PhD thesis, Texas A&M University, 2012.

CHAPTER 8

CARBON NANOTUBES FOR DRUG DELIVERY APPLICATIONS

JAHANVEE MITRA, G. K. P. SRILEKHA, NILESH WAGH*, and JAYA LAKKAKULA*

Amity Institute of Biotechnology, Amity University, Mumbai, Pune Expressway, Bhatan Post, Somathne, Panvel, Mumbai 410206, Maharashtra, India

**Corresponding author.*
E-mail: spencerjaya@gmail.com, jrlakkakula@mum.amity.edu

ABSTRACT

The use of nanomaterials especially carbon nanotubes (CNTs) in the field of cancer therapy is developing expeditiously. CNTs have been proved to be of high potential ever since their advent. Due to their variable physicochemical characteristics, they allow covalent and noncovalent addition of certain compounds for the development of novel drug delivery systems. These conjugated compounds tend to enhance the physical and chemical properties of both Single-walled carbon nanotubes and multi-walled carbon nanotubes making them the most effective nanomaterial for chemotherapeutic treatment of cancer. Various anticancer drugs like doxo-rubicin, paclitaxel, methotrexate, gemcitabine, platinum analogs, etc. were loaded onto the functionalized CNTs for targeted delivery of these drugs to the tumor site. They were found to be capable of repressing certain draw-backs like MDR (multidrug resistance), selectivity, target specificity, drug leakage and serious side effects of the drugs. It was also observed that the proliferation of tumor cells decreased drastically, moreover, the apoptosis of cancer cells was enhanced and extended due to the sustained release of drug(s) from the CNTs and toxic effects of functionalized CNTs. Thus, the concept of multifunctional hybrid–CNT was highly acknowledged by researchers and significant studies were conducted to obtain desired results. Therefore, this chapter describes the vivid aspects and properties of CNTs as drug delivery systems for the treatment of different types of cancer.

8.1 INTRODUCTION

The most significant disease known to be prevailing among the living entities is cancer, with around 1,762,450 cases and 606,880 deaths in the United States in 2019. Cancer is marked to be one of the most dangerous diseases in the world.[1] Cancer is a disease acquired by humans and animals that is caused due to uncontrolled proliferation of cells leading to the formation of tumors. Certain internal and external factors are responsible for the change in the programmed cell division pattern and pave the way to unconstrained mitosis. The genes inducing cancer are classified into two main categories: (1) protooncogenes (e.g., HER2) and (2) tumor suppressor genes (e.g., P53), overexpression and underexpression of respective genes promote tumor growth. Some of the different types of cancers are carcinoma, sarcoma, lymphoma, leukemia, melanoma, brain, and spinal cord tumors.[2]

Numerous studies were conducted concerning different ways for the treatment of cancer, few of which are chemotherapy, photothermal therapy, gene therapy, immunotherapy, and radiotherapy. This chapter mainly deals with chemotherapy in which drugs or chemical agents, such as doxorubicin, paclitaxel, cisplatin, etc. are used to kill tumor cells.[3] Although these methods produced satisfactory results, specific barriers prevented them from producing accurate and desired results. These barriers were normal cell cytotoxicity, a nonspecific site of action, and short-term drug effect, hence, to overcome these barriers, nanotechnology was introduced to cancer therapy techniques.[4]

Nanotechnology is the branch of science and engineering committed to the application of nanosized structures in different fields of sciences, primarily biomedical and healthcare. These nanostructures are of the size 1–100 nm and exhibit intrinsic properties that are advanced and reliable.[5] In oncology, nanostructures, such as nanospheres, nanocapsules, nanotubes, liposomes, dendrimers, etc. are used as carriers of chemotherapeutics.[6]

Out of all these nanostructures, carbon nanotubes (CNTs) are recognized to be substantially used for cancer treatment. CNTs are cylindrical nanostructures made out of rolled-up graphene sheets that have a diameter within the nanoscale range while the length can differ from centimeters to nanometers. These nanostructures exhibit extraordinary physical and chemical properties that caused striking interests in researchers around the globe.[7]

From the past few years, two types of CNTs are used in cancer therapy, that is, single-walled CNTs (SWCNTs), and multi-walled CNTs (MWCNTs) to deliver anticancer drugs to the tumor site. Drugs or genes are either loaded into the nanotubes or conjugated onto the walls and released in vitro/in vivo for the drug to reach the tumor cells.[8] The drug then enters the nuclei of the cells by rupturing its cell wall and DNA is fragmented leading to cell death or apoptosis.[9] Also, genes are delivered to the targeted site using CNTs, which on interacting with the DNA inhibits the uncontrolled mitosis of cells by silencing tumor-forming genes.[10] The reasons why CNTs play a potential role in apoptosis, are because of their sustained release and drug-loading capacity that can be changed by altering the length and diameter of the nanotubes, which eventually changes the surface area of the tubes, leading to highly efficient and accurate performance.[11]

However, there are certain circumstances in which the nonfunctionalized CNTs induce a high level of cytotoxicity that may cause apoptosis of normal cells also.[12] Hence, CNTs are conjugated with magnetic, organic, polymeric, metallic compounds, or quantum dots to control the cytotoxic effects, side effects of anticancer drugs, and targeted drug delivery. Conjugates, such as folic acid, gold nanoparticles, platinum, polyethylene glycol, chitosan, quantum dots, etc. are combined with CNTs which are known to be biocompatible.[13–15] In 2013, Moore et al. performed an experiment using polymer-functionalized MWCNTs to treat brain cancer. MWCNTs were combined with poly(lactide)-poly (ethylene glycol) (PLA-PEG) copolymer, loaded with paclitaxel, and used against U-87 cells. It was observed that the copolymer was able to enhance the high dosage effects of paclitaxel in MWCNTs, increase the therapeutic effect of paclitaxel, and eliminate the toxicity of normal cells.[16] Furthermore, in 2011, Wu et al. fabricated doxorubicin-loaded MWCNTs and cobalt ferrite (DOX/MWCNT–$CoFe_2O_4$) nanohybrid for the diagnosis and treatment of cervical cancer (HeLa cell line) in vitro. Results revealed elevated drug-loading capacity of MWCNT–$CoFe_2O_4$ and cytotoxicity against HeLa cells. Moreover, drug release was observed to be maximum at a low acidic environment.[17]

This chapter principally discusses the novel and modified strategies introduced by various researchers for the treatment of cancer using nanotechnology, which has aided biomedical fields over the past few years. It focusses on the advantages of carbon nanotube-based drug delivery systems (DDS), loaded with drugs and conjugated with different compounds, over conventional DDS. Moreover, it gives a detailed outline on the enhanced apoptosis and cell viability results obtained using modified carbon nanotube-based DDS, which can have a great impact on cancer treatment therapeutics in the immediate future (Table 8.1).

8.2 APPLICATION OF CARBON NANOTUBES IN DIFFERENT TYPES OF CANCER

8.2.1 CARBON NANOTUBES-BASED DELIVERY SYSTEM FOR LIVER CANCER TREATMENT

In recent years, CNTs have been proved to be of immense potential, specifically in the biomedical fields. Some of the recent avenues of CNTs for liver cancer treatment have been highlighted in this section. An experiment was conducted to make a comparative study of the efficiency of PEG applied to carboxylated CNTs of various lengths and biocompatibility for the drug delivery of doxorubicin against the HepG2 cells (liver cancer cells). The morphological characteristics of the PEG–MWCNTs were measured using TEM and the absorbance of the drug was monitored with UV-vis. It was observed that with an increase in the oxidation time (for 2, 4, 6, and 8 h) of different carboxylated multiwalled CNTs (CMWCNTs), the length was decreased and the roughness of surface increased. Dynamic light scattering (DLS) method was used to measure the hydrodynamic diameter and the studies explained that CNTs with shorter length and diameter favored cellular internalization. It was further reported that cytotoxicity of HepG2 cells was decreased with a decrease in length of CNTs and an increase in PEGylation degree. Thus, it was concluded that doxorubicin (DOX)-loaded PEG–CNTs with shorter length showed an enhanced inhibitory effect on HepG2 cells as compared with the free DOX.[18]

TABLE 8.1 Application of Different Types of Carbon Nanotubes Loaded with Anticancer Drug(s), for the Treatment of Various Types of Cancer.

Name of CNT	Type of CNT	Size	Characterization	Name of drug	Type of cancer	References
PEGylated CNT	MW	D (10–20 nm) L(5–15 μm)	TEM, FT-IR, TGA, DLS	Doxorubicin	Liver cancer	[17]
Water-soluble PEG Functionalized CNT	SW	D (1.3 nm) L(200 nm)	UV-vis, AFM			[28]
Estradiol PEG-appended CNT	MW	D (20–30 nm) L(1–5 μm)	SEM, TEM.TGA, FT-IR			[29]
					Breast cancer	
polylactic acid (PLA)/ polycthylcnc glycol (PEG)/multiwalled carbon nanotube	MW	D (30–70 nm), L (1–2 μm)	SEM		Lung cancer	[41]
Folate-conjugated PEG on Single-walled	SW	D (1–2 nm), L (less than 500 nm)	FT-IR, UV-vis, TEM, ZP, FL spec, CCD			[48]
Carbon nanotubes for						
PVPy–S–PEG folic acid-conjugated CNT	MW	D (8–15 nm) L (1–2 μm)	TEM, FT-IR, UV-vis			[50]
Poly (ethylene glycol) CNT	SW	D (1–1.2 nm) L (5–20 μm)	HNMR, GPC, UV-vis, FT-IR, TEM, TGA, DLS			[56]
					Cervical cancer	
PEGylated CNTs	MW	D (5–11 nm) L (900 nm)	SEM, TEM, TGA, FT-IR, Zeta potential		Bone cancer	[67]

TABLE 8.1 *(Continued)*

Name of CNT	Type of CNT	Size	Characterization	Name of drug	Type of cancer	References
PEG–-FA–MWCNTs	MW	D (1.4–100 nm), L (1–100 μm)	FT-IR, UV-vis spectrometry			[27]
PEI-SWCNTs linked to NGR	SW	D (1.2 nm), L (2–10 μm)	TEM, FT-IR, NMR, TGA			[26]
Glycopolymer– CNT	MW		HNMR, FT-IR, UV-vis			[23]
Polysaccharide–CNT	SW	D (1–2 nm) L (5–30 μm)	TEM, Zeta potential		Breast cancer	[60]
Folic acid-coupled CNT	SW	100 – 300 nm	HNMR, UV-vis, TEM, FT-IR, Zetasizer, PCA			[49]
Iron and folate CNT	MW	D (10–20 nm) L (0.5–1 μm)	TEM, XRD, FT-IR,		Cervical cancer	[45]
Γ-Fe_2O_3 multiwalled CNT	MW	D 20 nm	TEM, FT-IR, TGA, XRD, XPS		Liver cancer	[18]
Quantum dots conjugated with Fe_3O_4 CNT		D (6–20 nm)	UV-vis, TGA, VSM, TEM		Cervical cancer	[57]
CNT @ mesoporous silica	–	D 4 nm	TEM, TGA, Zeta potential, UV- vis, nitrogen adsorption/desorption analysis, confocal microscopy		Breast cancer	[33]

TABLE 8.1 *(Continued)*

Name of CNT	Type of CNT	Size	Characterization	Name of drug	Type of cancer	References
Mesoporous silica-grafted CNT	MW	D (20–30 nm) L (10–30 μm)	HNMR, TEM, TGA, XPS		Cervical cancer	[58]
TiO_2- Au-MWCNT	MW	D (30–50 nm) L (10–30 μm)	XRD, XPS, FESEM, TEM, UV- vis		Breast cancer	[37]
CNT–impregnated silk-based multifunctional hydrogel	SW	D (32–60 nm) L (500–1300 nm)	DLS, FT-IR, UV-vis, SEM, TEM, Swell ability		Lung cancer	[43]
Dendrimer-modified Multi-Walled Carbon Nanotubes	MW	D (30–70 nm), L (100 nm–2 μm)	HNMR, UV-vis, ZP, TEM			[47]
Polyampholyte-grafted single-walled CNT	SW	D (1–1.2 nm) L (5–20 μm)	HNMR, FT-IR, TGA, UV-vis, TEM		Cervical cancer	[55]
Chitosan–CNT	MW		TEM, DSC			[20]
Chitosan–CNT	MW		TEM	Doxorubicin and Rhodamine	Liver cancer	[19]
mildly oxidized large-inner diameter MWCNTs	MW	OD (30–60 nm), ID (20–50 nm)	UV-vis, SEM, HR-TEM, FT-IR	Doxorubicin, cisplatin		[30]
SWCNT	SW		TEM, UV-vis, Zeta potential	Doxorubicin and gadolinium-diethylenetriamine		[25]
PL–PEG–NH_2 functionalized SWNT	SW	D (60.4 ± 1.1 nm)	TEM, UV-vis, TGA,	Paclitaxel	Breast cancer	[34]
FA-CNT	SW	D (4–6 nm) L (0.7–10 μm)	FT-IR, SEM, TGA, UV-vis		Stomach cancer	[76]

TABLE 8.1 *(Continued)*

Name of CNT	Type of CNT	Size	Characterization	Name of drug	Type of cancer	References
Carboxyl–CNT	SW	D (1–2 nm) L (5–30 µm)	TEM, UV-vis		Lung cancer	[42]
Carboxyl–CNT	MW	D (20–30 nm) L (10–30 µm)	FT-IR, XRD, TEM, SEM	Gemcitabine	Breast cancer	[31]
PEG–CNT	SW		FT-IR, NMR, DSC, TEM, HNMR		Lung cancer	[40]
SWCNT and MWCNT	MW & SW	L (SWCNT: 1–2 nm) (MWCNT: 20–30 nm)	TEM, SEM, DLS, UV-Vis, ZP, surface tension	–	Breast cancer	[22]
Targeted SWCNT	SW	L(200 nm)	HNMR	Platinum-derived drugs	Prostate cancer	[63]
Platinum-based polymer-modified magnetic carbon nanotubes	MW	D (TEM – 5–15 nm) (DLS – 274 nm)	TEM, FT-IR, TGA, XRD, VSA, ICP-MS, FESEM		Cervical cancer	[52]
MWCNT	MW	L (2 µm)	TEM		Colorectal cancer	[73]
PEI–CNT	SW &MW	SW- 2–10 nm MW-20–40 nm	TEM, SEM	siRNA delivery	Cervical cancer	[54]
Cyclin A2 siRNA–f-SWNTs complexes.	SW	L (50–300 nm)	FT-IR, XPS		Leukemia	[70]
Octa ethylene glycol-linked SWCNT	SW	D (0.8–1.2 nm), L (100–1000 nm)	NMR, UV-vis, Raman spec.,	Tamoxifen	Breast cancer	[24]

TABLE 8.1 *(Continued)*

Name of CNT	Type of CNT	Size	Characterization	Name of drug	Type of cancer	References
Lipid-SWCNT	SW	D (1.5 nm) L (200 nm)	UV-vis spectra	Lipid drug		[32]
Single-walled carbon nanotube and graphene nanodelivery	SW	D (1.5–2 nm)	TEM, SEM, UV-vis	gambogic acid		[35]
Carboxyl-CNT	MW	D (1–3 nm)	TEM, TGA, Raman,	Mitoxantrone		[36]
MWCNT	MW	D (1–5 nm) L (180 nm)	DLS, TEM, SEM, TGA, UV-vis, FT-IR	Artemisinin		[38]
Carboxyl–CNT	MW	D (15–20 nm) L (50 μm)	FT-IR, SEM, TGA,	Letrozole		[39]
carboxylated multiwalled carbon nanotube	MW	D (2–100 nm)	FT-IR, TEM, TGA, Raman spec, SEM, UV-vis	Silibinin	Lung cancer	[44]
Multifunctional dendrimer-modified CNT	MW	D (30–70 nm) L (100 nm–2 μm)	UV-vis, TEM, TGA	Polyamidoamine	Cervical cancer	[46]
PCL/PVP core–shell nanofibers containing MWCNTs	MW	D (300–400 nm), L (20–30 μm)	FESEM, TEM, FT-IR	5-fluorouracil	Cervical cancer	[51]
PEGylated CNT	SW	D (5–6 nm)	UV-vis, TEM, AFM	Biotinylated amphicillin		[53]
Carboxylated CNT	SW	D (1–2 nm) L(1–3 μm)	FT-IR, XRD, DSC, Zeta potential, SEM, DLS	Formononetin		[59]
Carboxyl–CNT	MW	L(20–600 nm)	TEM, FT-IR, UV-vis, SEM, TGA, Zeta potential	Fluorescein isothiocyanate	Prostate cancer	[61]

TABLE 8.1 *(Continued)*

Name of CNT	Type of CNT	Size	Characterization	Name of drug	Type of cancer	References
SWCNT	SW	170.4 nm	Zeta potential, XRD, TEM, FT-IR, UV-vis	Curcumin		[62]
Carboxylated CNTs			Raman spec, TEM, FT-IR	Camptothecin		[64]
MWCNT	SW, DW	D (2 nm) L (100–200 nm)	SEM	IND		[65]
SWCNT	SW	D (0.7–1.3 nm)	TEM, UV-vis, FT-IR	Albumin based drug	Melanoma	[66]
P-Glycoprotein antibody functionalized CNT	SW		FT-IR, TEM, SEM	Antibody P-glycoprotein	Leukemia	[69]
Fe@MWCNT	MW	D (4 ± 25) L (50±30 μm)	SEM, TEM, Raman spec., M¨ossbauer spectroscopy	5 fluorouracil, purpurin, NIDIs	Colon cancer	[71]
Ir–loaded MWCNT	MW	ID (7.9–14)	TEM, IR spec., Raman spec., TGA, UV-vis,	Irinotecan	Bone cancer	[72]
RAFT-CNT	MW	D (5–15 nm)	FT-IR, HNMR, SEM, TEM, XRD, TGA	Capecitabine		[74]
Folate–chitosan CNT	MW		SEM, TEM, UV-vis	Irinotecan	Colon cancer	[75]
Biotin/FITC functionalized CNT	SW	D (3 nm) L (250 nm)	TEM, AFM, UV-vis, ATR-IR	Anticancer drug	Leukemia	[68]
Functionalized CNT	MW		SEM, UV-vis, FT-IR, ZP	Ruthenium polypyridyl complex	Liver cancer	[21]

CNT, Carbon nanotubes; MW, Multi-walled; SW, Single-walled; D, Diameter; L, Length; TEM, Transmission electron microscopy; SEM, Scanning electron microscopy; FT-IR, Fourier transform infrared; AFM, Atomic force microscopy; XPS, X-ray photoemission spectroscopy; XRD, X-ray diffraction; H-NMR, High-resolution nuclear magnetic resonance; ZP, Zeta potential; TGA, Thermogravimetric analysis; DLS, Dynamic light scattering; DSC, Differential scanning calorimeter; VSM, Vibrating sample magnetometer; ATR-IR, Attenuated total reflectance infrared; PEG, Polyethylene glycol; FA, Folic acid.

Similar to the previous experiment, functionalized MWCNTs with and without the deposition of γ-Fe_2O_3 on the surface of the tubes, loaded with doxorubicin drug was used for the treatment of liver cancer. The MWCNTs were functionalized with polyethyleneimine (PEI) and also linked with folic acid (FA) and PEG. The transmission electron microscopy (TEM) results showed that the diameter of the used MWCNTs was around 20 nm while the supernatant DOX absorbance was calculated using UV-vis spectroscopy. The DOX loading capacity on MWCNT/PEI–PEG–FA and MWCNT/γ-Fe_2O_3/ PEI–PEG–FA was calculated as 2.42 ± 0.16, and 2.38 ± 0.19 mg/g, respectively. After the conduction of N_2 adsorption/desorption, it was observed that surface area was decreased after the deposition of γ-Fe_2O_3. Ge et al. reported that the deposition of γ-Fe_2O_3 nanoparticles did not cause any significant change in the drug-loading capacity and also that doxorubicin could prevent cell proliferation in both cases. Thus, it was concluded that MWCNTs with or without the deposition showed a negligible difference in cell viability.[19]

Chemotherapy is the most widely used therapy for the treatment of cancer, but it may sometimes give undesirable outcomes and can fail to treat cancer. Thus, a new theory in which a combination of drugs was used was developed. In this regard, an experiment comprising hydrogels were combined with chitosan – MWCNTs and used for dual drug delivery. In this study, a combination of doxorubicin (DOX) and rhodamine (RB) was used as model drugs for the treatment of BEL-7402 (liver cancer cells) using the hydrogels MWCNTs which was monitored for the drug delivery using fluorescence imaging (in vivo). TEM determined the diameter of the CNTs/hydrogels to be 25 nm while DOX, RB absorbance was observed using UV-vis spectroscopy and fluorescence spectrometry. Cellular fluorescence imaging indicated that free DOX caused higher cytotoxicity (conc. in the range of 3.75–30 μg/mL) as compared with the DOX-CNTs/hydrogel and the intensity of fluorescence signals was high in free RB–DOX while weak in RB–DOX–CNTs/hydrogels (Fig. 8.1). Thus, it was concluded that DOX–RB–CNTs/hydrogels had a unique property of long-term sustained drug delivery and the technique of combined chemotherapy was found to be successful.[20]

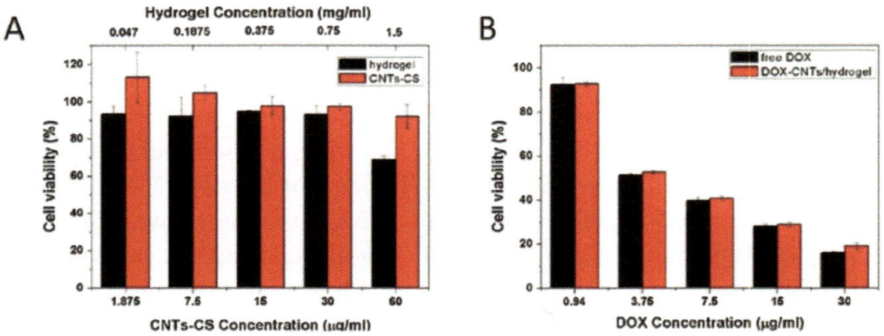

FIGURE 8.1 The toxicity of PCL–PEG–PCL hydrogel and CNTs–CS against BEL-7402 cells by the MTS assay (A). The cell viability of free DOX and the DOX-CNTs/hydrogel against BEL-7402 cells as measured using the MTS assay (B). The results are expressed as the mean ± SD (n=6).

Source: Reproduced with permission from Ref. [20]. © 2020 Elsevier.

Over the years, many techniques for drug delivery were developed but the main motive and challenge was controlled and sustained drug release. Hence, a significant strategy called NIR (Near-infrared) light-triggered drug delivery which focused mainly on the photothermal effects was developed. One such experiment comprising the strategy of chitosan (CS)-functionalized MWCNTs combined with thermosensitive hydrogel was introduced by Xia Dong et al. Due to their cross-linking structure, hydrogel CNTs showed a sustained release of drugs. This experiment focused on the release of anticancer drug doxorubicin by the photothermal hydrogel CNTs on the BEL-7402 (liver cancer cells). The morphological characteristics of hydrogel CNTs were measured using TEM. Differential scanning calorimetry (DSC) was used for thermal analysis of drug-loaded CNT/copolymers. It was observed that CNTs-CS can sustain higher irradiation and can increase the temperature by 15°C, whereas the CNTs could be increased only by 1°C with continuous irradiation. Also, the drug release rate was increased by 8–10% after undergoing NIR irradiation. These observations led to the conclusion that the release of DOX could be controlled by the use of NIR–laser irradiation and this method was successfully used on thermosensitive hydrogel CNTs.[21]

Cancer therapies face certain obstructions like multidrug resistance and radio resistance. To overcome these obstacles and produce lower drug dose, several experiments were performed to develop an efficient DDS. In one such experiment, radio sensitive CNTs were designed to antagonize the multidrug resistance in liver cancer cells (hepatocellular carcinoma). An anticancer drug named as ruthenium polypyridyl (RuPOP) having a high ability to induce cancer cell apoptosis was used in conjunction with MWCNTs. The morphological characteristics of MWCNTs were obtained using TEM and SEM, whereas FT-IR and UV-vis spectrophotometer was used to study the structure of drug-loaded MWCNTs (RuPOP-MWCNTs). It was observed that when the RuPOP–MWCNTs were treated with X-ray radiation, it enhanced the anticancer activity in R-HepG2 (liver cancer cell) than the group without X-ray irradiation. It was further reported that X-ray-irradiated RuPOP–CNTs inhibited the growth of R-HepG2 by promoting apoptosis while the RuPOP/CNTs without X-ray irradiation did not cause an increase in death rate. Hence, in this experiment, it was evident that radiochemotherapy using functionalized CNTs was an effective method to treat hepatocellular carcinoma and could also overcome cancer multidrug resistance.[22]

8.2.2 CARBON NANOTUBES-BASED DELIVERY SYSTEM FOR BREAST CANCER TREATMENT

Applicability of CNTs (SWCNTs and MWCNTs) has been found effective in treatment of breast carcinomas as well. This section depicts some of these achievements.

A comparative study was made between SWCNTs and MWCNTs to assess their properties like toxicity, therapeutic effects on MC4L2 cells (breast cancer cells). The size and characteristics of MWCNTs (20–30 nm length) and SWCNTs (1–2 nm length) were measured by TEM, SEM, and also the drug delivery was traced with UV-vis spectroscopy. Time and dose-dependent cell viability test was conducted on MC4L2 using SWCNT and MWCNT, it was observed that cell apoptosis was induced by both MWCNTs and SWCNTs and was time-dependent, with an increase in time the rate of apoptosis increased. The standard concentration for SWCNTs and MWCNTs was measured to be 50 and 400 μg/mL, respectively and an increase in dosage was toxic to organs. Thus, it was found that CNTs especially MWCNTs were highly promising and effective in lower dosages and could successfully treat breast cancer cells.[23]

Moreover, different carbon nanomaterials functionalized with different hydrophilic and biocompatible polymers either covalently or noncovalently were also tried to enhance their biocompatibility in the breast cancer treatment studies. In one such study, a comparative analysis was made between the covalent and noncovalent functionalization approach on glycoblock polymer conjugated with MWCNT. The anticancer drug DOX was loaded on the glycoblock copolymer- conjugated CNT for the treatment of breast cancer cells (MCF-7). The qualitative analysis of glycoblock structures was done by nuclear magnetic resonance spectroscopy-(NMR) analysis. The surface properties after the functionalization of CNTs were determined with zeta potential measurements. Further, through the MTT assay results, it was depicted that DOX-loaded glycoblock copolymer-conjugated CNTs showed more cytotoxicity toward the breast cancer cells as compared with the free DOX. In this study, the function of DOX-covalent-CNT and DOX-noncovalent-CNT was examined revealing that glycopolymer-conjugated CNT improved the Dox delivery to the breast cancer cells. Hence, based on the study, it was concluded that the hybrid–CNTs potentially delivered the DOX drug and increased anticancer efficacy.[24]

In a similar approach, SWCNTs conjugated with tamoxifen (TAM) and linked by octa ethylene glycol (OEG) were analyzed for the treatment of breast cancer. Carboxylic acid-functionalized SWCNTs were bound with tamoxifen and characterized with FT-IR, UV-vis spectrometer, Raman, and H-NMR spectroscopy. It was observed that the 0.8–1.2 nm wide and 100–1,000 nm long SWCNTs, on conjugation formed SWCNTs–CO–OEG–TAM showed absorbance at 275 nm on the UV-vis spectrometer. Additionally, the Raman spectrometer showed the G-band at ~1,578 cm^{-1} and RBM at ~150–250 cm^{-1}, similarly, FT-IR showed a peak at 1,741 cm^{-1} at the ester C=O stretch. Hence, the recognition and targeting properties of tamoxifen proved the potential of SWCNTs as a DDS with various characteristic properties.[25]

SWCNTs due to its distinct features and properties have gained considerable interest in combination therapy (chemophotothermal therapy) for cancer treatment. SWCNTs are known to be highly hydrophobic and insoluble in aqueous solution or common organic reagents, hence, it is a must for SWCNTs to increase its solubility for gaining biocompatibility. In one such experiments, the SWCNTs were modified with asparagine–glycine–arginine (NGR) by non- covalent approach and loaded with a combination of two drugs, namely, DOX and gadolinium-diethylenetriamine pentaacetic acid (Gd-DTPA) for the tumor-targeted delivery with MCF-7 cells (breast cancer cells). The size and length of DOX/NGR-SWCNTs/Gd-DTPA system were determined with TEM while the products and drug distribution were observed via UV-vis spectroscopy. It was observed in the MRI analysis of Gd-DTPA that signal intensity value increased with an increase in the concentration and the DOX/NGR–SWCNTs/Gd-DTPA system was successful in monitoring and delivering the drug to the tumor cells. The experiment further confirms Gd-DTPA as the most efficient MRI imaging contrast agent for the determination of tumoral site and DOX as the model anticancer drug. Thus, this multifunctional DDS for tumor imaging and treatment was found to be an efficient strategy for cancer therapy.[26]

In a similar study, cytotoxic effect of doxorubicin on MCF-7 cells was assessed in vitro using PEI-functionalized SWCNTs linked to Asn–Gly–Arg (NGR). TEM, FT-IR, NMR, and TGA were carried out to find the diameter (1.2 nm), length (2–10 μm), chemical composition, and temperature changes in the SWCNTs. Cell apoptosis assay was conducted between DOX, SWNTs-PEI, SWNT-PEI/NGR, and SWNTs–PEI/DOX/NGR and it was observed that SWNTs–PEI/DOX/NGR had the highest rate of cell apoptosis effect ($38.6 \pm 2.45\%$). Similarly, tumor growth inhibition in vivo tests revealed that at lower pH, interactivity between DOX and SWCNTs was decreased due to which the amount of DOX released was more. This experiment helped in understanding the potential application of carbon nanotubes in chemo and photothermal therapy.[27]

Surface-engineered CNTs are of high potential and possess unique physicochemical properties which make them ideal for targeted and controlled drug delivery. The stability of CNTs improves upon functionalization or modifying it with some chemicals. One such experiment was conducted to investigate the effect of functionalization on CNTs with the help of doxorubicin drug on MCF-7 cells (breast cancer cells). The FT-IR spectroscopy and UV-vis spectroscopy were used to examine the morphological characteristics of functionalized CNTs and the concentration of DOX drug, respectively. Sonam Sharma et al. observed that drug loading and entrapment efficiency of the drug DOX were enhanced after functionalization. Also, modified CNTs led to higher cell viability and reduced cell toxicity unlike free DOX in the MCF-7 breast cancer cells. The experiment depicted for the functionalized MWCNTs as the most potential DDS.[28]

Moreover, SWCNTs are known to be highly hydrophobic and thus to increase their efficiency they need to be functionalized which enhances their aqueous solubility. The SWCNTs are functionalized with PEG chains, for example, enable solubility of SWCNTs in aqueous solutions. An experiment introducing the concept of "functionalization partitioning" of SWCNTs was conducted with doxorubicin drug on breast cancer cells. The size and length of PEGylated SWCNTs were observed by TEM, and UV-vis spectroscopy was used to measure the absorbance of DOX-SWCNTs. Zhuang Liu et al. studied and observed various factors affecting doxorubicin loading on functionalized SWCNTs. The outcome of the experiment mentioned that the amount of DOX bound onto SWCNT increased with a decrease in pH, that is, at acidic pH and also the toxicity of DOX-loaded SWCNTs was more while it induced cancer cell death and cell apoptosis of MCF-7 cells. Thus, it was found that functionalized CNTs could adsorb much variety of molecules and also had increased the solubility which led to a new direction of targeted DDS.[29]

Furthermore, M. Das et al. conjugated PEG with 17β Estradiol (E_2) which in turn was conjoined with MWCNTs for the effective and long-term treatment of breast cancer. The female reproductive organs are mainly regulated by hormones wherein E_2 binds to the estrogen receptors for the transcription of genes which is usually overexpressed in malignant cells. In this study, Doxorubicin was used as a model anticancer drug along with E2–PEG–MWCNTs. The size and length of MWCNTs were analyzed using TEM and SEM. In the anticancer activity of DOX–PEG–MWCNTs, it was observed that the presence of E2 lowered the cytotoxicity of normal cells and a there was a decrease in growth inhibition by 10%. Through this study, it was evident for the development of an excellent and selective ER targeting system E2-PEG–MWCNT which successfully treated breast cancer.[30]

In another study for breast cancer treatment conducted by Yang et al., anticancer drugs DOX and cisplatin were used in conjunction with MWCNTs having large-inner diameter, PEG grafted and linked with folic acid. Through the TEM characterization, it was found that the inner and outer diameters of the MWCNTs were 20–50 and 30–60 nm, respectively. Further, the rough texture of the MWCNTs was observed by SEM and a peak at 490 nm of doxorubicin was also observed by UV-vis spectrometer. Since the DDS was pH-responsive, at pH under 6.5, it was observed and reported that the cytotoxicity effect of both PEG–MWNTs-FA-CDDS and

PEG–MWCNTs–CDDS–DOX systems on MCF-7 cells (breast cancer cell line) was higher and efficient. Hence, the dual DDS comprising of cisplatin and DOX using MWCNTs played an important role in controlling the cytotoxic activity.[31]

In a similar experiment, an anticancer drug gemcitabine-loaded FA-conjugated MWCNTs were used for the treatment of breast cancer cells (MCF-7). The size and length of CNTs were determined with the TEM and the surface charge of GEM/FA-NT was observed by Zeta potential. Usually, functionalization reduces the bundling, aggregation, and affects the size of CNTs but in this study, it was observed that after the CNTs were modified with FA ,the loading of GEM was easy and also there was no aggregation or bundling. The hemolytic toxicity by GEM/FA-NT was observed to be 8.23± 0.65 as compared with free GEM which was observed to be 17.34 ±0.56). The study conducted by Singh et al. finally reported that GEM/FA-NT caused effective apoptosis of the breast cancer cells in comparison to the free GEM. In conclusion, Gemcitabine (GEM) with the functionalized FA-CNTs is an effective drug targeting system for the complete cure of human breast cancer in the coming year.[32]

A report highlights the effectiveness of an anticancer drug molecule paclitaxel (PTX) conjugated with a lipid molecule docosanol for efficient drug loading onto the SWCNTs surface via hydrophobic interactions. The CNTs were further conjugated with FA for targeted drug delivery on human breast cancer cells. In this report, UV-vis spectra was used to measure the absorbance of lipid conjugated PTX molecules and TEM was used for determining the morphological characteristics of functionalized SWCNTs. This study conducted by Shao et al. using MCF-7 cells (breast cancer cells) showed that raw CNTs have high level of toxicity as compared with functionalized SWCNTs. It was also observed that cellular internalization and cell permeability of SWCNT increased after conjugation with the FA. The importance of this drug delivery approach was that after the loading of the drug on SWCNTs sidewalls, the CNT coating of sidewalls were left free for conjugation with other molecules and multifunctional drug delivery occurred. Thus, the lipid drug approach with FA-SWCNTs was successfully treated and inhibited the growth of breast cancer cells and was found to be an efficient drug delivery model.[33]

With the increasing advancement in the field of nanomedicine, the development of smart and activatable nanomaterials for controlled drug delivery has become very challenging. One such study was performed for the development of CNTs coated with hybrid mesoporous silica (MS) for the controlled and sustained delivery of DOX with the application of NIR laser (phototherapy). The CNTs were functionalized with isobutyramide (IBAM) and coated with HSA (human serum albumin), a nanosized macromolecular structure known to increase the drug-loading capacity of the antitumor drug DOX. The size and structure of CNTs before and after functionalization were measured by TEM and N_2 adsorption/desorption analysis, respectively. The absorbance of the drug DOX was determined by UV-vis spectroscopy. The cytotoxicity and drug-releasing capacity of CNT/MS/IBAM/HSA were observed to increase upon NIR laser treatment. This study conducted by Li et al. revealed that NIR-responsive DOX loaded on CNT/MS/IBAM/HSA showed an increase in cell cytotoxicity of breast cancer cells and also there was a controlled release of DOX due to the photothermal activity. Hence, it can be concluded that the use of these smart and highly potential nanomaterials can be of great interest for the treatment of cancer with the help of drugs like DOX.[34]

A similar strategy was used by Shao et al. who combined HSA with SWCNTs and loaded them with paclitaxel (PTX) to treat breast cancer. The SWCNTs had a diameter of 60.4 ± 1.1 nm as revealed from TEM and characterized by thermogravimetric analysis and UV-vis-NIR spectrometer to yield a PTX peak at 228 nm. In this study, the SWCNTs were first functionalized by PL–PEG–NH_2 (amphiphilic lipid polymer-phospholipid PEG mixed and then loaded with HSA and PTX, respectively, to form SWCNTs–HSA–PTX. SWCNTs–HSA–PTX were tested against MCF-7 cells (breast cancer cell line) and determined for the apoptosis or inhibition of cancer cell multiplication using TUNEL (TdT dUTP nick-end labeling) staining. Study revealed that the percentage of apoptotic cells was 8.5% for SWCNTs–HSA–PTX which was higher than that of free PTX and HSA-PTX alone. The study thus proved that SWCNTs are a reliable delivery system in the treatment of cancer.[35]

In another study involving Gambogic acid, a xanthonoid, known for its apoptosis-inducing property, was applied in cancer treatment. In this study, gambogic acid (GA) was loaded into SWCNTs to form a delivery system and result in the apoptosis of breast cancer cells (MCF-7). The diameter of the SWCNTs was determined by TEM (1.5–2 nm and samples of GA and GA+SWCNTs were tested against MCF-7 cells to study cell apoptosis as evident by flow cytometry and DNA fragmentation assays. This study conducted by Saeed et al. revealed that according to the flow cytometry and DNA fragmentation assay, GA+SWCNTs gave enhanced cytotoxicity outcomes than GA alone. The reason behind was speculated to be the sustained release of GA from SWCNTs, and on free release of GA, it was used up immediately decreasing its cytotoxic activity with time. GA was also used to treat other cancers such as pancreatic cancer, thereby proving its potential as an optimum drug for cancer treatment.[36]

A chemotherapeutic drug Mitoxantrone (MTO) is a widespread anti-cancer drug mainly used to treat breast cancer and leukemia. All well-known chemotherapeutic drugs show some or the other side effects but the mitoxantrone shows considerably less acute side effects. One such experiment was conducted to examine the effectiveness of MWCNT – MTO activity on the breast cancer cells (MDA231) and nonneoplastic fibroblast cells (NIH3T3) at a particular time and amount of dose. The dispersibility and morphology of MWCNT-MTO were measured by SEM and TEM, respectively. The cell viability and dispersibility were assessed at different concentrations for different incubation periods and it was observed that the cell viability for MDA231 cells (breast cancer cells) was dose- and time-dependent unlike the NH3T3 cells (nonneoplastic fibroblast). This study conducted by Giulia Risi et al. revealed that the adsorbed drug MTO showed positive effects on both the types of cells, and also a standard concentration of 0.4 μg/mL for MWCNT–MTO complex was determined .[37]

Nanomaterial-based drug delivery have been established on a large scale in the field of biomedicine because of its advantages like efficient drug loading, sustained, and controlled drug release. An experiment was conducted in which the MWCNTs (prepared using bark extract) were conjugated with titanium dioxide–gold nanoparticles, dispersed in ethylene glycol and water with a ratio of 1:1 and 1:3, for the efficient drug delivery of DOX to the MCF-7 and A549 cells. After coating MWCNT with TiO_2–Au, its physical and chemical interactions were determined using XPS. The morphological characters of TiO_2–Au-MWCNT were examined by SEM and the loading of drugs onto the MWCNT surface was characterized by HR-TEM. Morphological changes were observed in the cells after they were treated with DOX-loaded TiO_2–Au/MWCNTs, the cancer cells showed shrinkage and difference in shape ultimately leading to apoptosis. This experiment conducted by Vishwanathan karthika et al. demonstrated that DOX-loaded TiO_2–Au/MWCNTs were taken up readily by the MCF-7 and A549 cells and increased cytotoxicity as compared with the free DOX (Fig. 8.2). Consequently, this conjugated DDS was proved to be an efficient method for cancer therapy.[38]

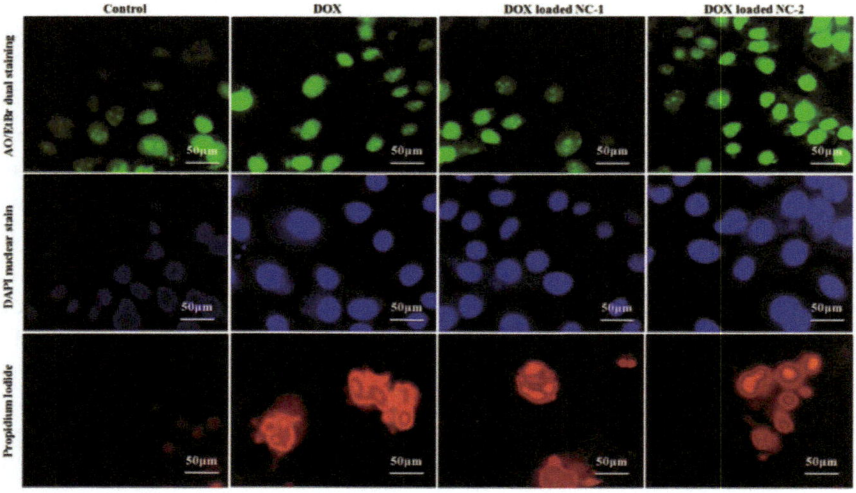

FIGURE 8.2 Cellular uptake and localization of DOX and DOX-loaded nanodrug carriers (NDC) against MCF-7 cancer cell observed by fluorescence microscopy.

Source: Reproduced with permission from Ref. [38]. © 2018 Elsevier.

Artemisinin (ART), is an antimalarial drug extracted from the Chinese herb *Artemisia annua L.* but recent studies have invented its potential for cancer treatment. It has a special property of endoperoxide bridge which can interact with the iron (Fe) molecule and cause cell damage and death. An experiment was conducted to develop a multifunctional drug targeting delivery system with the help of hyaluronic acid (HA)-derived MWCNTs, transferrin (Tf) as targeting ligand, and ART as the model drug for breast cancer treatment. In this study, the morphological characteristics were studied with the DLS (dynamic light scattering) and SEM microscopy whereas the drug-loading capacity of ART was measured with the help of UV-vis spectroscopy. It was observed that the MCF-7 cells could effectively take the ART drug with the delivery system HA-MWCNTs-Tfs-ART and the drug uptake was also faster with this HA-MWCNTs-Tf-ART-FITC system than without the Tf. This study conducted by Huijuan Zhang et al. depicted that HA-MWCNTs/Tf@ART was an effective DDS, moreover, they also showed that the rate of apoptosis increased after combining chemotherapy with photothermal therapy, that is, treating the system with irradiation. Thus, through this study, an efficient DDS with negligible toxicity and exceptional antitumor activity was developed.[39]

Letrozole is a nonsteroidal breast cancer drug and inhibits the growth of hormonally responsive breast cancer. In one such research involving Letrozole, the MWCNTs were modified with cyanuric chloride and 1,3-dihydroxybenzene (MWCNT-resorcinol) for the delivery and determination of Letrozole performance in breast cancer cells. The characteristics and surface morphology of the MWCNTs and functionalized MWCNTs were observed by FT-IR and SEM spectroscopy. After the addition of resorcinol (adsorbent), it was observed that the extraction rate of the drug was increased. Moreover, the adsorption of drugs varied with a change in pH with the highest amount of letrozole being extracted at acidic pH. The study conducted by Nodeh et al. stated that two major factors are responsible for the delivery of letrozole, that is, pH and the adsorbent used. Thus, the functionalization of MWCNTs with the adsorbent resorcinol was found to be an effective method and the effect of letrozole on breast cancer was found to be promising.[40]

8.2.3 CARBON NANOTUBES-BASED DELIVERY SYSTEM FOR LUNG CANCER TREATMENT

Lung cancer is one the most death-causing cancers all over the world and many researches are investigating for its cure. Nanotechnology has paved the way in this endeavor. This section depicts the applicability of CNTs in lung cancer treatment.

One such investigation on GEM, a hydrophilic anticancer drug commonly used for the treatment of lung, pancreatic, and colon cancer. It was used to develop an efficient DDS using functionalized SWCNTs to deliver GEM drugs for lung cancer treatment. In this study, FT-IR spectroscopy was used to observe the chemical structures and TEM microscopy was used to analyze the size and characteristics of SWCNTs and PEG–SWCNTs. After functionalizing the SWCNTs with PEG the drug-loading capacity of GEM was increased from 43.14 ± 1.82% (w/w) to 37.32 ± 2.73% while the cell cytotoxicity test results were observed to be positive for SWCNT-GEM and SWCNT–PEG–GEM. Specifically, the cytotoxicity increased by 19.66% and 11.1% for SWCNT-GEM and

SWCNT–PEG–GEM, respectively. The study conducted by Razzazan et al. reported that the functionalized SWCNTs with PEG led to an increase in cytotoxicity and also conjugation of GEM drug was enhanced. Hence, an effective DDS using GEM with decreased side effects and increased drug-loading efficiency was developed for the therapy of A549 cells (lung cancer cells).[41]

Since doxorubicin has been broadly used to treat cancer, Anaraki et al. used DOX-loaded polylactic acid (PLA)/PEG/MWCNTs nanofibrous scaffolds to treat lung cancer under in vitro conditions. The electrospinning method was chosen to produce PLA/PEG/MWCNTs nanofibrous scaffolds in which MWCNTs were added to different concentrations of N, N-dimethyl formamide and again added to PLA/PEG after stirring overnight at 25°C–27°C. The solution obtained was subjected to electrospinning, the obtained nanofibrous scaffolds were collected, loaded with DOX, and used against the A549 cell line. PLA/PEG/MWCNTs of diameter 30–70 nm and length 1–2 μm (SEM) were observed to have shown cancer cell growth suppression as evident through MTT assay. It was revealed that PLA/PEG/MWCNTs nanofibrous scaffolds were far more advanced in cancer cell apoptosis than freely released DOX due to its slow and sustained release time from the nanotubes. It was also noted that even after 72 h, the anticancer capability DOX-loaded PLA/PEG/MWCNTs nanofibrous scaffolds was not lost.[42]

Paclitaxel (PTX) is a chemotherapeutic drug mainly used for the treatment of lung, breast, and ovarian cancers. It was first derived from plants of the genus Taxus in the year 1963 and was found very efficient in the treatment of cancer. In one such study, A549 cells (lung cancer cells) were treated with a novel drug delivery technique in which SWCNTs attached with chitosan (CH) and hyaluronan (HA) to upgrade the biocompatibility and biodegradability. The size of unmodified SWCNTs was measured and found to be 20 nm, while the SWCNTs–CH, PTX-loaded SWCNTs–CH–HA were observed using TEM (Fig. 8.3). It was observed that the surface area and trapping ability of the SWCNTs were increased after modifying it with HA and CH and the drug release ability was found to be pH-dependent. Further, the LDH enzyme levels were measured in A549 cells and it was observed that PTX-loaded CNTs increased the LDH levels. The study conducted by Yu et al. reported that SWCNTs modified with CH and HA were highly promising DDS for the release of PTX which also induced higher toxicity toward A549 cells. Thus, this DDS was found to be effective one for anticancer therapy.[43]

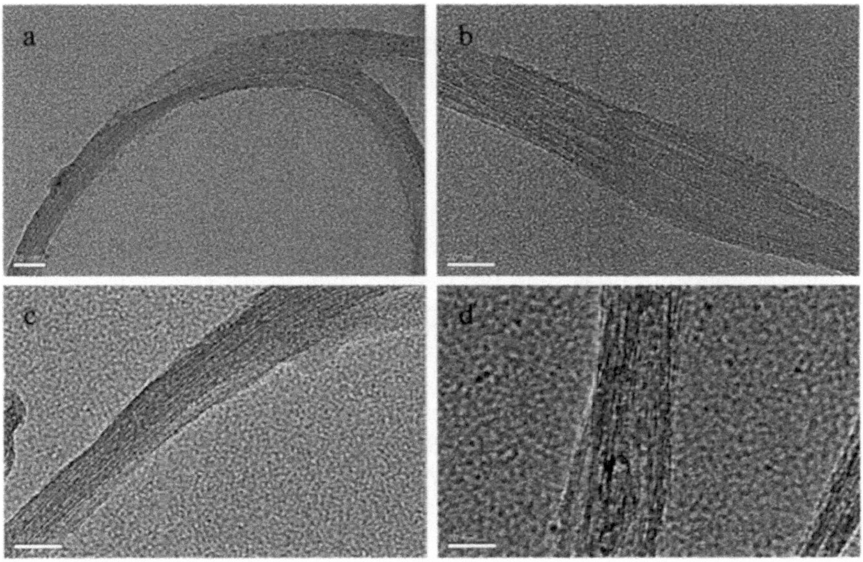

FIGURE 8.3 Transmission electron microscopy (TEM) images of (a) single-walled (SWNTs), (b) chitosan-modified SWNTs (SWNTs–CHI), (c) chitosan- and hyaluronan-modified SWNTs (SWNTs–CHI–HA), and (d) CHI- and HA-modified SWNTS loaded with paclitaxel (SWNTs–CHI–HA–PTX).

Source: Reproduced with permission from Ref. [43]. © 2016 Elsevier.

In a similar approach, SWCNTs were modified with FA and loaded with a silk protein as well as DOX for the treatment of lung cancer. In this study, the length and diameter of SWCNTs were observed using TEM, and the modified SWCNTs were analyzed using FT-IR. Silk hydrogel is supposed to be nonresponsive toward NIR light but, it was observed that the DOX release rate increased with the presence of silk protein after undergoing irradiation. Moreover, the LDH release which acts as a signal for cell apoptosis, it was evident that the LDH amount increased due to SWCNT-FA/DOX. The study performed by Ankita Gangrade et al. reported that modifying the drug release by SWCNTs–FA with silk protein was responsive only after phototherapy and thus, it was proved to be an effective drug delivery method for cancer treatment.[44]

Silibinin (SB), a chemotherapeutic, is not a widely known anticancer drug, yet it has been useful in treating all types of cancers, one of which is lung cancer. In vitro delivery of SB using pH-responsive MWCNTs was performed by Tan et al. for the treatment of lung cancer. These MWCNTs had a diameter of 2–100 nm (TEM), and in the case of SB-MWCNTs, the G-band and D-band peak was not noticeable (Raman spectrometer), the absorption band shifted from 1636 to 1630 cm^{-1} (FT-IR), weight loss of 52.8% was observed at 1000°C (TGA). MWCNTs were conjugated with SB (SB-MWCNTs) and used against the A549 cell line, and the cytotoxicity of these cells was determined by proliferation assay. Through this study, it was reported that SB-MWCNTs showed enhanced cytotoxicity at lower concentrations than the free SB due to sustained release of the drug from the nanotubes. Yet again, Tan et al. helped in proving CNTs as a superior delivery system for cancer treatment.[45]

8.2.4 CARBON NANOTUBES-BASED DELIVERY SYSTEM FOR CERVICAL CANCER TREATMENT

Over the years, the CNTs are being used as drug carriers and are proved to be of high interest in the field of biomedical and biological fields. A variety of CNTs are being developed with the help of oxidants, surfactants, polymers, and other biomolecules for enhancement of the drug delivery method. This section highlights their potential as anticancer agents in cervical cancer.

In this regards, one such experiment conducted by Li et al. reveals on development of a difunctionalized MWCNTs for the delivery of DOX into the Hela cells (cervical cancer cells). The oxidized MWCNTs were modified with folate and iron molecules while their characterization was done by TEM, XRD, and FT-IR spectrum. It was observed that the DOX loading capacity of FA-MWCNT@Fe differed with and without exposure to the NIR light. After the exposure to NIR light, the rate of DOX released was increased from 23% to 59% and the highest amount of cytotoxicity was induced into the HeLa cells (cervical cancer cells). Hence, DOX an effective DDS was developed with FA and Fe which could recognize cancer cells and also killed them in a specified manner.[46]

A new approach for the improvement of the CNTs with poly(amidoamine) (PAMAM) dendrimers with fluorescein isothiocyanate (FI) and FA was developed for the cancer treatment and imagining. Dendrimers (G5·NHAc-FI-FA) are monodispersed highly branched macromolecules, they are nanoscale structures that can be attached to different anticancer drugs for its application in cancer treatment. In this approach, the MWCNTs were acid treated and acetylated before modification with the dendrimers, FA and FI. TEM was used for the identification of morphological characteristics of MWCNTs while H-NMR spectroscopy confirmed the modification of MWCNTs with the dendrimers. The cell viability results showed that KB cells (sub-cell line for cervical cells), when treated with MWCNT/G5–NHAc–FI-FA were observed to be more biocompatible and also the cellular uptake was dose-dependent. This approach reported by Xiangyang Shi et al. showed that FA and FI conjugation on the dendrimer came up with enhanced targeting and imaging proficiency. Hence, this approach of modifying MWCNTs with targeting ligands and imaging molecules was a success and also speculated about them in various other biological systems for imaging and targeting a disease.[47]

In a similar manner, FI and FA-modified PAMAM dendrimers were used to modify MWCNTs, eventually forming MWCNTs/G5. NHAc-FI-FA was loaded with DOX and characterized using H-NMR, UV-vis spectrometer, and TEM which revealed about the diameter 30–70 nm and length 100 nm–2 µm of MWCNTs. The therapeutic effect of dendrimer-modified MWCNTs was studied against KB cells (cervical cancer cell line) which revealed that DOX-loaded multifunctional dendrimer-modified MWCNTs caused changes in the morphology of the cancer cells. This was followed by the cytotoxicity of cancer cells leading to their apoptosis that was notably higher than freely released DOX. This study was conducted by Wen et al. used dendrimers to produce a unique DDS that proved to be effective for treating cancer cells in vitro without any complications or drawbacks.[48]

Similarly, another experiment exploiting DOX and PEG was conducted by Niu et al. in which PEG conjugated with folic acid (PEG–FA) was linked to SWCNTs and loaded with DOX to treat cervical cancer (HeLa cells) in vitro. As evident from TEM characterization, a diameter and length of 1–2 nm and less than 500 nm respectively, and high loading efficiency of 149.3 ± 4.1%, the SWCNTs were also characterized using FT-IR, UV-vis spectrometer (peak of DOX at about 490 nm) and fluorescence spectrometer. The cell viability after treatment of drugs was carried out by WST-1 assay through which it was revealed that FA was responsible for targeted delivery of DOX that was released into the cells at low pH. It was also reported that DOX/PEG–FA/SWNTs system entered cancer cells by clathrin-mediated endocytosis and caused apoptosis without affecting normal cells. This experiment demonstrated that the novel DDS has many advantages over conventional delivery systems with accuracy, precision, and high probability of positive results.[49]

Prodrugs are pharmacologically inactive derivatives of drugs that can be converted into their active form through chemical reactions. The application of the anticancer prodrug system into CNTs for the treatment of cervical cancer was introduced by Fan et al. The prodrug system consisted of characteristic features, such as pH-triggered drug release, active targeting, and photodynamic therapeutic (PDT) effect. In one of such experiment, FA coupled and hydrazone-linked MWCNTs were loaded with DOX and characterized by H-NMR, UV-vis spectrometer, TEM, and FT-IR. Through TEM studies, an average size of 100 – 300 nm was evident. Further, the DOX–hydrazone–MWCNTs–FA's cell viability, PDT, and in vitro chemotherapy were studied. This study conducted by Fan et al. reported that under low pH conditions, folate receptor (FR)-positive HeLa cells are taken over by the prodrug system resulting in targeted drug release, thereby yielding lower viability. MTT assay carried out to check the effect of blue light irradiation on cancerous cells in the presence of DOX–hydrazone–MWCNTs–FA revealed that the cell death increased upon blue light illumination, which on further combining with PDT and chemotherapy, the cell death was more prominent.[50]

In a recent approach, an efficient bifunctional system with both chemo-photothermal cancer treatment was developed. The MWCNTs were coated with poly (N-vinyl pyrrole)(PVPy) and folic acid terminated polyethylene glycol-thiol (FA–PEG–SH) to magnify the properties like biocompatibility, dispersibility, and increased circulation time of the drug into the blood. The length and diameter of MWCNTs length 1–2 μm, diameter 8–15 nm was observed by TEM and the functionalized MWNT–PVPy–S–PEG–FA was specified by FT-IR spectroscopy. DOX loading and its absorbance were measured by UV-vis spectroscopy. The photothermal ability of the modified MWCNTs was observed to be dependent on the NIR laser concentrations, an increase in concentration increased the irradiation effect and thus killed more cancer cells. The cytotoxicity test was conducted onto the HeLa cells cervical cancer cells and it was observed that FA combined MWCNTs caused more cytotoxicity to the cancer cells than the free MWCNTs. The study performed by Daquan Wang et al. reported that coating PVPy and FA onto MWCNTs effectively enhanced the photothermal property and drug-loading capacity respectively. Thus, this bifunctional DDS was a success and could effectively kill the cancer cells.[51]

On the other hand, CNTs are not only used for the delivery of anticancer drugs but also exhibit anticancer properties themselves. To prove this point, Nasari et al. used poly (ε-caprolactone) (PCL)/poly (N-vinyl-2-pyrrolidone) (PVP) core–shell nanofibers containing MWCNTs and 5-fluorouracil (5FU) against HeLa cells. Field emission scanning electron microscope (FESEM) revealed the uniform texture of the core–shell nanofibers (prepared by electrospinning method) of diameter 300–400 nm and MWCNTs of diameter 17 nm (TEM). Along with characterization, tensile properties, hydrophilicity assessment, and biodegradability of nanofibers mat were studied that provided information regarding the PVP amount which was directly proportional to the amount of drug release. The results confirmed the total drug release duration ranging from 312 to 528 h, high efficiency of drug-loaded nanofibers on HeLa cells, and nontoxicity of MWCNTs or nanofibers. This study proved the efficiency of MWCNTs and anticancer drug (5FU) when used in the field of healthcare in cancer treatment.[52]

Platinum, a chemical element, is also known to have exhibited anticancer properties over various types of cancers, such as breast, cervical, ovarian, etc. To improve its cytotoxicity properties, Rezaei et al. performed an experiment in which MWCNTs mixed with $FeSO_4.7H_2O/FeCl_3$ (FO), polycitric acid (PC), PEG, and loaded with Pt(II) complex, were used to treat HeLa cells. The MWCNTs/FO@PC/Pt (II)-b-PEG magnetic nanovehicle of diameter 5–15 nm (DLS) and a negative charge of −21.54 ± 1.1 mV (ZP) was tested on the basis of cellular uptake and cytotoxicity assay. The cellular uptake (using FITC label) and cytotoxicity (MTT) assay revealed that the amount of cell penetration was more due to the acidic conditions wherein the release of platinum drug was quick. Moreover, at low concentration of platinum, the toxicity was observed to be high in HeLa cells.[53]

Another novel approach was introduced by Brahmachari et al. which deals with noncovalently water dispersed biotinylated amphiphile–SWCNTs nanoconjugate. The terminal ends (-OH, -COOH, and $-NH_2$) of L-tyrosine were introduced with C-16 hydrophobic segment, PEG hydrophilic fragment, and biotin in order to produce neutral amphiphile. This neutral amphiphile was useful in dispersing SWCNTs in water (in measured amounts), cell proliferation assay, and cytotoxicity of cancer cells. Likewise, biotin was responsible for the transportation of loaded cargo to the cancer cells. Characterization of SWCNTs was carried out using UV-vis spectrometer, atomic force microscope (AFM), and TEM which confirmed the diameter of SWCNTs as 5–6 nm. Toward the end, PEG was also conjugated to SWCNT (PEG–SWCNT)-amphiphile to assess the cell viability of HeLa cells. Firstly, an oligonucleotide (Cy3-Olg) was loaded onto the surface of the conjugate, and after obtaining the desired results, DOX was loaded similarly. It was observed that at high concentration of SWCNT, almost 60% of HeLa cells were killed which proves that cell apoptosis was completely dependent on the concentration of SWCNT and DOX together.[54]

Despite drug delivery, CNTs are also widely used as carriers for gene delivery and they need to be functionalized for binding of plasmid DNAs, small interfering RNAs (siRNA), or other chemical compounds for delivery into various cancer cell lines. Because of their high potential in gene/drug delivery, an experiment was conducted to check the efficiency of PEI polyethylenimine functionalized MWCNTs for the delivery of siRNAs into the human cervical cancer cells. The size and length of PEI aminated MWCNTs were measured by TEM and SEM microscopy. It was observed that PEI functionalization led to the positive charge on the surface of MWCNTs and thus the binding of siRNA electrostatically was easy due to its negative charge, and the cell viability of the HeLa cells was decreased with an increase in the concentration of PEI–NH–CNTs. This study conducted by Huang et al. further reported that siRNA loading and delivery were highly effective with the PEI–NH–MWCNTs. This study was found to be highly promising for eliminating carcinogens and viral vectors and was a safe method for gene therapy.[55]

On the other hand, new strategies involving the production of SWCNTs (diameter 1–1.2 nm and length 5–20 μm (TEM), grafted with polyampholyte (PMT) via green synthesis and loaded with DOX (SWCNTs/PMT–DOX) was introduced to check its cytotoxicity against HeLa cells. The experiment was initiated with the synthesis of polyampholyte alternating monomers from maleic anhydride–thiolactone adduct monomer (Ma-Tla) (using water) which was then grafted over SWCNTs forming SWCNTs–PMT. These SWCNTs–PMT composites were then loaded with

DOX and were characterized by H-NMR, FT-IR, UV-vis spectrometry, and TGA. Different concentrations of SWCNTs–PMT–DOX (0.1, 0.5, 1, 5, 10, and 15 μg/mL) and free DOX were tested on HeLa cells that resulted in concentration-dependent cytotoxicity. Concisely reported by Phan et al. found that on increasing the concentration of SWCNTs/PMT–DOX, cell proliferation or cell viability decreased accordingly. When free dox was released, cytotoxicity was more due to more availability of DOX to cancer cells, but the effect did not last as long as that of SWCNTs/PMT–DOX.[56]

Similar to the previous experiment, Cao et al. performed an experiment using SWCNTs grafted with green and direct functionalized PEG-grafted furfuryl-grafted-polystyrene-alt-maleic anhydride and loaded with DOX (SWCNTs/PSMF–PEG–DOX). The SWCNTs were characterized for diameter 1–1.2 nm and length 5–20 μm (DLS) using H-NMR, gel permeation chromatography (GPC), UV-vis spectrometer, FT-IR, TEM, and TGA. Consecutively, cytotoxicity assay was performed on HeLa cell line in vitro which revealed that the cytotoxicity and cell viability were entirely dependent on the concentration of SWCNTs/PSMF–PEG–DOX. Also, the CNT-based DDS showed sustained released hence, the cytotoxicity lasted longer than that of freely released DOX.[57]

Since magnetic nanoparticles have been functional in biomedical fields, Chen et al. came up with a novel approach for the imaging and in vitro treatment of cancer involving magnetic nanoparticles and quantum dots conjugates. CNTs of diameter 6–20 nm (TEM) were incorporated with Fe_3O_4 nanoparticles and then conjugated with SiO_2-coated quantum dots (made using CdTe nanocrystals). This conjugate was then coated with transferrin and loaded with doxorubicin forming DOX–Fe_3O_4–CNT–HQDs–Trf system. They were further characterized using UV-vis spectrometer, TGA, and VSM. MTT assay was conducted to check the biocompatibility between DOX–Fe3O4–CNT–HQDs–Trf and cell viability of HeLa cells. It was observed and reported that DOX molecules entered the nuclei of cancer cells within 1 h of incubation which indicates that due to the presence of CNTs in the conjugate, cellular uptake was enhanced efficiently. Fluorescent dye was used to check the status of cancer cells because of which imaging and cytotoxicity was observed easily.[58]

Many experiments were conducted using CNTs -based DDS, however, some resulted in drug leakage which caused inefficiency in cell apoptosis. Hence, to overcome this issue, Zhang et al. came up with a strategy involving the use of tri-stimuli responsive MWCNTs coated with MS graft multifunctional copolymer. MWCNTs of diameter 20–30 nm and length 10–30 μm (TEM) were coated with poly (N-isopropyl acrylamide-block-poly (2-4-formylbenzoyloxyethyl methacrylate) copolymer through disulfide cysteamine bonds forming MWCNT-MSN-s-s-g-PNIPAM-b-PFBEMA and tested against HeLa cells. This nanostructure (MWCNTs–MSN–s–s–g–PNIPAM–b–PFBEMA) was then loaded with DOX and characterized by H-NMR, XPS, and TGA. Results revealed that at optimum temperature and low pH, DOX release was higher. MTT assay revealed that when the concentration of MWCNTs-MSN-s-s-g-PNIPAM–b–PFBEMA/DOX was increased to 10 μg/mL, the cell proliferation decreased from 76.8 to 3.2% which proved the effective cytotoxicity of MWCNTs–MSN–s–s–g–PNIPAM–b–PFBEMA/DOX.[59]

Cyclodextrin is a cyclic oligosaccharide which is used in various fields, such as healthcare or biomedical, agriculture, and many other industries. Since cyclodextrin is one of the most cost-effective drugs carriers, Liu et al. fabricated hydroxypropyl-β-cyclodextrin-grafted carboxylated SWCNTs (SWCNTs–COOH) to form cyclodextrin-modified SWCNTs (SWCNTs–CD). This nanoconjugate containing SWCNTs of diameter 1–2 nm and length 1–3 μm (DLS) was loaded with formononetin (FNM), and characterization was carried out using SEM, FT-IR, XRD, DSC, and DLS. Also, the dispersity of SWCNTs–CD and SWCNTs–COOH in water was observed (Fig. 8.4). Firstly, the encapsulation efficiency of SWCNTs–CD was calculated using HPLC which was recorded to be 88.66 ± 3.13 %, and then the in vitro cytotoxicity of HeLa cells was observed. It was reported that SWCNTs–CD showed sustained release of FNM and even at lower concentrations, SWCNTs–CD–FNM demonstrated higher cytotoxicity than freely released FNM.[60]

An another novel experiment was conducted to form a novel DDS by adding the chitosan (CHI) layer onto the SWCNTs and hyaluronan (HA) as the target ligand for efficient delivery of DOX with HeLa cells. The morphological structure was observed by TEM and the diameter was found to be 20 nm. It was observed that modification of SWCNTs with CHI and HA increased the drug-loading capacity and the release of drug which was time- and pH-dependent. Additionally, the fluorescence test results showed that the SWCNT–CHI–DOX did not cause much cytotoxicity. Yunfei Mo et al. reported through this study that the chitosan layered SWCNTs effectively killed HeLa cells and had comparatively lower cytotoxicity. Thus, this DDS was highly promising and had fewer side effects.[61]

FIGURE 8.4 The dispersibility of SWCNTs–COOH (A) and CD-SWCNTs (B) in water after 10 min.

Source: Reproduced with permission from Ref. [60]. © 2018 Elsevier.

8.2.5 CARBON NANOTUBES-BASED DELIVERY SYSTEM FOR PROSTATE CANCER TREATMENT

The nanostructured materials need to be integrated with molecular imaging and therapy technologies since cancer therapy cannot be successful without tracking and visualizing the tumor lesions. In this experiment, the MWCNTs were cut short and modified with PEI (polyethylenimine) for further conjugation to fluorescein isothiocyanate (FITC) and prostate stem cell antigen (PSCA) monoclonal antibody (mAb), added with Doxorubicin loading. The TEM and SEM microscopy were used to measure the length and diameter of short carboxylated MWCNTs (Fig. 8.5), whereas the absorption of MWCNTs after grafting of PEI was observed by FT-IR spectroscopy. It was observed in the results of UV-vis spectroscopy that FITC was conjugated and showed fluorescence. The cytotoxicity of CNT–PEI(FITC)-mAb on the PC-3 (prostate cancer cells) was increasing with increased concentrations. A hemolysis test was conducted, and it was observed that CNT-PEI(FITC)-mAb had negligible hemolytic activity. Wu et al. reported that CNT-PEI(FITC)-mAb can be used for targeted delivery of drugs and can suppress tumor growth. Consequently, a bifunctional and smart platform was developed for the application in cancer diagnosis and therapy.[62]

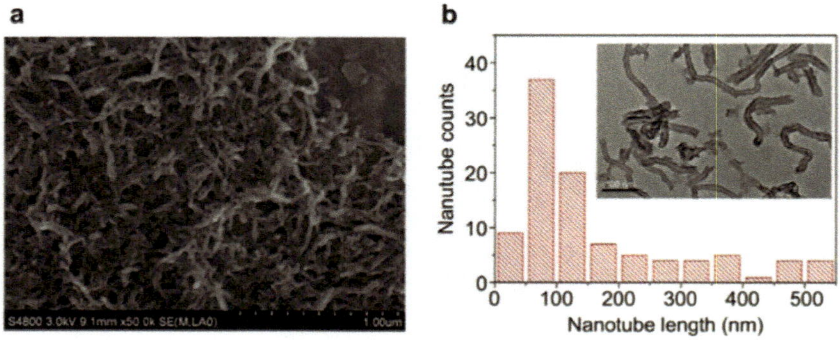

FIGURE 8.5 SEM image (a) and length distribution histogram (b) of MWCNT-COOH. The inset in (b): TEM image.

Source: Reproduced with permission from Ref. [62]. © 2014 Elsevier.

Curcumin is a natural and commonly used anticancer drug with highly advantageous properties like multitarget and broad-spectrum activity, chemopreventive nature, and ability to overcome multidrug resistance. But it has some obstructions which need to be controlled like low-aqueous solubility, light sensitivity, instability, and poor bioavailability. In this study, SWCNTs were used as carriers for curcumin to overcome their obstructions and effective delivery for the treatment of prostate cancer cells (PC-3). UV-vis spectroscopy was used for the measurement of SWCNTs – Cur concentrations and TEM microscopy was used for the identification of SWCNT aggregates. In the curcumin release test, it was observed that curcumin could be well released after dissolving it and by single SWCNTs than the bundles. Also, the SWCNT-Cur delivery system did not cause much cytotoxicity in the PC-3 cells and was thus noncovalently functionalized with PC and PVP for enhanced biocompatibility and higher cytotoxicity to the PC-3 cells. Haixia Li et al. reported that functionalization of the SWCNTs for the release of high concentrations of curcumin was necessary and could also increase the solubility with aqueous media for advanced treatment against tumor cells.[63]

Carboplatin is a low-molecular weight platinum anticancer drug most commonly used for the chemotherapeutic treatment of cancer. The platinum-derived anticancer drugs usually have short blood circulation time which lowers its DNA binding and tumor uptake capacity. So, the carboplatin was conjugated with the folate molecule (FA). As we know, SWCNTs are of high interest for delivery of drugs, the folate-conjugated platinum (IV) complex was loaded on the surface of amine-functionalized SWCNTs and a smart delivery system was constructed. MTT assay was used to study the potential of SWCNTs–NH_2–Pt(IV) to detect and destroy the PC-3 cells and uterine epithelial cancer cells and it was observed that cis-DDP could successfully reach the DNA in the nucleus and caused cytotoxicity of the cells. Shanta Dhar et al. demonstrated the ability of folate (FA) and Pt (IV) conjugated SWCNTs for entry into nuclear DNA and also the formation of cross-links with the DNA molecule in the presence of a monoclonal antibody R-C18. This experiment demonstrates the potential of a CNT-based delivery system in biomedical applications in the immediate future.[64]

8.2.6 CARBON NANOTUBES-BASED DELIVERY SYSTEM FOR SKIN CANCER TREATMENT

Angiogenesis and metastasis are two major properties of cancer and Integrin αvβ3 is a part of cancer cell lines that promotes the above two properties for the growth of the tumor. Treating Integrin αvβ3 with some drugs can be advantageous for the therapy of cancer. In this experiment, cyclic arginylglycylaspartic acid (RGD) peptide was conjugated with carboxylic CNTs with the drug camptothecin (CPT) for the treatment of A375 cells (melanoma cancer cells). The characterization of CPT-fCNTs-RGD was done by FT-IR and Raman spectroscopy. The concentration of CPT in A375 cells upon incubation with CPT@fCNTs-RGD, intravenous and intraperitoneal injection of CPT@fCNT-RGD, and tumor-bearing mice injected with CPT@fCNT-RGD was noted (Fig. 8.6). A delivery experiment to check the ability of CPT@fCNT-RGD onto 2D and 3D A375 cells were conducted and it was observed that the cell viability was decreased. The cellular uptake of the CPT drug was more than 50% of the total released. Koh et al. determined that CPT encapsulated fCNT- RGD showed a positive effect on the αvβ3-expressing cells and caused apoptosis. In conclusion, the CPT-loaded fCNTs-RGD was an identical DDS and had great potential against cancer cells.[65]

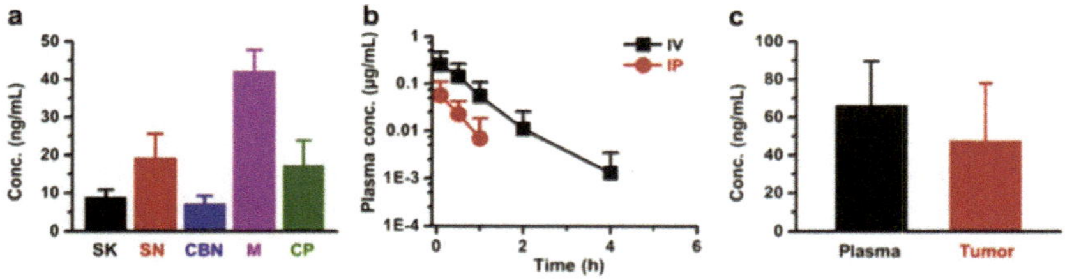

FIGURE 8.6 (a) Concentration of CPT detected in each A375 fraction after incubation with CPT–fCNT–RGD. (b) Plasma concentration of CPT after intravenous and intraperitoneal injection of CPT–fCNT–RGD. (c) Concentration of CPT detected on tumor-bearing mice after injection with CPT–fCNT–RGD. CP, cytoplasmic extract; M, plasma, mitochondria, and endoplasmic reticulum/golgi membrane extract; SN, soluble nuclear extract; CBN, chromatic bound nuclear extract; SK, cytoskeletal extract; IV, intravenous; IP, intraperitoneal.

Source: Reproduced with permission from Ref. [65]. © 2014 Elsevier.

Melanoma or skin cancer is a type of cancer caused due to the abnormal growth of the epidermal cells, and the use of nanoparticles for its treatment has gained particular attention in the field of research. Despite its distinct properties, the use of CNTs has led to skin irritations after they entered into the epidermis. In this study, the properties of CNTs for drug delivery into the skin and penetration enhancement were introduced. The morphological characteristics of the MWCNTs were studied with AFM (atomic force microscopy). IND a hydrophobic drug was used and adsorbed on the surface of functionalized MWCNTs and the rate of adsorption was measured by HPLC. Skin does not allow penetration of any kind of substances but exceptions exist like pores, shunts, or lesions which have a size range and thus nanosized tubes were capable of penetrating through the skin and led to the delivery of IND due to its hydrophobic nature. Degim et al. reported that MWCNTs were identical for the adsorption and desorption of the drug and its penetration through the transdermal layer. CNTs are the most accepted adsorptive material having high surface area and are also capable of penetrating through the skin for the delivery of hydrophobic drugs.[66]

CNTs are proved to be highly advantageous especially the SWCNTs due to their single-layer sheet of graphene. However, CNTs are susceptible to aggregation and hence needs to be modified covalently or noncovalently to increase their biocompatibility. In this experiment, the combined effect of chemo and phototherapy was studied with EB (Evans Blue) functionalized SWCNT and albumin derived drug for therapeutic purposes. TEM and AFM microscopy were used for the measurement of length and diameter of SWCNTs, whereas UV-vis spectroscopy was used to monitor the drug loading and delivery. PTX an anticancer drug was used with the SWCNTs/EB/albumin complex and the drug-loading efficiency onto MDA-MB 435 cells (melanoma cells) was observed to increase with an increase in temperature of irradiation. Cytotoxicity of melanoma cells in the presence of NIR laser light was increased and SWCNT/EB/albumin/PTX complex caused more cell death. Liwen Zhang et al. submitted a report stating that SWCNT/EB/albumin/PTX complex-mediated chemo/phototherapy is a highly promising system and could diagnose and kill MDA-MB 435 cells. The EB-modified SWCNTs could very well load PTX drugs and was an excellent strategy to treat cancer.[67]

8.2.7 CARBON NANOTUBES-BASED DELIVERY SYSTEM FOR BONE CANCER TREATMENT

CNTs are found to be of great interest in oncology studies and in the past years have been used on a large scale for the treatment of cancer. In this study, different types of CNTs were designed like MWCNTs, PEGylated MWCNTs, and bone-targeting (BT) alendronate-conjugated MWCNTs (BT--PEG–MWCNTs) and a comparative study of their efficiency was made with the help of Doxorubicin (DOX). The different types of functionalized CNTs were characterized with the help of SEM, TEM, TGA, and FT-IR microscopy (Fig. 8.7). In the results, it was observed that the use of single MWCNTs was safer than a bundle of MWCNTs. Also, the drug-loading efficiency of PEG–dMWCNTs and BT–PEG–dMWCNTs were effective at low dosage, they showed a decrease in tumor burden of the Burkitt's lymphoma. Thus, Falank et al. depicted that a minimal difference was observed between PEG–dMWCNTs and BT–PEG–dMWCNTs and their efficiency in drug delivery. The bone-targeting system developed was found to be highly promising for treating disease related to bone or bone marrow and could load many therapeutic drugs and biomolecules.[68]

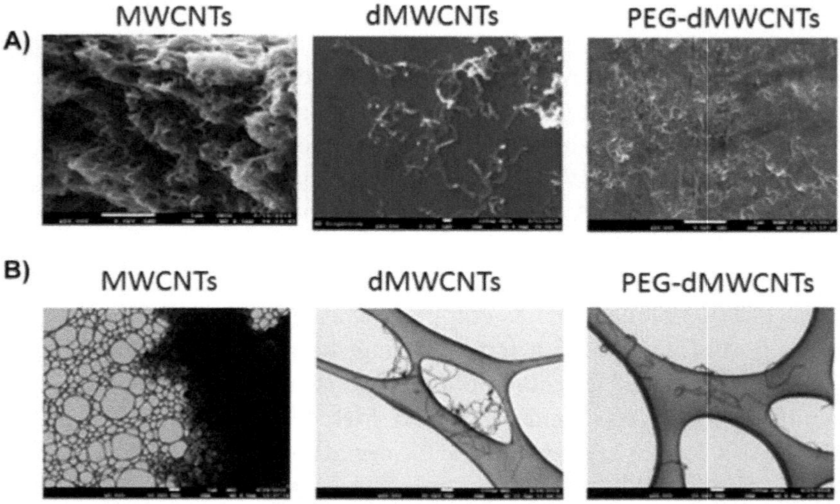

FIGURE 8.7 Physical characterization of dMWCNTs. (A) Representative SEM micrographs of entangled MWCNTs aggregated tubes and individualized dMWCNTs and PEG–dMWCNTs are discrete and substantially free of contaminants (silicon wafer seen in backdrop). B) TEM analysis proves they have an outer diameter of 10.74 +/− 2.46 (SD) nm and an inner diameter of 4.959 +/−1.07 (SD).

Source: Reproduced with permission from Ref. [68]. © 2019 Elsevier.

The conjugation of SWCNT with anticancer drug conjugate is highly desirable and can be regarded as the most coherent strategy for targeted drug delivery and minimal cell toxicity. The tumor-targeting drug conjugates mainly include vitamins, polypeptides, fatty acids, antibodies, etc. and have many advantages like they can cleave the cell for drug entry and are found to be highly cytotoxic to cancer cells. In this experiment, an effective DDS was designed with the help of SWCNTs and an anticancer agent, its cytotoxicity was assessed onto L1210 cells (leukemia cell line). The size and other characteristics of functionalized SWCNTs were measured by TEM, and the UV-vis spectroscopy was utilized to observe the fluorescence activity and drug-loading capacity. Jingyi Chen et al. studied that cell internalization was temperature-dependent and consequently increased with an increase in temperature and the cytotoxicity of the biotin-SWCNT-based DDS was observed to increase by 87.6 ±4.2 nm. The biotin-conjugated SWCNT was proved a high potential as a DDS for treating cancer and was superior to the free DDS.[69]

Multidrug resistance (MDR) is a major drawback related to chemotherapy which leads to an increased dosage of the anticancer drug. An experiment was conducted in which an antibody P-gp was functionalized with SWCNTs (water-soluble) against the K562 cells (leukemia) aiming at the development of a target-specific DDS to overcome MDR. The loading and release of the DOX were observed by FT-IR spectroscopy and the characteristics of antibody-conjugated SWCNTs were measured by SEM and TEM. It was observed that Antibody P (Ap) increased the surface area of the CNTs, thus the loading capacity was increased and also that the Ap-SWCNTs efficiently targeted and entered the leukemia cells through endocytosis. Li et al. reported that the DDS system developed with Ap–SWCNTs successfully delivered the drugs and caused a high amount of cell cytotoxicity onto K562R cells, which could overpower the efflux of DOX by MDR. The best and highly demanding antibody-conjugated SWCNTs was established mainly to overcome the MDR and also to enhance the targeted delivery of DOX.[70]

Earlier, it was discovered that cyclin A_2 is a protein responsible for DNA replication, transcription, and regulating the cell cycle. On suppressing cyclin A_2 by delivering siRNA, the proliferation of cancer cells can be controlled and apoptosis can be induced. Wang et al. applied the same strategy in their experiment by using functionalized single-walled CNTs mixed with siRNA (f-SWCNTs–siRNA), made out of a complementary strand of that of cyclin A_2 producing gene. F-SWCNTs–siRNA of length 50–300 nm were analyzed by FT-IR and XPS and deployed for the treatment of chronic myelogenous leukemia cells (K562). Wang et al. reported that siRNA delivered by f-SWCNT was successful in suppressing cyclin A_2 producing genes and eventually breaking the cycle of cancer cell division. This was one of the new strategies brought up by researchers in order to treat leukemia which is predicted to be used extensively in the coming future.[71]

8.2.8 CARBON NANOTUBES-BASED DELIVERY SYSTEM FOR COLON CANCER TREATMENT

In some cases, instead of encapsulating a drug into the delivery system, various magnetic elements are loaded and the drug is instead combined to the surface of the delivery system. This was experimentally demonstrated by

Boncel et al. in which ferromagnetic iron-loaded MWCNTs (Fe-MWCNTs) were combined with 5-fluorouracil/purpurin/1,8-naphthalimide DNA intercalators (NIDIs) and used against HCT116+ cell lines (colon cancer) in vitro. Fe-MWCNTs (synthesized by chemical vapor deposition method) was characterized using SEM (vertically aligned), TEM (diameter –4 ± 25 and length –50 ± 30 μm), and Raman spectrometer. Throughout the experiment, it was observed that MWCNTs, with low concentrations of drugs, played an important role in inducing apoptosis in HCT116+ cancer cells (MTS assay). It was reported that maximum cytotoxicity was observed in 5-fluorouracil or purpurin-conjugated Fe-MWCNTs, while NIDI-conjugated Fe-MWCNTs were viable against Caco-2 cells (human epithelial colorectal adenocarcinoma cells). This delivery system paved the way for in vivo treatment of cancer with higher toxicity levels and enhanced apoptosis induction.[72]

Irinotecan is another well-known anticancer drug that prevents cancer cells from dividing by blocking DNA transcription because of which cell apoptosis occurs. In this study, MWCNTs with a large-inner diameter from 7.9 to 14 nm (TEM) were loaded with irinotecan (Ir) and exploited for the treatment of colorectal cancer cells in vitro. The MWCNTs (synthesized by chemical vapor deposition) were washed with irinotecan, and later it was confirmed (using infrared spectroscopy) that there were no traces of irinotecan of the outer shell of the MWCNTs. Other than TEM and infrared spectroscopy, Raman spectrometer, TGA, and UV-vis spectrometer was also used for characterization. Tripisciano et al. gave a detailed outline in which it was noted that at high pH (above 6) and in the presence of light, irinotecan was unstable and did not work efficiently against colorectal cancer cells. But when the conditions were favorable concerning irinotecan's properties, apoptosis was successfully inducted into the cancer cells.[73]

CNTs have the tendency to absorb infrared radiations and deliver drugs as a carrier because of which they can be used for photothermal therapy while drug delivery (chemotherapy). The following experiment was conducted by LeVi-Polyachenko et al. in which MWCNTs (synthesized by chemical vapor deposition) were used to treat colorectal cancer in vitro. MWCNTs of length 2 μm (TEM) were loaded with oxaliplatin (MWCNTs-OX) and used against RKO and HCT 116 cell lines. Studies revealed that cell viability decreased significantly on introducing nanotubes as well as nanotubes and laser. Also it was noted that CNT-based photochemotherapy was faster than hyperthermic chemotherapy. This experiment proved the efficacy of CNTs as an optimum choice for photothermal and chemotherapy in biomedical fields.[74]

Reverse addition–fragmentation chain transfer is a polymerization process in which an active radical is added to one of the ends of a dormant chain. In the following work, nanocomposite of reversible addition–fragmentation chain transfer-CNTs -hydrogel loaded with capecitabine (RAFT/CNT/Hydrogel) was prepared by living radical polymerization for the treatment of stomach cancer in vitro and in vivo. Hosseinzadeh et al. used CNTs of diameter 5–15 nm (TEM), pentanoic acid as a chemical agent for chain transfer, and acrylic acid and acrylamide copolymerized through RAFT polymerization. Characterization of RAFT/CNT/Hydrogel nanocomposite was carried out by FT-IR, XRD, TGA/DTG, SEM, and HNMR. Results indicated that capecitabine release took place efficiently at low pH which states that the system demonstrated pH-sensitive behavior and viability against stomach cancer. This experiment paved the way for hydrogel nanocomposites as anticancer DDS by obtaining desired results.[75]

Microspheres are spherical hollow structures used for loading cargo and delivering them to a targeted site. These structures have various biomedical applications, one of which is cancer. MWCNTs conjugated with irinotecan, folic acid, and chitosan (through noncovalent bonding) were encapsulated (oil-in-oil technique) in microspheres (MWCNTs/FA-CS/Ir). The particle size and surface morphology of MWCNTs/FA-CS/Ir were analyzed by TEM, SEM, and Uv-vis spectrometer. Firstly, the in vitro drug release rate was analyzed which revealed that irinotecan showed sustained release from the composite. Consequently, the cell viability of HT-29 cells (colorectal cancer) was determined by MTT assay, in which it was found that on increasing the concentration of MWCNTs, cell viability decreased and maximum cytotoxicity of HT-29 cells was observed when MWCNTs/FA-CS/Ir was used.[76]

Functionalization of the CNTs either covalently or noncovalently induces biocompatibility and also facilitates a high amount of drug loading by increasing the surface area. SWCNTs were conjugated with FA and an anticancer chemotherapeutic drug PTX targeting mainly the MKN45 cells (colon cancer cells). The size and morphological characteristics of FA–SWCNTs–PTX was analyzed by TEM, SEM, FT-IR spectroscopy. It was observed that the properties of PTX drugs were improved like the solubility and the selectivity nature after conjugation with the FA- SWCNTs. The role of FA was observed to induce cell proliferation and its water-soluble nature causes an increase in solubility of the DDS. Tavakolifard et al. reported the successful development of a functionalized CNT delivery system and its high potential in the nanobiotechnology field.[77]

8.3 CONCLUSION

The main focus of this chapter is on the significance and efficiency of carbon nanotube-based DDS for cancer therapy. This chapter gives an outline of the different experiments performed by various researchers using two types of CNTs, that is, MWCNTs and SWCNTs. CNTs were functionalized using various chemical compounds, such as folic acid, quantum dots, chitosan, PEG, PVP, hyaluronic acid, etc. which enhanced the physical and chemical properties of CNTs. Moreover, certain obstacles like solubility, multidrug resistance, the formation of aggregates, drug leakage, etc. were also eliminated upon functionalizing. With the help of anticancer drugs, namely, doxorubicin, gemcitabine, paclitaxel, curcumin, and genes were loaded onto functionalized CNTs for the chemotherapeutic treatment of different types of cancer like lung cancer, breast cancer, cervical cancer, prostate cancer, skin cancer, colon cancer, and bone cancer. It was observed that the drug-loading capacity and cytotoxicity effect were heightened after the functionalization of CNTs also, a decrement in the cell viability or proliferation of cancer cells was noted that proved CNT-based DDS to be highly promising and productive. Furthermore, CNTs were also proficient in reducing the side effects caused due to high dosage of anticancer drugs. Although these experiments have paved the way for CNTs to be recognized as a safe, efficient, and feasible way of in vitro and in vivo treatment of cancer, these systems are not widely recognized and utilized by industries all around the world. However, nanotechnology has proven to be a potential technology with significant advantages over other technologies, which makes it more acceptable and safer for the people.

KEYWORDS

- **single-walled**
- **cancer**
- **size**
- **functionalized**
- **anticancer drugs**
- **cytotoxicity**
- **apoptosis**
- **cell viability**

REFERENCES

1. Siegel, R. L.; Miller, K. D.; Jemal, A. Cancer Statistics, 2019. *CA: Cancer J. Clin.* **2019,** *69* (1), 7–34.
2. Franks, L. M.; Teich, N. M., Eds. *Introduction to the Cellular and Molecular Biology of Cancer*; Oxford University Press, USA, 1997.
3. Pucci, C.; Martinelli, C.; Ciofani, G. Innovative Approaches for Cancer Treatment: Current Perspectives and New Challenges. *Ecancermedicalscience* **2019,** 13.
4. Hartshorn, C. M.; Bradbury, M. S.; Lanza, G. M.; Nel, A. E.; Rao, J.; Wang, A. Z.; Wiesner, U. B.; Yang, L.; Grodzinski, P. Nanotechnology Strategies to Advance Outcomes in Clinical Cancer Care. *ACS Nano* Jan 23, **2018,** *12* (1), 24–43.
5. Gopalakrishnan, S.; Narendar, S. Introductory Concepts of Wave Propagation Analysis in Structures. In *Wave Propagation in Nanostructures*; Springer: Cham, 2019; pp 19–29.
6. Navya, P. N.; Kaphle, A.; Srinivas, S. P.; Bhargava, S. K.; Rotello, V. M.; Daima, H. K. Current Trends and Challenges in Cancer Management and Therapy Using Designer Nanomaterials. *Nano Convergence* Dec 1, **2019,** *6* (1), 23.
7. Dai, H. Carbon Nanotubes: Synthesis, Integration, and Properties. *Acc. Chem. Res.* **2002,** *35* (12), 1035–1044.
8. Dineshkumar, B.; Krishnakumar, K.; Bhatt, A. R.; Paul, D.; Cherian, J.; John, A.; Suresh, S. Single-Walled and Multi-Walled Carbon Nanotubes Based Drug Delivery System: Cancer Therapy: A Review. *Indian J. Cancer* Jul 1, **2015,** *52* (3), 262.
9. Weaver, Beth A. A.; Cleveland, D. W. Decoding the Links between Mitosis, Cancer, and Chemotherapy: The Mitotic Checkpoint, Adaptation, and Cell Death. *Cancer Cell* **2005,** *8* (1), 7–12.
10. Kam, N. W. S.; Liu, Z.; Dai, H. Functionalization of Carbon Nanotubes via Cleavable Disulfide Bonds for Efficient Intracellular Delivery of siRNA and Potent Gene Silencing. *J. Am. Chem. Soc.* **2005,** *127*, 12492–12493.

11. Ji, S. R.; Liu, C.; Zhang, B.; Yang, F.; Xu, J.; Long, J.; Jin, C.; Fu, D. L.; Ni, Q. X.; Yu, X. J. Carbon Nanotubes in Cancer Diagnosis and Therapy. *Biochim. et Biophys. Acta (BBA)-Rev. Cancer* Aug 1, **2010,** *1806* (1), 29–35.
12. Kesharwani, P.; Mishra, V.; Jain, N. K. Validating the Anticancer Potential of Carbon Nanotube-Based Therapeutics through Cell Line Testing. *Drug Discov. Today* **2015,** *20* (9), 1049–1060.
13. Mahajan, S.; Patharkar, A.; Kuche, K.; Maheshwari, R.; Deb, P. K.; Kalia, K.; Tekade, R. K. Functionalized Carbon Nanotubes as Emerging Delivery System for the Treatment of Cancer. *International J. Pharm.* Sep 5, **2018,** *548* (1), 540–558.
14. Tan, A.; Yildirimer, L.; Rajadas, J.; De La Peña, H.; Pastorin, G.; Seifalian, A. Quantum Dots and Carbon Nanotubes in Oncology: A Review on Emerging Theranostic Applications in Nanomedicine. *Nanomedicine* Aug **2011,** *6* (6), 1101–1114.
15. Soldano, C. Hybrid Metal-Based Carbon Nanotubes: Novel Platform for Multifunctional Applications. *Prog. Mater. Sci.* **2015,** *69*, 183–212.
16. Moore, T. L.; Pitzer, J. E.; Podila, R.; Wang, X.; Lewis, R. L.; Grimes, S. W.; Wilson, J. R.; Skjervold, E.; Brown, J. M.; Rao, A.; Alexis, F. Multifunctional Polymer-Coated Carbon Nanotubes for Safe Drug Delivery. *Particle Particle Syst. Charact.* Apr **2013,** *30* (4), 365–373.
17. Wu, H.; Liu, G.; Wang, X.; Zhang, J.; Chen, Y.; Shi, J.; Yang, H.; Hu, H.; Yang, S. Solvothermal Synthesis of Cobalt Ferrite Nanoparticles Loaded on Multiwalled Carbon Nanotubes for Magnetic Resonance Imaging and Drug Delivery. *Acta Biomater.* Sep 1, **2011,** *7* (9), 3496–3504.
18. Zhao, X.; Tian, K.; Zhou, T.; Jia, X.; Li, J.; Liu, P. PEGylated Multi-Walled Carbon Nanotubes as versatile Vector for Tumor-Specific Intracellular Triggered Release with Enhanced Anti-Cancer Efficiency: Optimization of Length and PEGylation Degree. *Colloids Surf. B Biointerf.* Aug 1, **2018,** *168*, 43–49.
19. Ge, X.; Fu, M.; Niu, X.; Kong, X. Atomic Layer Deposition of γ-Fe_2O_3 Nanoparticles on Multi-Wall Carbon Nanotubes for Magnetic Drug Delivery and Liver Cancer Treatment. *Ceram. Int.* Jul 21, **2020.**
20. Wei, C.; Dong, X.; Zhang, Y.; Liang, J.; Yang, A.; Zhu, D.; Liu, T.; Kong, D.; Lv, F. Simultaneous Fluorescence Imaging Monitoring of the Programmed Release of Dual Drugs from a Hydrogel-Carbon Nanotube Delivery System. *Sens. Actuat. B Chem.* Nov 10, **2018,** *273*, 264–275.
21. Dong, X.; Wei, C.; Liang, J.; Liu, T.; Kong, D.; Lv, F. Thermosensitive Hydrogel Loaded with Chitosan-Carbon Nanotubes for Near Infrared Light Triggered Drug Delivery. *Colloids Surf. B: Biointerf.* Jul 1, **2017,** *154*, 253–262.
22. Wang, N.; Feng, Y.; Zeng, L.; Zhao, Z.; Chen, T. Functionalized Multiwalled Carbon Nanotubes as Carriers of Ruthenium Complexes to Antagonize Cancer Multidrug Resistance and Radioresistance. *ACS Appl. Materials Interf.* Jul 15, **2015,** *7* (27), 14933–14945.
23. Kavosi, A.; Noei, S. H.; Madani, S.; Khalighfard, S.; Khodayari, S.; Khodayari, H.; Mirzaei, M.; Kalhori, M. R.; Yavarian, M.; Alizadeh, A. M.; Falahati, M. The Toxicity and Therapeutic Effects of Single-and Multi-Wall Carbon Nanotubes on Mice Breast Cancer. *Sci. Rep.* May 30, **2018,** *8* (1), 1–2.

24. Ozgen, P. S.; Atasoy, S.; Kurt, B. Z.; Durmus, Z.; Yigit, G.; Dag, A. Glycopolymer Decorated Multiwalled Carbon Nanotubes for Dual Targeted Breast Cancer Therapy. *J. Mater. Chem. B* **2020,** *8* (15), 3123–3137.
25. Nelson, D. J.; Kumar, R. Characterization of a Tamoxifen-Tethered Single-Walled Carbon Nanotube Conjugate by Using NMR Spectroscopy. *Analyt. Bioanalyt. Chem.* Aug 1, **2012,** *404* (3), 771–776.
26. Yan, C.; Chen, C.; Hou, L.; Zhang, H.; Che, Y.; Qi, Y.; Zhang, X.; Cheng, J.; Zhang, Z. Single-Walled Carbon Nanotube-Loaded Doxorubicin and Gd-DTPA for Targeted Drug Delivery and Magnetic Resonance Imaging. *J. Drug Target.* Feb 7, **2017,** *25* (2), 163–171.
27. Wang, L.; Shi, J.; Jia, X.; Liu, R.; Wang, H.; Wang, Z.; Li, L.; Zhang, J.; Zhang, C.; Zhang, Z. NIR-/pH-Responsive Drug Delivery of Functionalized Single-Walled Carbon Nanotubes for Potential Application in Cancer Chemo-Photothermal Therapy. *Pharm. Res.* Nov 1, **2013,** *30* (11), 2757–2771.
28. Sharma, S.; Mehra, N. K.; Jain, K.; Jain, N. K. Effect of Functionalization on Drug Delivery Potential of Carbon Nanotubes. *Artif. Cells Nanomed. Biotechnol.* Nov 16, **2016,** *44* (8), 1851–1860.
29. Liu, Z.; Sun, X.; Nakayama-Ratchford, N.; Dai, H. Supramolecular Chemistry on Water-Soluble Carbon Nanotubes for Drug Loading and Delivery. *ACS Nano* Aug 14, **2007,** *1* (1), 50–56.
30. Das, M.; Singh, R. P.; Datir, S. R.; Jain, S. Intranuclear Drug Delivery and Effective in Vivo Cancer Therapy via Estradiol–PEG-Appended Multiwalled Carbon Nanotubes. *Mol. Pharm.* Sep 3, **2013,** *10* (9), 3404–3416.
31. Yang, T.; Wu, Z.; Wang, P.; Mu, T.; Qin, H.; Zhu, Z.; Wang, J.; Sui, L. A Large-Inner-Diameter Multi-Walled Carbon Nanotube-Based Dual-Drug Delivery System with pH-Sensitive Release Properties. *J. Mater. Sci. Mater. Med.* Jul **2017,** *28* (7), 1–3.
32. Singh, R.; Mehra, N. K.; Jain, V.; Jain, N. K. Gemcitabine-Loaded Smart Carbon Nanotubes for Effective Targeting to Cancer Cells. *J. Drug Target.* Jul 1, **2013,** *21* (6), 581–592.
33. Shao, W.; Paul, A.; Zhao, B.; Lee, C.; Rodes, L.; Prakash, S. Carbon Nanotube Lipid Drug Approach for Targeted Delivery of a Chemotherapy Drug in A Human Breast Cancer Xenograft Animal Model. *Biomaterials* Dec 1, **2013,** *34* (38), 10109–10119.
34. Li, B.; Harlepp, S.; Gensbittel, V.; Wells, C. J.; Bringel, O.; Goetz, J. G.; Begin-Colin, S.; Tasso, M.; Begin, D.; Mertz, D. Near Infra-Red Light Responsive Carbon Nanotubes@ Mesoporous Silica for Photothermia and Drug Delivery to Cancer Cells. *Mater. Today Chem.* Sep 1, **2020,** *17*, 100308.
35. Shao, W.; Paul, A.; Rodes, L.; Prakash, S. A New Carbon Nanotube-Based Breast Cancer Drug Delivery System: Preparation and in Vitro Analysis Using Paclitaxel. *Cell Biochem. Biophys.* Apr 1, **2015,** *71* (3), 1405–1414
36. Saeed, L. M.; Mahmood, M.; Pyrek, S. J.; Fahmi, T.; Xu, Y.; Mustafa, T.; Nima, Z. A.; Bratton S. M.; Casciano, D.; Dervishi, E.; Radominska-Pandya, A. Single-Walled Carbon Nanotube and Graphene Nanodelivery of Gambogic Acid Increases Its Cytotoxicity in Breast and Pancreatic Cancer Cells. *J. Appl. Toxicol.* Nov **2014,** *34* (11), 1188–1199.
37. Risi, G.; Bloise, N.; Merli, D.; Icaro-Cornaglia, A.; Profumo, A.; Fagnoni, M.; Quartarone, E.; Imbriani, M.; Visai, L. In Vitro Study of Multiwall Carbon Nanotubes

(MWCNTs) with Adsorbed Mitoxantrone (MTO) as a Drug Delivery System to Treat Breast Cancer. *RSC Adv.* **2014,** *4* (36), 18683–18693.
38. Karthika, V.; Kaleeswarran, P.; Gopinath, K.; Arumugam, A.; Govindarajan, M.; Alharbi, N. S.; Khaled, J. M.; Al-anbr, M. N.; Benelli, G. Biocompatible Properties of Nano-Drug Carriers Using TiO_2-Au Embedded on Multiwall Carbon Nanotubes for Targeted Drug Delivery. *Mater. Sci. Eng. C.* Sep 1, **2018,** *90*, 589–601.
39. Zhang, H.; Ji, Y.; Chen, Q.; Jiao, X.; Hou, L.; Zhu, X.; Zhang, Z. Enhancement of Cytotoxicity of Artemisinin toward Cancer Cells by Transferrin-Mediated Carbon Nanotubes Nanoparticles. *J. Drug Target.* Jul 3, **2015,** *23* (6), 552–567.
40. Mohammadi Nodeh, M. K.; Rahemi Haghighi, M.; Soltani, S.; Rashidi Nodeh, H. Release and Extraction of Letrozole in Blood Plasma Using Resorcinol Functionalized Multi-Walled Carbon Nanotube Coupled with High-Performance Liquid Chromatography. *J. Liquid Chromatogr. Relat. Technol.* Mar 16, **2018,** *41* (5), 239–245.
41. Razzazan, A.; Atyabi, F.; Kazemi, B.; Dinarvand, R. In Vivo Drug Delivery of Gemcitabine with PEGylated Single-Walled Carbon Nanotubes. *Mater. Sci. Eng. C* May 1, **2016,** *62*, 614–625.
42. Anaraki, N. A.; Rad, L. R.; Irani, M.; Haririan, I. Fabrication of PLA/PEG/MWCNT Electrospun Nanofibrous Scaffolds for Anticancer Drug Delivery. *J. Appl. Polym. Sci.* Oct 17, **2014,** *3* (132), n-a.
43. Yu, B.; Tan, L.; Zheng, R.; Tan, H.; Zheng, L. Targeted Delivery and Controlled Release of Paclitaxel for the Treatment of Lung Cancer Using Single-Walled Carbon Nanotubes. *Mater. Sci. Eng. C* Nov 1, **2016,** *68*, 579–584.
44. Gangrade, A.; Mandal, B. B. Injectable Carbon Nanotube Impregnated Silk Based Multifunctional Hydrogel for Localized Targeted and On-Demand Anticancer Drug Delivery. *ACS Biomater. Sci. Eng.* Apr 3, **2019,** *5* (5), 2365–2381.
45. Tan, J. M.; Karthivashan, G.; Arulselvan, P.; Fakurazi, S.; Hussein, M. Z. Characterization and in Vitro Sustained Release of Silibinin from pH Responsive Carbon Nanotube-Based Drug Delivery System. *J. Nanomater.* Jan 1, **2014,** *2014*.
46. Li, R.; Wu, R. A.; Zhao, L.; Hu, Z.; Guo, S.; Pan, X.; Zou, H. Folate and Iron Difunctionalized Multiwall Carbon Nanotubes as Dual-Targeted Drug Nanocarrier to Cancer Cells. *Carbon* Apr 1, **2011,** *49* (5), 1797–1805.
47. Shi, X.; Wang, S. H.; Shen, M.; Antwerp, M. E.; Chen, X.; Li, C.; Petersen, E. J.; Huang, Q.; Weber, Jr. W. J.; Baker, Jr. J. R. Multifunctional Dendrimer-Modified Multiwalled Carbon Nanotubes: Synthesis, Characterization, and in Vitro Cancer Cell Targeting and Imaging. *Biomacromolecules* Jul 13, **2009,** *10* (7), 1744–1750.
48. Wen, S.; Liu, H.; Cai, H.; Shen, M.; Shi, X. Targeted and pH-Responsive Delivery of Doxorubicin to Cancer Cells Using Multifunctional Dendrimer-Modified Multi-Walled Carbon Nanotubes. *Adv. Healthcare Mater.* Sep **2013,** *2* (9), 1267–1276.
49. Niu, L.; Meng, L.; Lu, Q. Folate-Conjugated PEG on Single Walled Carbon Nanotubes for Targeting Delivery of Doxorubicin to Cancer Cells. *Macromol. Biosci.* Jun **2013,** *13* (6), 735–744.
50. Fan, J.; Zeng, F.; Xu, J.; Wu, S. Targeted Anti-Cancer Prodrug Based on Carbon Nanotube with Photodynamic Therapeutic Effect and pH-Triggered Drug Release. *J. Nanoparticle Res.* Sep 1, **2013,** *15* (9), 1911.

51. Wang, D.; Ren, Y.; Shao, Y.; Yu, D.; Meng, L. Facile Preparation of Doxorubicin-Loaded and Folic Acid-Conjugated Carbon Nanotubes @ Poly (N-Vinyl Pyrrole) for Targeted Synergistic Chemo–Photothermal Cancer Treatment. *Bioconjugate Chem.* Nov 15, **2017,** *28* (11), 2815–2822.
52. Nasari, M.; Semnani, D.; Hadjianfar, M.; Amanpour, S. Poly (E-Caprolactone)/Poly (N-Vinyl-2-Pyrrolidone) Core–Shell Nanofibers Loaded by Multi-Walled Carbon Nanotubes and 5-Fluorouracil: An Anticancer Drug Delivery System. J. Mater. Sci. **2020,** *55*, 10185–10201.
53. Tabatabaei Rezaei, S. J.; Hesami, A.; Khorramabadi, H.; Amani, V.; Malekzadeh, A. M.; Ramazani, A.; Niknejad, H. Pt (II) Complexes Immobilized on Polymer-Modified Magnetic Carbon Nanotubes as a New Platinum Drug Delivery System. *Appl. Organometal. Chem.* Jul **2018,** *32* (7), e4401.
54. Brahmachari, S.; Ghosh, M.; Dutta, S.; Das, P. K. Biotinylated Amphiphile-Single Walled Carbon Nanotube Conjugate for Target-Specific Delivery to Cancer Cells. *J. Mater. Chem. B* **2014,** *2* (9), 1160–1173.
55. Huang, Y. P.; Lin, I. J.; Chen, C. C.; Hsu, Y. C.; Chang, C. C.; Lee, M. J. Delivery of Small Interfering RNAs in Human Cervical Cancer Cells by Polyethylenimine-Functionalized Carbon Nanotubes. *Nanoscale Res. Lett.* Dec **2013,** *8* (1), 1–1.
56. Phan, Q. T.; Patil, M. P.; Tu, T. T.; Le, C. M.; Kim, G. D.; Lim, K. T. Polyampholyte-Grafted Single Walled Carbon Nanotubes Prepared via a Green Process for Anticancer Drug Delivery Application. *Polymer* Mar 2, **2020,** 122340.
57. Cao, X. T.; Patil, M. P.; Phan, Q. T.; Le, C. M.; Ahn, B. H.; Kim, G. D.; Lim, K. T. Green and Direct Functionalization of Poly (Ethylene Glycol) Grafted Polymers onto Single Walled Carbon Nanotubes: Effective Nanocarrier for Doxorubicin Delivery. *J. Indust. Eng. Chem.* Mar 25, **2020,** *83*, 173–180.
58. Chaban, V. V.; Savchenko, T. I.; Kovalenko, S. M.; Prezhdo, O. V. Heat-Driven Release of a Drug Molecule from Carbon Nanotubes: A Molecular Dynamics Study. *J. Phys. Chem. B* Oct 28, **2010,** *114* (42), 13481–13486.
59. Zhang, R. Q.; Liu, Z. Q.; Luo, Y. L.; Xu, F.; Chen, Y. S. Tri-Stimuli Responsive Carbon Nanotubes Covered by Mesoporous Silica Graft Copolymer Multifunctional Materials for Intracellular Drug Delivery. *J. Indust. Eng. Chem.* Dec 25, **2019,** *80*, 431–443.
60. Liu X, Xu D, Liao C, Fang Y, Guo B. Development of a Promising Drug Delivery for Formononetin: Cyclodextrin-Modified Single-Walled Carbon Nanotubes. *J. Drug Deliv. Sci. Technol.* Feb 1, **2018,** *43*, 461–468.
61. Mo, Y.; Wang, H.; Liu, J.; Lan, Y.; Guo, R.; Zhang, Y.; Xue, W.; Zhang, Y. Controlled Release and Targeted Delivery to Cancer Cells of Doxorubicin from Polysaccharide-Functionalised Single-Walled Carbon Nanotubes. *J. Mater. Chem B* **2015,** *3* (9), 1846–1855.
62. Wu, H.; Shi, H.; Zhang, H.; Wang, X.; Yang, Y.; Yu, C.; Hao, C.; Du, J.; Hu, H.; Yang, S. Prostate Stem Cell Antigen Antibody-Conjugated Multiwalled Carbon Nanotubes for Targeted Ultrasound Imaging and Drug Delivery. *Biomaterials* Jul 1, **2014,** *35* (20), 5369–5380.
63. Li, H.; Zhang, N.; Hao, Y.; Wang, Y.; Jia, S.; Zhang, H.; Zhang, Y.; Zhang, Z. Formulation of Curcumin Delivery with Functionalized Single-Walled Carbon

Nanotubes: Characteristics and Anticancer Effects in Vitro. *Drug Deliv.* Aug 1, **2014,** *21* (5), 379–387.

64. Dhar, S.; Liu, Z.; Thomale, J.; Dai, H.; Lippard, S. J. Targeted Single-Wall Carbon Nanotube-Mediated Pt (IV) Prodrug Delivery Using Folate as a Homing Device. *J. Am. Chem. Soc.* Aug 27, **2008,** *130* (34), 11467–11476.
65. Koh, B.; Park, S. B.; Yoon, E.; Yoo, H. M.; Lee, D.; Heo, J. N.; Ahn, S. αVβ3-Targeted Delivery of Camptothecin-Encapsulated Carbon Nanotube-Cyclic RGD in 2D and 3D Cancer Cell Culture. *J. Pharm. Sci.* Nov 1, **2019,** *108* (11), 3704–3712.
66. Degim, I. T.; Burgess, D. J.; Papadimitrakopoulos, F. Carbon Nanotubes for Transdermal Drug Delivery. *J. Microencapsulation* Dec 1, **2010,** *27* (8), 669–681.
67. Zhang, L.; Rong, P.; Chen, M.; Gao, S.; Zhu, L. A Novel Single Walled Carbon Nanotube (SWCNT) Functionalization Agent Facilitating in Vivo Combined Chemo/Thermo Therapy. *Nanoscale* **2015,** *7* (39), 16204–16213.
68. Falank, C.; Tasset, A. W.; Farrell, M.; Harris, S.; Everill, P.; Marinkovic, M.; Reagan, M. R. Development of Medical-Grade, Discrete, Multi-Walled Carbon Nanotubes as Drug Delivery Molecules to Enhance the Treatment of Hematological Malignancies. *Nanomed. Nanotechnol. Biol. Med.* Aug 1, **2019,** *20*, 102025.
69. Chen, J.; Chen, S.; Zhao, X.; Kuznetsova, L. V.; Wong, S. S.; Ojima, I. Functionalized Single-Walled Carbon Nanotubes as Rationally Designed Vehicles for Tumor-Targeted Drug Delivery. *J. Am. Chem. Soc.* Dec 10, **2008,** *130* (49), 16778–16785.
70. Li, R.; Wu, R. A.; Zhao, L.; Wu, M.; Yang, L.; Zou, H. P-Glycoprotein Antibody Functionalized Carbon Nanotube Overcomes the Multidrug Resistance of Human Leukemia Cells. *ACS Nano* Mar 23, **2010,** *4* (3), 1399–1408.
71. Wang, X.; Ren, J.; Qu, X. Targeted RNA Interference of Cyclin A2 Mediated by Functionalized Single-Walled Carbon Nanotubes Induces Proliferation Arrest and Apoptosis in Chronic Myelogenous Leukemia K562 Cells. *ChemMedChem: Chem. Enab. Drug Discov.* Jun 16, **2008,** *3* (6), 940–945.
72. Boncel, S.; Pluta, A.; Skonieczna, M.; Gondela, A.; Maciejewska, B.; Herman, A. P.; Jędrysiak, R. G.; Budniok, S.; Komędera, K.; Błachowski, A.; Walczak, K. Z. Hybrids of Iron-Filled Multiwall Carbon Nanotubes and Anticancer Agents as Potential Magnetic Drug Delivery Systems: In Vitro Studies against Human Melanoma, Colon Carcinoma, and Colon Adenocarcinoma. *J. Nanomater.* Jan 1, **2017,** *2017.*
73. Tripisciano, C.; Rümmeli, M. H.; Chen, X.; Borowiak-Palen, E. Multi-Wall Carbon Nanotubes–a Vehicle for Targeted Irinotecan Drug Delivery. *Phys. Status Solidi (B)* Dec **2010,** *247* (11–12), 2673–2677.
74. Levi-Polyachenko, N. H.; Merkel, E. J.; Jones, B. T.; Carroll, D. L.; Stewart, IV J. H. Rapid Photothermal Intracellular Drug Delivery Using Multiwalled Carbon Nanotubes. *Mol. Pharm.* Aug 3, **2009,** *6* (4), 1092–1099.
75. Hosseinzadeh, A.; Pashaei, S.; Hosseinzadeh, S.; Namazi, H. Surface Modification of Multiwalled Carbon Nanotubes via Surface RAFT Copolymerization Method and Capecitabine-Loaded Anticancer Hydrogel for Controlled Drug Delivery in Stomach. *Polym.-Plastics Technol. Mater.* Jun 1, **2020,** 1–0.
76. Zhou, M.; Peng, Z.; Liao, S.; Li, P.; Li, S. Design of Microencapsulated Carbon Nanotube-Based Microspheres and Its Application in Colon Targeted Drug Delivery. *Drug Deliv.* Mar 1, **2014,** *21* (2), 101–109.

77. Tavakolifard, S.; Biazar, E.; Pourshamsian, K.; Moslemin, M. H. Synthesis and Evaluation of Single-Wall Carbon Nanotube-Paclitaxel-Folic Acid Conjugate as an Anti-Cancer Targeting Agent. *Artif. Cells Nanomed. Biotechnol.* Jul 3, **2016,** *44* (5), 1247–1253.

CHAPTER 9

CARBON NANOTUBE ELECTRONICS

C. SWETHA*

Department of Chemistry, St. Thomas College, Ranni, Kerala, India

E-mail: swethachengat@gmail.com

Nanotechnology is the idea that we can create devices and machines all the way down to the nanometer scale, which is a billionth of a meter, about half the width of a human DNA molecule.

—Paul McEuen

ABSTRACT

A hollow tube with diameter in nanometer dimensions has revolutionized many areas of technology, such as electronics, optics, nanotechnology, nanomedicine, sensors, membranes, displays, smart textiles, photovoltaics, energy conversion devices, etc. The unique electronic property of carbon nanotubes makes them a suitable candidate to replace silicon. Further miniaturization of electronics (keeping up with Moore's law) will largely rely on carbon nanotubes. Silicon as channels in traditional FET (Field Effect Transistors) can be replaced by a single carbon nanotube or an array of carbon nanotubes. Application of carbon nanotubes in microelectronics can pave the way to light-weight, flexible, and stretchable smart integrated circuits and switching devices. In this chapter, common synthetic approaches, such as arc-discharge method, laser ablation, and chemical vapor deposition are discussed along with classification and characteristic properties, supremely electronic properties, of carbon nanotubes. The need for silicon replacement in the electronic industry is overviewed. Major challenges of using carbon nanotubes in electronic devices, such as sorting and alignment are addressed and among various potential applications, diodes, FET, logic circuits, and interconnects are discussed.

9.1 INTRODUCTION

Group 14 elements have great significance in industry and nature. The valence configuration of the family is ns^2np^2 and +4 is the dominant oxidation state. Carbon, the lightest member, is the most prominent. It is the building block of life and the backbone of Organic Chemistry. The name carbon comes from the Latin word "carbo" meaning "coal." The element is represented by the symbol C and has the atomic number 6. In nature, carbon exists as carbon dioxide in the atmosphere and in dissolved state in natural water and also as insoluble calcium and magnesium carbonates. Carbon has the unique ability to bind together in different ways which further lead to the existence of different allotropes. In the electronegativity scale, carbon falls in the midrange, about 2.5. Stable covalent bonds can be formed only when the difference in electronegativity between atoms is not more than two. This enables carbon atoms to form strong and stable covalent bonds with many atoms of lower and higher electronegativities. No other element in the periodic table has the capability to form such a variety of strong bonds with itself.[1] Interestingly, different allotropic forms exhibit different physical properties. Diamond and graphite are the purest forms of carbon and are obtained by mining. Coke and lamp black are the less pure forms. Fullerene, another allotrope of carbon was discovered in the 1980s leading to the emergence of a new field within the Inorganic Chemistry of carbon. Fullerene research has further led to the discovery of carbon nanotubes (CNTs). They are concentric cylindrical tubes of graphene sheet (one single layer of graphite with honeycomb lattice of sp^2 carbon atoms) which may or may not be closed at ends with fullerene-like caps.[2]

First image of carbon nanotube was produced by Roger Bacon in 1959. In the 1980s, application for a patent was made by Howard Tennant for a method of production of carbon nanotube. By 1990, Richard Smalley realized that big enough buckyballs become carbon cylinders. CNTs were first identified in 1991 by Sumio Iijima by making use of his sophisticated high-resolution transmission electron microscopic techniques. The credit for the discovery of CNTs is vested on Sumio Iijima. He successfully established that CNTs are composed of concentric graphene tubes and have a helical atomic arrangement. His findings were published in Nature, "Helical microtubules of graphitic carbon" in 1991. He made use of an arc-discharge evaporation method similar to that of fullerene synthesis. The growth of needles was found at the negative end of the electrode used for arc-discharge method.[3] In this publication, he reported the presence of 2–20 layered multiwalled CNTs. The subsequent publication from the same author in 1993 confirmed the existence of single-walled CNTs.[4] After Iijima's first publication, enormous research has been carried out in this area and interestingly the highest number of publications is based on the electronic properties of CNTs.[3]

9.2 SYNTHESIS

Major challenges in the synthesis of CNTs are—mass production, control over structural and electronic properties, controlled orientation and placement of nanotube over the substrate and understanding the mechanism of process involved. Several methods of synthesis are reported for both single-walled and multiwalled CNTs (SWCNTs and MWCNTs). All the methods make use of a carbon source, catalyst, and sufficient energy. The first recognized method for the synthesis of both types of CNTs is the arc-discharge synthesis. This is similar to fullerene synthesis by Kratschmer-Huffman method. This method was used by Iijima for the synthesis of MWCNTs. Iijima and Bethune found that for SWCNTs synthesis, addition of a metal catalyst to anode is necessary. Iron:carbon anode in methane:argon environment is used by Iijima while cobalt:carbon anode with helium environment is used by Bethune. Another common method is laser ablation synthesis. This method enabled the first large-scale production of SWCNTs in gram quantities. In this technique, a pulsed or continuous wave laser can be used. This method makes use of metal-impregnated graphite target as anode to produce SWCNTs, similar to arc-discharge technique. If pure graphite is used instead, fullerenes and MWCNTs are obtained. Chemical vapor deposition (CVD), high-pressure carbon monoxide synthesis and flame synthesis comes under the category of thermal synthesis. This is considered as medium temperature synthesis as the hot zone of the reaction never crosses 1200°C.[4] Characteristic features of the common methods are discussed below.

9.2.1 ARC DISCHARGE

Two graphite rods placed a few millimeters away are connected to a power supply. Carbon vaporizes at 100 amperes and hot plasma is formed. Low-pressure inert gas condition is used here. Pure graphite acts as the carbon source and 30–90% yield is obtained. Both SWCNTs and MWCNTs can be prepared using this technique. SWCNTs may show few structural defects and MWCNTs are obtained without a catalyst. The synthesized carbon nanotube needs to be further purified which is a major disadvantage of this technique.

9.2.2 LASER ABLATION

Here, graphite is blasted with intense laser pulses. CNTs are formed from the carbon gas produced. The synthesis is carried out with the condition of argon/nitrogen at 500 torr. A total of 70% of the yield is reported. Similar to arc-discharge method, this method also uses graphite as the carbon source. Compared with arc-discharge method, SWCNTs obtained by this method are of higher quality with narrow diameter distribution. Because it requires expensive lasers for this technique, this technique is expensive.

9.2.3 CHEMICAL VAPOR DEPOSITION

Carbon bearing gas such as methane is slowly added into an oven (which is preheated to high temperature) and the substrate is placed in the oven. Carbon atoms are formed upon decomposition of gas which recombines to form CNTs. The yield obtained by this method is 20–100%. Fossil-based hydrocarbon and botanical hydrocarbon are used as the carbon source. Compared with other techniques, this method is cost-effective. This technique can be made use for industrial production due to that easiness in scaling up and also due to the long, diameter-controlled pure CNTs are obtained. Presence of defects is the major challenge faced in CVD. Many modifications of this technique have been reported. The most promising ones include high pressure catalytic decomposition of carbon monoxide (HiPCO), methane CVD, carbon monoxide CVD, alcohol CVD, plasma enhanced CVD.

9.3 CHARACTERISTIC PROPERTIES AND APPLICATIONS

Carbon–carbon covalent bonding and seamless hexagonal network architecture makes carbon nanotube the strongest and most flexible molecular material. Compared with steel with the same diameter, the tensile strength of carbon nanotube is approximately hundred times greater. The Young's modulus of CNT is 1.8 TPa. It possesses elastic properties and can easily withstand a bend or compression. Another creditable property of CNTs is their light weightedness. For multiwalled nanotubes density is 1.8 g/cm^3 and for single-walled nanotubes it is 0.8 g/cm^3. CNTs are resistant toward chemical attack and are difficult to oxidize. The surface area is of the order 10–20 m^2/g. The thermal conductivity of CNT is higher than that of silver and with decreasing diameter of the tube, its value increases. Compared with metals, conduction of electricity is better in CNTs. Electrons traveling in CNTs obey the rules of quantum mechanics. This quantum movement is called ballistic transport, that is it behaves like a wave in a smooth channel without scattering.[4,5] Metallic and semiconducting properties are exhibited by CNTs. While some nanotubes have conductivities higher than that of Cu, others behave similar to silicon.

These unique properties enable a number of applications for CNTs. Due to their electronic properties, they are used in transistors, gas sensors, high frequency diodes, light-emitting diodes and in efficient photovoltaic devices.[6–9] CNTs are used in nonelectromechanical systems, in thin conducting sheets, and in composite materials due to their high strength and low density.[10–12] High aspect ratio of CNTs is made use in scanning probe microscope tips and flat panel display field emitters.[13,14] Hydrogen storage in fuel cells and capacitors make use of the large surface-to-volume ratio of CNTs.[15,16]

9.4 CLASSIFICATION OF CARBON NANOTUBES

There are two modes of classification for CNTs. In the first method, it is classified into SWCNT and MWCNT (Fig. 9.1). Multiwalled are further classified into Russian doll and Parchment type. The Russian doll consists of concentric tubes with approximately 3.4 A° interlayer separation. In the parchment type, a single graphene sheet is rolled into a parchment.

SWCNT requires a catalyst for synthesis and bulk synthesis is comparatively difficult. It usually contains around 10 atoms along the circumference. Thickness of the tube is only one atom. Large aspect ratio, that is, length-to-diameter ratio enables SWCNTs to be considered as one-dimensional structure. They are easy to characterize and can be easily twisted.

MWCNTs can be produced without the aid of a catalyst and can be produced in bulk. They cannot be easily twisted and defects are less especially when arc-discharge method is used. The name MWCNT is restricted to nanostructures with diameter less than 15 nm. Structures with higher diameter are called nanofibers, which are not single molecules but strands of layered graphite sheets.[17]

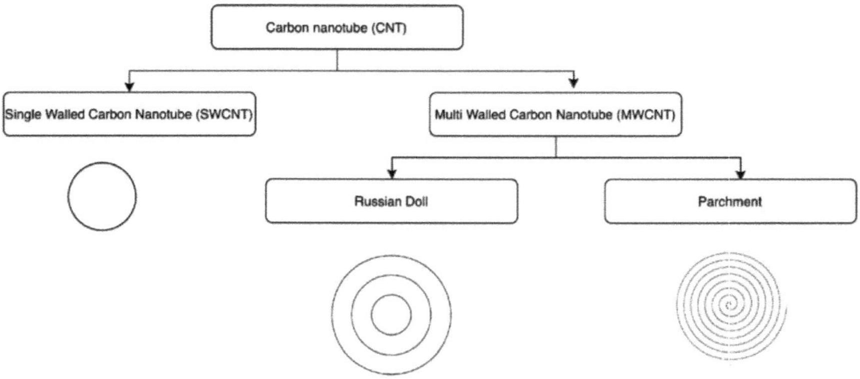

FIGURE 9.1 Classification of CNTs.

The second mode of classification is based on chirality (handedness) (Fig. 9.2). Based on this, CNTs are classified into chiral (handed) and achiral (not handed). In the chiral type, the mirror plane is absent and in the achiral, mirror plane is present. The achiral type can further be subdivided into zig-zag and armchair CNTs. Zig-zag and armchair type are named after the pattern of atoms found at the circumference of the tube.

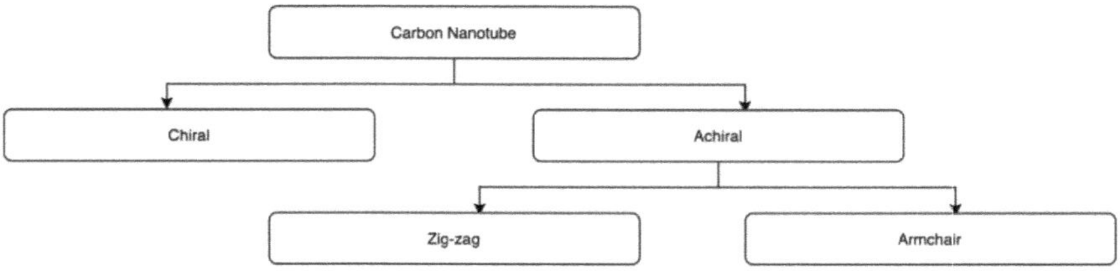

FIGURE 9.2 Classification of CNTs (based on chirality).

SWCNT geometry can be explained with the help of two integers (n,m) which denotes the relative position of a pair of atoms on a graphene strip, which when rolled onto each other form a tube. In the lattice sense, equivalent atoms are obtained when we consider alternate atoms and not near neighbors. Two vectors separated by 60°C are selected as the basis vectors, a_1 and a_2. Carbon nanotube is obtained when a lattice equivalent atom overlaps the origin. Now the chiral vector (or wrapping vector or circumferential vector) characteristic of the SWCNT and corresponding to the edge of graphene strip can be represented as:

$C_h = na_1 + ma_2$, with indices (n,m) which are positive integers with $n \geq m \geq 0$.

Zig-zag type is obtained when the circumferential vector lies along one of the two basis vectors. For the armchair nanotube, circumferential vectors lie along the direction exactly between the two basis vectors, that is $n = m$. When $n \neq m$ chiral nanotubes are obtained. Chiral angle, represented by θ, is the angle which the chiral vector makes with one of the basis vectors.

(n,0)	$m = 0$	$\theta = 0°$	zig-zag
(n,m)	$n = m$	$\theta = 30°$	armchair
(n,m)	$m \neq 0$	$\theta = 0°–30°$	chiral

Interestingly, the optical, mechanical, and electronic properties of a CNT depend on chirality, that is n and m values.[18,19] In the case of armchair and zigzag nanotubes, top and bottom honeycomb lattices are always parallel to the tube axis. Carbon–carbon bonds on opposite sides of the hexagon are perpendicular to the tube axis in the armchair while it is parallel in case of zig-zag. All other confirmations belong to chiral structures where carbon–carbon bonds lie at an angle to the tube axis.[20] Pictorial representation (side-view) of the CNTs (armchair, zigzag and chiral) are given in Figure 9.3. Origin and lattice equivalent atom for each type on a graphene sheet is also shown.

9.5 ELECTRONIC PROPERTIES

CNT is formed by superpositioning a lattice point with origin (0,0). The chiral vector determines the direction of rolling. The diameter of the tube obtained is given by the formula,

$$d = \left(a\sqrt{m^2 + mn + n^2} \div \Pi \right)$$

Where “a” represents the lattice constant and is given by $1.42 \times \sqrt{3}$ A^0

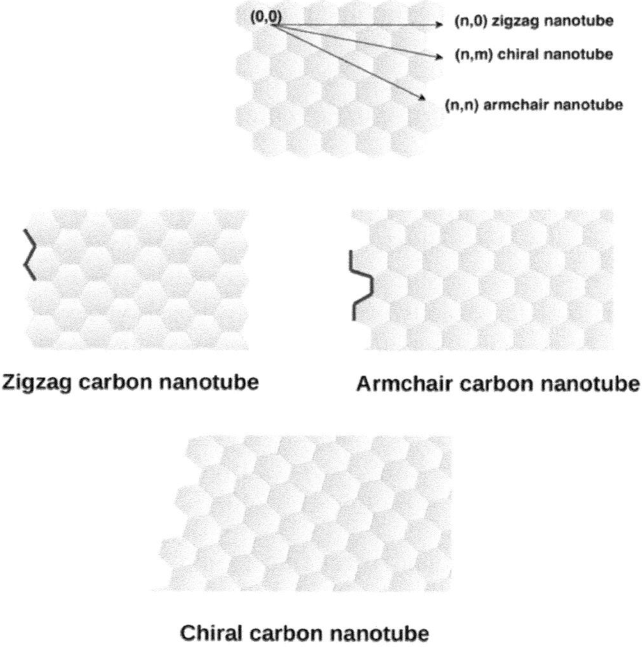

FIGURE 9.3 Pictorial representation of zigzag, armchair, and chiral carbon nanotubes.

The chiral vector is given by $C = ma_1 + na_2$

The chiral angle $\theta = \arctan \dfrac{-\sqrt{3}\, n}{2m + n}$

From theoretical studies, it is found that all armchair SWCNT are metallic while one-third of zig-zag and chiral tubes also belong to the metallic category. The remaining ones are semiconducting. Metallic conduction occurs when n-m = 3q, where q is an integer.[21,22] This observation of metallic behavior even in 1 nm diameter CNT is not a trivial matter as Peierls instability occurs in most one-dimensional materials. This leads to an opening of the energy gap and transition from metal-to-insulator character. This is due to reduction in energy by doubling the unit cell.

Due to strong carbon–carbon bonds and cylindrical arrangement of atoms, CNTs are immune to these.[4]

Electronic local density in individual SWNTs can be determined with the help of scanning tunneling spectroscopy (STS) and it was shown that based on small variations in chiral angle or diameter, SWNTs can act as metals or semiconductors. Quantum confinement of electrons normal to the CNT axis is responsible for the unique electronic properties exhibited. Propagation of electrons occurs only along the axis and the standing waves produced around the circumference of the CNT determine the number of one-dimensional conduction and valence bands.[20] In semiconducting tubes, it is interesting to note that the energy gap increases as the diameter of the nanotube decreases. Also, MWCNTs tend to be metallic as there is low probability for exclusive semiconducting nested nanotubes.[19]

Method of preparation largely determines the transport response among bulk samples of MWNTs. Degree of crystallinity and geometric differences, such as defects, chirality, and diameter determines the electronic response. It has been shown that bundles of nanotubes behave similar to metals with resistivities in the range of 0.34×10^{-4} to 1×10^{-4} ohm cm.[23]

Similarity of SWNTs to quantum wires was proved in 1997 with the help of transport measurements on individual SWNTs with 1 nm diameter. In this, electrical conduction occurs via quantum mechanically coherent discrete electron states that are well separated.[24] Also, on application of strong magnetic fields, semiconducting nanotubes, both single-walled and multiwalled nanotubes act as metals, were also predicted based on theoretical calculations.[25]

Superconductivity in CNTs was first demonstrated in 1999 by Kasumov. It was reported that SWNTs, below 1 K, exhibit superconductivity when connected between two superconducting electrodes. Variations in electronic properties of SWNTs in the presence of external magnetic fields and temperature were also studied.[26] After 2 years superconductivity at 0.55 K was reported by the same group by using nonmetallic pads to contact SWNT bundles.[27]

9.6 NEED FOR REPLACEMENT OF SILICON AND SILICON-BASED COMPONENTS

Silicon, another member of the carbon family shows interesting properties and has huge applications in industries. The conductivity of silicon can be varied over several orders of magnitude. Silicon is the most widely used semiconductor due to various factors. It is available in its elemental form and can be converted to tough insulating oxide for isolation and protection and has controlled the conduction properties by making use of various impurity atoms.

The celebrated Moore's law, stated by Gordon E Moore, co-founder of Intel, predicts that the rate of advancement in processor technology doubles every 18–24 months or in other words the number of transistors on a processor doubles every 18 months. This observation was given in his article, "Cramming more components onto integrated circuits" published in 1965. During this time, the number of transistors in a single chip was just 32, which has now increased to half a billion. This shows the significance and exactness of the prediction made by this visionary. Moore's law was given a mathematical form by Meindle.[28]

$$N = F^{-2} D^2 P_E$$

where N represents the number of transistors per chip, F is the minimum feature size, D, the chip area and P_E gives the transistor packaging efficiency measured per minimum feature area.

The demands of the transistor industry, such as smaller, faster, and advanced computing can be met by scaling down transistors. Computer processor world makes use of transistors as switches that allow or block the path of electrons. To be more specific, they are called field effect transistors (FET). Advances in silicon technology in the past few decades mainly rely on scaling down of metal oxide semiconductor FET (MOSFET). Scaling of devices can be better understood with the help of scaling factor, κ. Change in device parameters during scaling will be as follows:

Channel length:	L/κ
Channel width:	w/κ
Oxide thickness:	t_{ox}/k
Supply voltage:	V/κ
Threshold voltage:	V_{th}/k
Number of devices fitted on a chip:	$N * k^2$
Transition time:	t_{trans}/k
Power consumption:	P/k^2

Nonscalable parameters include silicon band gap, E_g and subthreshold slope, S. These scaling requirements are not immutable. One can focus on few easily scalable parameters such as oxide thickness.

It is difficult to make bulk silicon devices with gate length less than 20 nm. Trigate FET or 3D FET or FinFET is used now. In this, silicon is made into a thin fin-like structure and the channel is covered on all three sides by the gate. Dramatic mobility loss and threshold variation may occur when channel width is reduced to less than 5 nm, to enable gate length below 10 nm in FINFET. Another option is to convert FinFET to a nanowire with gate all around geometry. But there are uncertainties in performance and desired integration densities when scaling Si devices and circuits in this diameter range.[29]

Now, Silicon-based FET has reached their physical limit and have problems with wire connections, heat dissipation, etc. Moreover, when t_{ox} reaching approximately, 1 nm, gate to channel leakage current increases sharply due to tunneling.[4] It is also important to note other problems related to silicon chip manufacturing and applications in this context. Millions of gallons of water and high temperature are required during manufacture of silicon and silicon-based components. Large amount of toxic solvents, such as xylene and toluene is also used. At high temperatures, silicon chips can malfunction. Direction of charge in silicon wafers may change due to the impurities added during doping.

CNTs are suitable candidates to replace silicon. They have interesting electrical and electronic properties, excellent heat conduction, high thermal and chemical stabilities, high mechanical strength and small diameter. Based on the tube diameter, geometry, and chirality, it can be metallic or semiconducting. They can be easily placed in a variety of substrates, have high field effect mobilities, better electronic characteristics and have low operating voltages.[1,30]

The band gap of semiconducting CNT is inversely related to its tube diameter.

$$E(g) \approx 0.84\ (ev) \div d\ (nm)$$

Fermi level position with respect to band edges determines the conduction level in semiconducting tubes. The conductivity can be modified by chemical or electrostatic doping. Conduction level of metallic tubes can be achieved in case of highly doped ones. Devices with switching behavior can be fabricated using CNT due to the strong dependence of conductivity on electrochemical potential. When the Fermi level is in the band gap, CNT acts as an insulator and when the Fermi level moves into valence or conduction band due to the potential on a capacitively coupled gate, it can conduct current. This kind of device is similar to FET. MOSFET can be replicated by making use of CNT in the channel in place of silicon.[4] Schematic diagrams of conventional MOSFET and carbon nanotube field effect transistor (CNTFET) are shown in Figure 9.4.

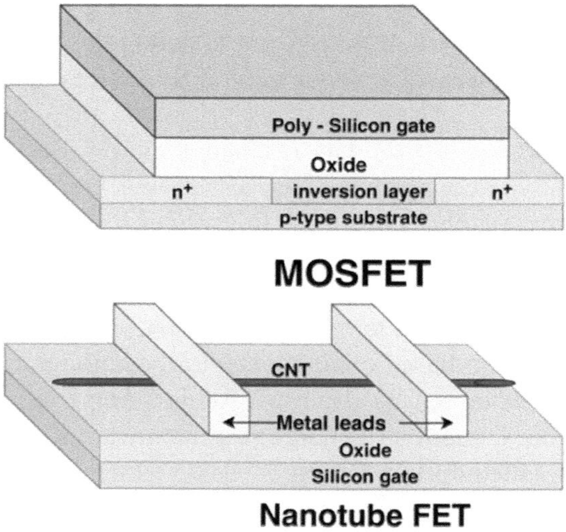

FIGURE 9.4 Conventional MOSFET and carbon nanotube FET.

9.7 NANOELECTRONIC DEVICES

High aspect ratio, structural stability and semiconducting character exhibited by two-thirds of CNTs enable its application in a wide range of nanoscale electronic devices from two terminal rectifiers to FET. The major challenges faced in CNT-based devices are as follows:

1. Purity of CNT (semiconducting versus metallic).
2. Allowed range of CNT diameter.
3. Placement of CNT on substrate with precision.

We can discuss these factors in detail.

9.7.1 SORTING OF CNT

There are three methods for sorting CNTs. They are preferential synthesis, selective destruction and post-synthesis sorting. Designing a process for the synthesis of a particular type of CNT is challenging. The attempted techniques include the use of appropriate CNT seed which is further elongated by sequential addition of building blocks. Half-a-buckyball cap or cycloparaphenylene can act as a seed which can produce semi-open or fully-open CNTs by the addition of monomers. Another approach is cloning, where a parent CNT is cut into shorter segments and made into templates which are elongated by CVD by using iron as catalyst. Orientation and diameter was preserved in this technique. Other methods include engineering a low-cost catalyst system offering high selectivity and activity. Careful selection of reaction promoters can also aid in preferential synthesis.

Selective destruction is mainly used for differentiating metallic and semiconducting CNTs. Available methods under this technique include electrical, light-assisted, plasma-assisted, and microwave-assisted breakdown and functionalization. Enrichment of semiconducting character of the sample can be attained by selective removal of metallic SWCNTs by current-induced oxidation. Organic film coating on CNTs can improve this method. High-power long-arc xenon lamp and laser irradiation can increase semiconducting CNT content in the sample. The method can be made faster by pretreating the sample with 4-bromine benzenediazonium salt which selectively binds to metallic CNTs. When suspended in hydrogen peroxide selective breakdown of semiconducting CNTs can also be achieved using light irradiation. Variation in parameters such as wavelength or time of treatment can accomplish selective removal of CNT of certain diameter. Improvement in diameter distribution and semiconductor character can be achieved by making use of methane and hydrogen plasma. Higher carbon–carbon bond strain in smaller diameter CNT leads to its faster removal. Microwave radiation also shows similar effects. To differentiate metallic and semiconducting CNTs, functionalization or selective side-wall chemistry can be employed. This followed by electrophoresis or chromatography techniques can sort the nanotubes. Functionalization allows a particular CNT to be more soluble in a desired solvent. Introduction of charged or neutral functional groups can alter the mobility of CNTs which can be made use of in electrophoresis technique. Most studied approach under this category is diazotization of metallic CNTs due to the availability of electrons near Fermi level.

Three variations in electrophoresis method are available, which employ electric field to sort CNT based on their mobilities. These include free electrophoresis (matrix less separation), capillary electrophoresis (carried out in capillaries with less than 1 mm diameter), and gel electrophoresis (make use of a matrix mostly agarose, agar or polyacrylamide). Smaller diameter CNTs reach the collector electrode first as they are more mobile. Two types of CNT population are obtained after sonication, short bigger diameter CNT, and long smaller diameter CNT. Here, the short bigger diameter nanotube is preferentially deposited, as molecular size plays a bigger role. Another modification of the technique is employment of a nonuniform electric field called dielectrophoresis. Coupling of CNT with a marker can improve its mobility in the electric field further enhancing the efficiency of the technique.

Separation of CNT mixture based on buoyant density difference is called ultracentrifugation or density-gradient centrifugation. Resolution based on length, diameter, electrical character, enantiomer form and chiral angle can be achieved using this technique. Components of a complex CNT mixture separate into discrete bands in a density-gradient medium when subjected to high centripetal force based on their buoyant density. This is due to variation in CNT structure when dispersed in water assisted by surfactants leads to changes in their hydration layer. The surfactant determines the shape of the hydration layer and type of sorting. Here, it is important to note that water-filled open CNTs and empty end-capped CNTs have different buoyant density and can influence the separation efficiency.

Chromatography is another available technique for separation. Separation can be done using controlled pore glass, agarose gel using sodium deoxycholate as eluent, sephacryl gel, etc. Temperature can influence the course of separation, and stepwise fractionation by connecting chromatographic columns in series at different temperatures can be achieved.

CNTs of particular chirality, metallic character, diameter and length show preferential binding to certain polymers and copolymers. This is made use of in the polymer isolation technique. DNA is the most interesting tailorable polymer used for this technique.[31]

9.7.2 CNT ALIGNMENT TECHNIQUES

Alignment of CNT is important as most of the properties vary along the tube axis and normal to it. We obtain randomly oriented CNTs during mass production. Two approaches are identified for alignment, one is growing the tube along a particular direction during synthesis and the other one is post-growth processing of random bundles. Common methods to evaluate the alignment of CNTs are atomic force microscopy, scanning tunneling microscopy and scanning electron microscopy. Optical absorption, photoluminescence or Raman scattering methods are followed when the aligned CNTs are buried in a dispersant.

Alignment during synthesis is achieved by covering the substrate with a metal catalyst and growing the tubes. Bending down of CNT is prevented by high density of catalyst on the substrate and nanotubes are grown normally to the substrate. Another approach is to apply an electric field during the growth normally to the substrate. This allows control over orientation and structure by changing the direction and strength of the applied field. Templates with nanochannels can be made use of for vertical alignment and diameter can be varied with pore size. Suitable templates for vertical alignment include mesoporous silica, alumina, liquid–crystal matrices, aluminophosphate, or zeolites. For horizontal alignment, this vertically aligned nanotube forest can be swept with a knife-edge smearing CNT along the substrate. A better approach for horizontal alignment is to grow the nanotube along the substrate rather than normally. CNT will bend with a low density catalyst and alignment is achieved with the help of specially oriented sapphire. Major drawback of this alignment during growth is that the nanotubes obtained have metal catalyst nanoparticles with it. We have to further remove these metal nanoparticles by acids or annealing. Electron beam or lithography setup can also be made use of for the same purpose. But this could be an advantage if we require functionalized CNTs such as those required in magnetic field controlled devices. Here, CNTs with iron, cobalt or nickel could be an added advantage.

Application of external forces or fields over isolated CNTs is required for post-growth alignment. A number of techniques have been reported along these lines. Dispersion of CNT in a suitable solvent is one such technique, but we will have to use a small amount of surfactant to facilitate the dispersion. This can form micelles around CNT and improve dispersing ability. Solution of polymers can also be used. Removal of dispersant is the major problem faced in this technique. Most popular technique for research purposes is to stretch the polymer CNT film after softening by dipping into water or alcohol and clamping the opposite edges. But this is not suitable for industrial applications. Elastic field of the polymer matrix reorient CNTs during stretching. An extreme case of this stretching is fracture where aligned nanotubes are obtained in the fracture gap formed by overstretching. Uniaxial pressure of about 10 K Bar can align nanotubes in a polymer matrix. Friction alignment is another approach where a blade or set of parallel grooves is unidirectionally rubbed with softened CNT-embedded polymer film. But, this can lead to damaged film. Nanotube solution when filtered through a porous membrane will align the tubes. Electrospinning or fiber drawing approach can also be made use of. Gas flow with linear velocity can align the nanotube during deposition over a substrate. This method can easily be automated. Capillary forces will align the tubes when a substrate is pulled from CNT solution. This is called Langmuir-Blodgett technique. When a substrate located inside a magnet is used for casting, the nanotubes will align to the direction of the magnetic field. Strong fields are required for alignment, this can be reduced by making use of magnetic nanoparticles. Interdigitated electrodes in a substrate produce good quality CNTs by electrophoresis. An AC electric field is required for alignment. This technique has an added advantage that metallic and semiconducting CNTs can be separated as metallic nanotubes alone are attached to electrodes and semiconducting ones can be rinsed away. When liquid crystal substance is mixed in the dispersing solution of CNTs, alignment can be achieved by electric field, magnetic field, light or mechanical forces.[32]

9.7.3 DIODES

Elbow-like bends are occasionally observed in nanotubes. Armchair and zigzag tubes could be joined with a pentagonal and heptagonal ring on the outer and inner side of the elbow, respectively. This enables the synthesis of intramolecular metal–metal, metal–semiconductor and semiconductor–semiconductor junctions. These nanotube junctions are found with nanotube prepared by arc-evaporation method. When metallic and semiconducting tubes are connected, heterojunctions are formed. High-energy electrons will flow downhill from semiconducting to metallic side, and the other way travel is not possible. This property can be made use of in many electronic devices.[33] One such application is a rectifying diode. This was first reported by Collins in 1998.[34] Rectifying behavior of elbow connection was experimentally proven a year later by Cees Dekker.[35] Nanotube diodes can also be produced by doping. For this, 50% of semiconducting SWNT was doped with potassium making it n-type due to the donation of electrons from potassium atoms. The undoped part behaves as a p-type semiconductor. Thus, a p-n junction diode, similar to silicon diode is fabricated. This was reported by Hongjie Dai in 2000[36] who used a SWNT of 2 nm diameter covered half way with 340 mm thick polymethylmethacrylate. Doping was carried out in vacuum by heating potassium source electrically. The potassium atom gets adsorbed on the uncovered part of the nanotube making it n-type.

9.7.4 FIELD EFFECT TRANSISTORS

The current progress seen in electronic technology can be mainly attributed to size scaling of devices, especially MOSFET. As this scaling down has almost reached its physical and theoretical limit, a search for new competitive technologies has been started. This issue can be faced from two aspects. One method is to adopt totally new concepts such as two-terminal molecular devices, spintronics, quantum computing, etc. But, this will not be compatible with present applications. The second approach is to make use of alternative materials in the three-terminal transistor concept. Here, we have a suitable candidate, SWCNTs. CNTFETs have shown superior characteristics, but many issues need to be solved for its implementation.[19] Speed, scalability and power are the most important characteristics of any transistor. One-dimensional nature enables ballistic transport (meaning a free path is longer than the path the carrier travels) in CNTs. Thus, only negligible resistance is there and it can achieve speed of terahertz or more. Power wastage is also minimum in the case of CNTs.

In FET, a semiconducting channel connects the two metal electrodes, source, and drain. A thin insulator film separates the third electrode gate from the channel. Flow of charge between source and drain is controlled by the gate. In conventional devices, silicon constitutes the channel while in CNTFET, an individual semiconducting SWNT is used. Such a device was first fabricated in 1998 by Dekker group.[37] The optimizers of CNTFET technology are gate length, CNT pitch and CNT number per device. The contact resistance between metal electrodes and CNTs need to be reduced. Increased tuning efficiency is preferred between gate and CNT channel. CNT channel length has to be minimized and optimum device structure needs to be developed. Complementary devices, n-type and p-type are used in both CNFET and MOSFET.

a. p-type
   In this, conduction of holes takes place. Conduction of electrons does not take place even at high positive gate voltages. Doping is not required for p-type CNTFET as they naturally become so on exposure to air. Fermi level at the contacts shifts closer to the valence band because of the oxygen in the air. As a result, holes easily tunnel through as they see a small barrier while electrons do not tunnel as they see a large barrier. From the transfer characteristics, it is evident that when the gate voltage is positive, conduction is too low to turn on the transistor and when the gate voltage is negative, the transistor is turned on, and a considerable increase in drain current is observed.
b. n-type
   Here, conduction is due to electrons. The nanotube needs to be converted and sealed to prevent turning back to p-type. There are two methods for conversion, such as annealing and doping. In annealing, the oxygen absorbed on exposure to air is driven out. This is carried out at 450°C in the nitrogen atmosphere. Removal of oxygen shifts the Fermi level to the conduction band and the barrier is reduced for electrons. This increases the barrier for holes and it cannot tunnel through. Electron donors such as potassium are used in doping. Doping causes the threshold voltage to shift downwards. Fermi level does not move but rather it is pinned down by the doping. Nanotubes need to be heavily doped to make n-type transistors. Transfer characteristics are opposite to that of p-type. When gate voltage is negative, small off current occurs and considerable increase is observed when it turns positive.
c. Ambipolar device
   It can conduct both holes and electrons. During annealing for conversion of p-type to n-type, an intermediate phase occurs which can be termed as an ambipolar device. Partial removal of

adsorbed oxygen moves the Fermi level between conduction and valence bands. This results in small barriers for both electrons and holes which enables them to conduct in nanotube. The conduction is based on the polarity of the gate voltage.

9.7.4.1 STRUCTURE-BASED CLASSIFICATION

9.7.4.1.1 Back-Gated CNFET

Initial design of transistors consists of two gold electrodes as source and drain and the CNT was draped over these. Silicon oxide was used to separate the gate and the nanotube (Fig. 9.5). The transistor was switched on or off by the electric field from the gate. Drawbacks of these designs include exposure of CNT to the air making it a p-type and thickness of the gate oxide need to be approximately 100 nm necessitating high voltages to penetrate and switch the transistor. Moreover, as the same gate is shared by all transistors, and switching of all transistors occurs at the same time. This creates problems in large circuits or chips. Improvement in performance can be made by reducing the insulator thickness.

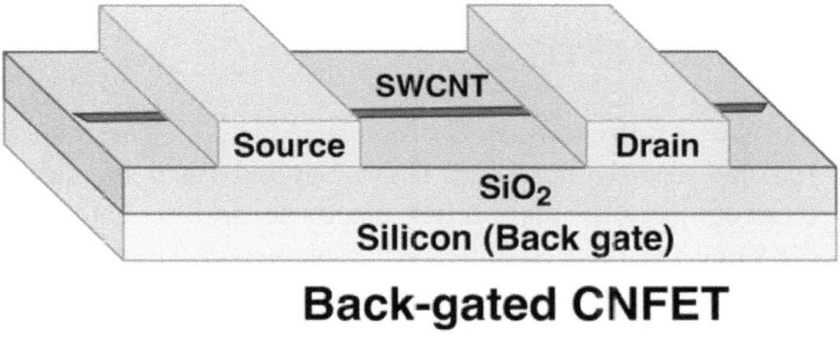

FIGURE 9.5 Schematic diagram of back-gated CNFET.

9.7.4.1.2 Top-Gated CNFET

An improvement in the design was made by placement of the gate electrode on top of the nanotube (Fig. 9.6). This provides covering to the nanotube and prevents air exposure. Gate oxide being thinner, voltage requirement is also minimized. This design can be made use in high frequency operations with slight modifications. Another method for top gate design makes use of an electrolyte for the gate. Higher capacitance is achieved by the electrolyte solution, thus enhancing the performance. Here, the liquid electrolyte needs to be sealed to prevent leakage and air diffusion which complicates the manufacturing procedure. Compared with back-gated CNFET, the contact resistance is reduced, threshold voltage is lowered and high-drive current as well as transconductance is observed in top-gated CNFET.

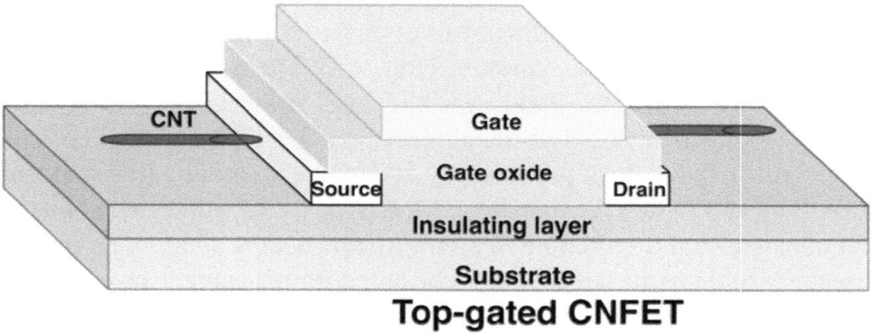

FIGURE 9.6 Schematic diagram of top-gated CNFET.

9.7.4.1.3 Vertical CNFET

This consists of a surround-gated structure (Fig. 9.7). The transistor size can be reduced to the diameter of CNT. Nanotube is connected to drain (upper electrode), source and bottom electrode, and the gate electrode is wrapped around the tube. A transistor element with single vertical CNT is formed at every cross point of source and drain electrodes. The hole diameter of gate oxide determines the number of CNT in the transistor. Higher packing densities are achieved by arranging source and drain areas on top of each other.

9.7.4.2 OPERATION-BASED CLASSIFICATION

In CNFET, the source terminal supplies electrons and the drain terminal collects the electrons supplied. That is the current flows from drain to source terminal. The current intensity in the transistor channel is controlled by the gate terminal.

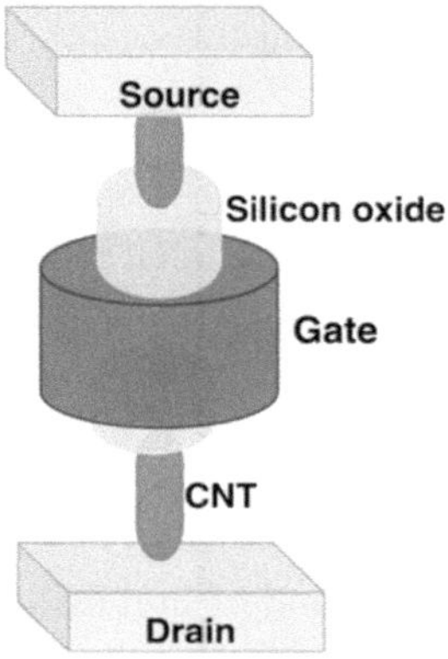

FIGURE 9.7 Schematic diagram of vertical CNFET.

9.7.4.2.1 Schottky-Barrier CNFET

A potential barrier exists at every contact between metal and semiconductor. This is known as the Schottky barrier. In this device, source and drain terminals are metals and the terminal contact between metal and nanotube has a Schottky barrier. Working principle is based on direct tunneling through this barrier at source–channel junction. Transconductance is gate voltage-dependent as the barrier width is controlled by this voltage. With increase in gate bias, barrier width decreases which further increase in quantum mechanical tunneling and current flow in the transistor channel. Occurrence of ambipolar conduction at low gate oxide thickness is a major setback. This can lead to exponential increase in leakage current. Increase in gate oxide thickness reduces the leakage current as well as improves the performance. Asymmetric gate oxide is another alternative solution. Other drawbacks include limited channel length leading to increased source to drain tunneling and inability to place gate and source close by as this leads to increased parasitic capacitance. The schematic diagram of the design is given in Figure 9.8.

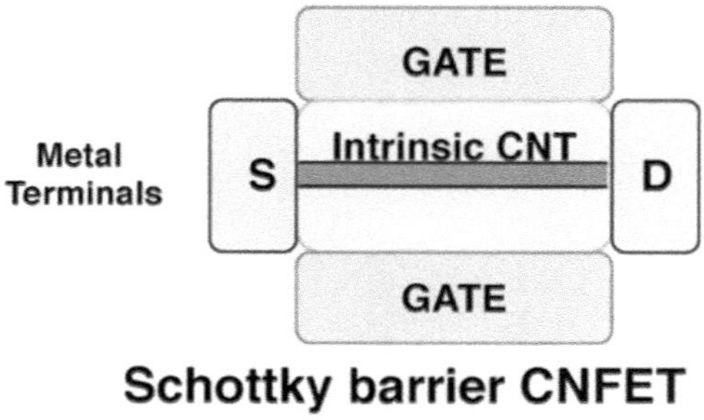

FIGURE 9.8 Schematic diagram of Schottky barrier CNFET.

9.7.4.2.2 MOSFET-Like CNFET

Heavily doped terminals in place of metals are characteristic of this type of device (Fig. 9.9). Modulation of barrier height with gate voltage is the basic principle used here. Absence of ambipolar conduction, longer channel length limit, reduced parasitic capacitance and faster operation sets higher priority to this design.

FIGURE 9.9 Schematic diagram of MOSFET-like CNFET.

9.7.5 LOGIC CIRCUITS

n- and p-type FETs are required in logic circuits. NOT gate (inverter) is the first nanotube-based logic gate. It requires just an ambipolar CNFET or both n- and p-type CNFET. Inverters constructed from complementary n- and p-type CNFET are slower than those using ambipolar CNFET.

Moreover, thresholds in ambipolar devices can be tuned even after fabrication. Requirement of higher input voltage increases power consumption in n- and p- type CNFET. Another design is an intrananotube inverter which uses a single nanotube. Its construction is similar to internanotube n- and p- type inverters, but in intratype, only one nanotube is draped across the contacts.

Doping and annealing techniques are used by Avouris for getting n-type behavior in short sections of nanotube bundles. Silicon oxide, gold electrodes, and a single nanotube bundle was placed one over the other in such a way that 2 p-type CNFETs are produced in series. The whole setup was covered by PMMA (Poly (methyl methacrylate)) and a part of the nanotube was exposed by electron beam lithography. The exposed portion was turned into n-CNTFET by potassium vapors while the other part remains p-type. The device thus obtained operates as NOT gate.[38]

NOR gate (opposite of OR gate) is fabricated by placing two n-type CNFETs in parallel. For NAND gate CNFET need to be placed in series. OR gate make use of second generation geometry and a p-type CNFET with two gates.

9.7.6 INTERCONNECTS

Individual gates in circuits and chips need to be connected which are done using metals in conventional circuits. Problems such as increased resistivity and power consumption delay and poor tolerance to electromigration are common occurrences in copper and aluminum interconnects. Search for novel interconnects technologies has led to optical, radio frequency (capacitive or inductive) coupling, antennas interconnect, superconductor interconnects, CNTs, and graphene materials. Better electrical properties of metallic CNTs enable its application in interconnects. Compared with metal, CNTs will offer lower resistance due to ballistic transport. Reduced time delay due to reduced scattering is an added advantage. Extraordinary mechanical strength, high thermal conductivity, low inductance, and saturation of skin effect favor them in three-dimensional integration and high frequency electronics. Selective growth of metallic CNTs, chirality control, orientation control, and formation of low resistance CNT–metal contacts are some of the issues that have to be addressed here.[39]

9.8 CONCLUSION

Unique properties of CNTs enable their application from molecular electronics to a diverse field as medicine. Many current and future demands of various fields can incite further research in this ever-growing field. Currently, many fundamental problems inhibit their application, but these issues do not outshine the popularity received by CNTs from the days of their discovery. Some properties, methods of synthesis, classification and applications particularly in the field of electronics are discussed here. But this is just a glimpse of the wide world of opportunities that lay ahead and it is fascinating to see the evolutions and the amazing applications that will eventually emerge from the research.

KEYWORDS

- **nanoelectronics**
- **carbon nanotubes**
- **transistors**
- **carbon nanotube field effect transistor**
- **carbon nanotube synthesis**
- **electronic properties**

REFERENCES

1. Booker, R. D.; Boysen, E. *Nanotechnology for Dummies*; John Wiley & Sons, 2005.
2. Shriver, D. F.; Atkins, P. W.; Langford, C. H. Inorganic Chemistry; W. H. **1990,** 131–132.
3. Harris, Peter J. F.; Harris, P. J. F. *Carbon Nanotube Science: Synthesis, Properties and Applications*; Cambridge University Press, 2009.
4. Popov, V. N. Carbon Nanotubes: Properties and Application. *Mater. Sci. Eng.: R: Rep.* **2004,** *43* (3), 61–102.
5. Pradeep, T. *NANO: The Essentials-Understanding Nanoscience and Nanotechnology*; McGraw Hill, 2007.
6. Suehiro, J.; Zhou, G.; Hara, M. Fabrication of A Carbon Nanotube-Based Gas Sensor Using Dielectrophoresis and Its Application for Ammonia Detection by Impedance Spectroscopy. *J. Phys. D Appl. Phys.* **2003,** *36* (21), L109.

7. Manohara, H. M. et al. Carbon Nanotube Schottky Diodes Using Ti– Schottky and Pt– Ohmic Contacts for High Frequency Applications. *Nano Lett.* **2005,** *5* (7), 1469–1474.
8. Mueller, T. et al. Efficient Narrow-Band Light Emission from a Single Carbon Nanotube p–n Diode. *Nat. Nanotechnol.* **2010,** *5* (1), 27–31.
9. Gabor, N. M. et al. Extremely Efficient Multiple Electron-Hole Pair Generation in Carbon Nanotube Photodiodes. *Science* **2009,** *325* (5946), 1367–1371.
10. Peng, B.; Ding, L.; Guo, Z. Resonant Modelling of Two Types of Tunable Carbon Nanotube Electromechanical Oscillators. *Micro Nano Lett.* **2010,** *5* (6), 365–369.
11. Zhang, M. et al. Strong, Transparent, Multifunctional, Carbon Nanotube Sheets. *Science* **2005,** *309* (5738), 1215–1219.
12. Schadler, L. S.; Giannaris, S. C.; Ajayan, P. M. Load Transfer in Carbon Nanotube Epoxy Composites. *Appl. Phys. Lett.* **1998,** *73* (26), 3842–3844.
13. Hafner, J. H. et al. High-Yield Assembly of Individual Single-Walled Carbon Nanotube Tips for Scanning Probe Microscopies. *J. Phys. Chem. B* **2001,** *105* (4), 743–746.
14. Jang, Y-T. et al. Fabrication and Characteristics of Field Emitter Using Carbon Nanotubes Directly Grown by Thermal Chemical Vapor Deposition. *Thin Solid Films* **2003,** *436* (2), 298–302.
15. Deshpande, R. et al. Hydrogen Adsorption in Single-Walled and Multi-Walled Carbon Nanotubes Grown in a Hot-Wire CVD (Cat-CVD) Reactor. *Thin Solid Films* **2006,** *501* (1–2), 224–226.
16. Kimizuka, O. et al. Electrochemical Doping of Pure Single-Walled Carbon Nanotubes Used as Supercapacitor Electrodes. *Carbon* **2008,** *46* (14), 1999–2001.
17. Saifuddin, N.; Raziah, A. Z.; Junizah, A. R. Carbon Nanotubes: A Review on Structure and Their Interaction with Proteins. *J. Chem.* **2013,** *2013*.
18. Léonard, F. *Physics of Carbon Nanotube Devices*; William Andrew, 2008.
19. Chen, C.; Zhang, Y. *Nanowelded Carbon Nanotubes- From Field-Effect Transistors to Solar Microcells;* Springer, 2009; pp 1434–4904.
20. Terrones, M. Science and Technology of the Twenty-First Century: Synthesis, Properties, and Applications of Carbon Nanotubes. *Annu. Rev. Mater. Res.* **2003,** *33* (1), 419–501.
21. Saito, R. et al. Electronic Structure of Chiral Graphene Tubules. *Appl. Phys. Lett.* **1992,** *60* (18), 2204–2206.
22. Saito, R. et al. Electronic Structure of Graphene Tubules Based on C 60. *Phys. Rev. B* **1992,** *46* (3), 1804.
23. Thess, A. et al. Crystalline Ropes of Metallic Carbon Nanotubes. *Science* **1996,** *273* (5274), 483–487.
24. Tans, S. J. et al. Individual Single-Wall Carbon Nanotubes as Quantum Wires. *Nature* **1997,** *386* (6624), 474–477.
25. Bailey, S. W. D. et al. Giant Magneto-Conductance in Twisted Carbon Nanotubes. *EPL (Europhysics Letters)* **2002,** *59* (1), 75.
26. Kasumov, A. Y. et al. Supercurrents through Single-Walled Carbon Nanotubes. *Science* **1999,** *284* (5419), 1508–1511.
27. Madou, Marc J. *Fundamentals of Microfabrication and Nanotechnology, Three-Volume Set*; CRC Press, 2018.

28. Meindl, J. D. Theoretical, Practical and Analogical Limits in ULSI. In *1983 International Electron Devices Meeting*; IEEE, 1983.
29. Tulevski, G. S. et al. Toward High-Performance Digital Logic Technology with Carbon Nanotubes. *ACS Nano* **2014,** *8* (9), 8730–8745.
30. Viswanathan, B. *Nano Mater*. **2009**. 978-81-7319-936-3.
31. Janas, D. Towards Mono Chiral Carbon Nanotubes: A Review of Progress in the Sorting of Single-Walled Carbon Nanotubes. *Mater. Chem. Fronti.* **2018,** *2* (1), 36–63.
32. Iakoubovskii, K. Techniques of Aligning Carbon Nanotubes. *Centr. Eur. J. Phys* **2009,** *7* (4), 645–653.
33. Harris, P. J. F. Carbon Nanotubes and Related Structures: New Materials for the Twenty-First Century. **2004,** 415–415.
34. Collins, P.; Zettl, A.; Bando, H.; Thess, A.; Smalley, R. E. Nanotube Nanodevice. *Science* **1997,** 100–102.
35. Yao, Y. et al. Temperature-Mediated Growth of Single-Walled Carbon-Nanotube Intramolecular Junctions. *Nat. Mater.* **2007,** *6* (4), 283–286.
36. Zhou, C. et al. Modulated Chemical Doping of Individual Carbon Nanotubes. *Science* **2000,** *290* (5496), 1552–1555.
37. Tans, S. J.; Verschueren, A. R. M.; Dekker, C. Room-Temperature Transistor Based on a Single Carbon Nanotube. *Nature* **1998,** *393* (6680), 49–52.
38. Raychowdhury, A. et al. Carbon Nanotube Field-Effect Transistors for High-Performance Digital Circuits—DC Analysis and Modeling toward Optimum Transistor Structure. *IEEE Trans. Electron Devices* **2006,** *53* (11), 2711–2717.
39. Srivastava, A.; Liu, X. H.; Banadaki, Y. M. Overview of Carbon Nanotube Interconnects. In *Carbon Nanotubes for Interconnects*; Springer: Cham, 2017; pp 37–80.

CHAPTER 10

CARBON NANOTUBES—A PATHWAY TOWARD GREEN APPLICATIONS

K. B. BHAVITHA[1,2,*] and SRINIVASARAO YARAGALLA[3,*]

[1]*Department of Physics, St Teresas's College, Ernakulam 682011, Kerala, India*

[2]*International and Inter University Centre for Nanoscience and Nanotechnology, Mahatma Gandhi University, Kottayam 686560, Kerala, India*

[3]*Smart Materials, Istituto Italiano di Tecnologia, Genova 16163, Italy*

[*]*Corresponding author. E-mail: Srinivasarao Yaragalla: yaragallasrinu@gmail.com; bhavithakb@gmail.com*

ABSTRACT

The unique and tuneable properties of carbon nanotubes enable them to identify and address new technological as well as environmental challenges. This chapter critically addresses a wide range of green applications of carbon nanotubes such as green technologies and environmental applications: water and air pollution prevention strategies, biotechnological applications, environmental sensors, renewable energy applications, super capacitors, and environmental friendly composites. Here, we also discuss the large-scale production stratagies, potential hazards, and future opportunities for the green technological and environmental applications.

10.1 INTRODUCTION

Among all carbon-based nanoparticles, carbon nanotubes (CNTs), due to its unique and tuneable physical, chemical, and electrical properties, have up-raised and drawn profound interest toward the nanotechnology fields since its discovery in 1991 by Iijima[1] during his preparation for fullerenes. It has been extensively explored[2,3] for the technological applications though there is a tough competition with two-dimensional graphene[4–6] and its composites.[7–14] Ideally, CNTs are one-dimensional counterpart of fullerene, with a shape of cylindrically rolled graphene sheet having a few nanometres in diameter and few micrometers in length and closed at the ends with fullerene spheres.[15] The sidewalls of CNTs have sp^2 hybridization, the same as grapheme, and, at the ends, the sp^3 hybridization of fullerene spheres causes less chemical reactivity compared to graphene and fullerene. CNTs can be single-walled (SWCNTs) or multi-walled (MWCNTs). The quantum confinement in one dimension in accordance with their tubular arrangement makes CNTs distinct and avail them unique electronic properties such as metallic, semi metallic, or semiconducting performance depending upon the orientation of its hexagonal lattice. Instead, MWCNTs exhibits semiconducting nature having concentric cylindrically shaped graphene sheets rolled together around a central axis with 0.34 nm interlayer distance.[16] Several synthesis strategies are reported for the preparation of CNTs of which chemical vapor deposition (CVD),[17] laser ablation,[18] vapor phase deposition,[19] electrochemical synthesis,[20] and pyrolysis of organic precursor molecules are some of the commonly used ones.[21]

CNTs have drawn great interest among the researchers and their preparation, properties, functionalization, mechanism of formation, chemical modifications, and applications are widely been investigated and reported.[22–28] The investigations on the environmental applications and implications are the new path ways going on with immense interest and importance.[29–31] Contamination of environment due to the toxic effect of nanomaterials is of high concern among the environmentalists around the world. The tuneable chemical, physical, electrical, thermal, and optical properties of CNTs correlating to its structural (size, shape, surface area, and high-aspect ratio) and its hybridization state is the answer for most of the environmental challenges associated with health and safety. New innovations are already being carried out for the use of CNTs for drug delivery, removal of contaminants, and for water and air purification.[32–35]

The basic properties of carbon nanotubes are outlined in the Table 10.1 and, in this chapter, we mainly focus on their applications in the field of green technologies and its environmental implications. We included the major applications such as water and air pollution prevention strategies, biotechnological applications, environmental sensors, renewable energy applications, super capacitors, and environmental friendly composites.

TABLE 10.1 Theoretical and Experimental Properties of CNTs.[36,37]

Properties	SWCNTs	MWCNTs
Specific gravity	0.8 g/cm^3	1.8 g/cm^3
Elastic modulus	~ 1 TPa	~ 0.3–1 TPa
Strength	50–500 GPa	10–60 GPa
Resistivity	5–50 μΩ cm	5–50 μΩ cm
Thermal conductivity	3000 W/m/K	3000 W/m/K
Thermal stability	2800°C (in vacuum)	2800°C (in vacuum)
Specific surface area	~ 400–900 m^2/g	~ 200–400 m^2/g

10.2 CARBON NANOTUBES AS AN EFFICIENT SORBENT FOR WASTE WATER TREATMENT

Activated carbon, zeolites, and resins act as the major sink of contaminants for most of the natural and engineered environmental sources for the past few decades. But the capacity of conventional sinks are limited by large sizes, less-activated sites, less-activation energy of bonds, slow kinetics, non-equilibrium, and less transfer rate of sorbent surface. Carbon nanotubes having high-aspect ratio and controllable surface chemistry can easily overcome these limitations of conventional sorbents. Both organic and inorganic contaminants can be adsorbed to CNT surfaces depending upon the functional moieties attached to it. CNTs, with its intrinsic nonpolar and hydrophobic nature, can be used to adsorb organic contaminants efficiently through physical interaction. Whereas, functionalized CNTs are used for the polar sorbate. Selective functionalization has proved CNTs to be high-efficient sorbents than conventional sorbents especially for micropollutants, low-concentration pollutants, and polar compounds with low molecular weight. The pH value of the solution also plays a major role in the sorbent activity in which CNTs proved to be effective in a wide range of pH values. Moreover, the availability performance of CNT for a number of sorption/desorption cycles are found to be higher than the conventional sorbents like activated carbon (Table 10.2).

TABLE 10.2 Recent Developments of CNTs as Sorbents for Waste Water Treatments.

Adsorbents	Pollutants	Highlights	Refs.
MWCNT	N-methylcarbamate insecticides	MWCNTs were used as solid-phase extraction sorbents combined with liquid chromatography–electrospray ionization–mass spectrometry detection for the removal of N-methylcarbamate insecticides from diverse surface water samples. It was found that the adsorption capacity of MWCNTs was larger than activated carbon. The analysis of tap and surface waters are successfully analyzed through this method.	[38]
MWCNT	Phenol	HPLC technique was used to analyze the amount of phenol before and after treatment. The efficiency of removal was found to be 85.54%.	[39]
CNF-CNT Composite	Micropollutant uptake	The achieved micropollutant uptake comparable to granular activated carbon by integrating CNTs into CNFs and realizing a macroporous composite. The optimal formulation realized was polyacrylonitrile 8 wt.%, CNT 2 wt.%, and phthalic acid 2.4 wt.% (40% relative humidity; 280°C stabilization), which provided better strength and performance.	[40]
PS-CNT	Oil-spill clean up	It has been found that fabricated super-hydrophobic-oleophilic PS-CNTs sorbent has potential of reusability by the electro-spinning method. The fluorination and covalent modification CNTs improved the dispersibility and interfacial interaction with PS, resulting in aligned CNTs inside the porous fiber structure. The maximum oil sorption capacity of the PS-CNTs was found to be higher by 42–70% than that of the PS sorbent without CNTs.	[41]

TABLE 10.2 *(Continued)*

Adsorbents	Pollutants	Highlights	Refs.
Chitosan/ CNT	Mercury	The results demonstrated the removal of mercury from industrial wastes using CS/CNTs beads prepared using protected crosslinking via the reaction of the beads with Hg(II) as the protector. This work showed that beads fabricated using the protected crosslinking method removed 2.5 times more Hg(II) from water than beads made with ordinary crosslinking.	[42]
CNT-biochar	Dye sorption	The fabricated CNT-coated biochars by dip-coating biomass in carboxyl-functionalized CNT solutions and the good sorption ability toward methylene blue (MB) was found (6.2 mg/g).	[43]
MWCNT–zirconia	Arsenic	The prepared MWCNT-ZrO_2 with 4.85% zirconia had an adsorption capacity of 2000 μg/g and 5000 μg/g for As (III) and As (V), respectively, and is independent of pH. The composite was successful in meeting the drinking water standard levels of 10 μg/L.	[44]
CNT yarn	Nitroaromatic compound, 2,4-dinitrotoluene (DNT)	CNTY was used to treat waste water polluted with 2,4-dinitrotoluene (DNT). The authors of this work had fitted the adsorption isotherm of DNT onto CNTY by the Freundlich isotherm with a Freundlich constant, K_F, of 55.0 mg/g (L/mg)1/n and a Freundlich exponent, 1/n, of 0.737.	[45]

Despite saving as effective sorbent, CNTs are also used as filters for water filtration with specific selectivity. Due to its hydrophobic nature, it proves to be an excellent material for water transport with speedy and less-friction way due to its minimal interaction with water. Efficient composite membranes with CNTs are reported by many studies recently. Membranes of polymer blend with CNTs as water-filtration membranes are also a current scenario of studies with high-end results. In addition to contaminant removal, recent studies have also shown that the CNT-membrane filters are also effective to remove pathogens such as bacteria, virus, and protozoa from contaminated water. The schematic representation of CNT-hybrid filter is shown in the Figure 10.1.

Even though toxicity of nanomaterials is a greatest concern for the environmentalist worldwide, the antimicrobial activity of CNTs is also revealed in the recent studies. The antibacterial activity of CNTs toward gram-positive and gram-negative bacteria are also revealed in the recent studies. The exact mechanism of the antibacterial properties is explicit and the ongoing studies are in this direction. The removal and inactivation of bacteria and virus endeavored them to be used against bacterial colonization and biofilm development in water systems. Researchers are going to use them effectively as medical implant equipments and other inner surfaces (Table 10.3).

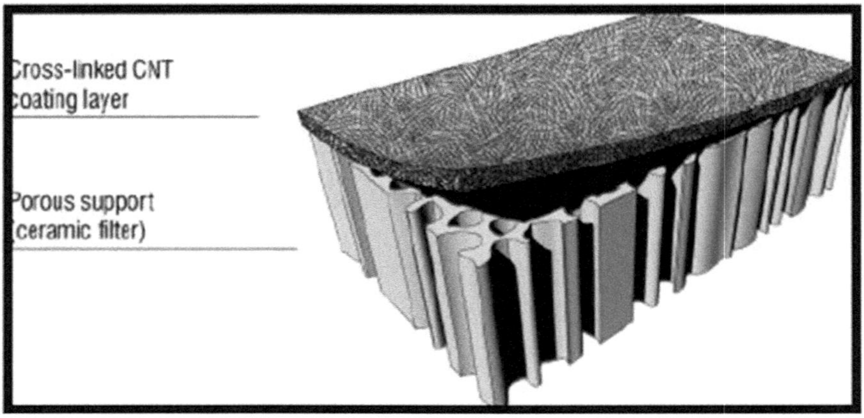

FIGURE 10.1 Schematic representation of CNT hybrid filter.

Source: Reported with reprint permission from Ref. [46]. © American Chemical Society, 2008.

TABLE 10.3 Recent Developments of CNTs as Filters for Waste Water Treatments.

Material	Filterate	Comments	Refs.
Carbon nanotube sponges	Nanoparticles and dye molecules	The three-dimensional interconnected porous structure formed by entangled nanotubes can trap nanoparticles and molecules by physisorption without the need for chemical functionalization. The sponge filters are potential environmental materials for water treatment.	[47]
TiO_2/MWCNT arrays	*Escherichia coli* O157:H7	Photocatalysis-assisted water filtration: Using TiO_2-coated vertically aligned multi-walled carbon nanotube array for removal of *E. coli* O157:H7. The photocatalytic-killing rate constant for TiO_2-ceramic and MWCNT/TiO_2-ceramic under fluorescent light was found be 1.45×10^{-2} min^{-1} and 2.23×10^{-2} min^{-1}, respectively.	[48]
Electrochemical MWNT microfilter	Viruses (MS2) and bacteria (*E. coli*)	It has been noticed that the concomitant electrolysis along with filtration could significantly remove influent bacteria and viruses. They found that applying 2 and 3 V for 30 s post-filtration removed greater than 75% of the bacteria and greater than 99.6% of the viruses.	[49]
Multi-walled carbon nanotube (MWNT)	MS2 bacteriophage virus	It was found that MWNT filter removed MS2 bacteriophage virus by 1.5 and 3 log higher than SWNT filter with comparable loading (0.3 mg/cm^2) of CNTs. The larger removal was endorsed to the formation of uniform CNT-filter matrix, which removes viruses by depth filteration.	[50]

TABLE 10.3 *(Continued)*

Material	Filterate	Comments	Refs.
Stainless steel-CNT meshes	Water-in-oil emulsions	SS-CNT meshes allowed repellsion of water and permeation of oil due to their superhydrophobic and superoleophilic property. They also showed better than 80% efficiency compared to other membranes for water-in-oil emulsions.	[51]

10.3 CARBON NANOTUBES FOR AIR POLLUTION SENSOR

Air pollution is a large-scale environmental challenge, which can be effectively monitored and tracked by using CNT-based environmental sensor by utilizing the electronic properties of CNTs. CNT has the potential to replace traditional carbon electrode sensors due to its exceptional properties like high electrical conductivities, mechanical stability, chemical stability, and functionalization abilities. CNTs are also found to be used in biosensing platform for the detection of microbial pathogens. Many studies revealed that the detection limit with CNT biosensor is even lower than the range of 10^{-18} to 10^{-21} M, which proves its use in precise applications (Table 10.4).

TABLE 10.4 Recent Developments of CNTs for Air Pollution Sensor.

Material	Air Pollutants	Comments	Refs.
SWNTs were decorated with metal nanoparticles	H_2, CH_4, CO, and H_2S gases	The fabricated-isolated SWNT networks on a single chip on to which various metals such as Pd, Pt, Rh, and Au are electroplated. These electronic small sensor arrays with low power are used for the detection of toxic gases for personal safety and air pollution monitoring.	[52]

TABLE 10.4 *(Continued)*

Material	Air Pollutants	Comments	Refs.
(CNTs) on Si_3N_4/Si substrates provided with Pt electrodes	NO_2	The prepared CNT/Si_3N_4/Si films with good electrical response with high reproducibility. They propose the sensor for the air-quality monitoring due to its low resistance value upon exposing to NO_2 gas and its response to even low concentrations of 10 ppb NO_2.	[35]
Antimony–carbon nanotube–tin oxide (Sb–CNT–SnO_2) thin film	Indoor air pollutants-formaldehydeammonia, and toluene	The authors of this work had performed the gas-sensing measurements of SnO_2, Sb–SnO_2, and Sb–CNT–SnO_2 thin films and found to have better sensitivity and fast response time for Sb–CNT–SnO_2 thin films. The better sensing property was mainly attributed due to the growth-guiding and energy-buffering effects caused by CNTs.	[53]
SWCNTs	Ozone gas	The fabricated sensor exhibits high sensitivity to ozone gas at concentration as low as 50 ppb and fast response time.	[54]
Pt- and Pd-nanocluster functonalized MWCNTs	NO_2, H_2S, NH_3, CO	The authors compared unfunctionalized MWCNTs and Pt and Pd-nanoclusters functionalized MWCNTs for various gas sensing and the later was to have better sensing performance and proposes it for the air pollution environmental monitoring.	[55]
SWCNTs/SnO_2	NO_2	Investigation was carried out on the blank SnO_2 and SWCNTs/SnO_2 sensors for the sensing of NO_2 gas and the later one was found to be a better sensor with high sensitivity and recovery property, which showed that doping with SWCNTs enhances the sensitivity of hybrid sensors.	[56]

TABLE 10.4 *(Continued)*

Material	Air Pollutants	Comments	Refs.
Indium-tin oxide(ITO)-SWCNT	NH_3	The prepared ITO-SWCNT composite exhibited high sensitivity toward ammonia gas and reached a detection limit of 13 ppb, which proves it to be a better candidate for environmental ammonia monitoring.	[57]

10.4 CARBON NANOTUBES FOR POLLUTION PREVENTION THROUGH BIOLOGICAL APPLICATIONS

Carbon nanotubes are resourcefully used for the preparation of biofuel cells, which are generally classified as enzymatic biofuel cells (EFC) and microbial fuel cells (MFC). In these fuel cells, electric power is generated due to bio-catalytic activity in which it is due to microbial catabolic activities and enzyme/protein catalysis for EFC and MFC, respectively. Due to their lower performance capacity, MFC are not used for much practical applications. Many studies are ongoing to improve the performance of MFC using CNT electrodes for improving power generation utilizing the CNTs having high conductivity and large surface area.

Short lifetime, less power production, and poor enzyme stabilities are the major concerns for EFC. The introduction of CNT into the field has brought good platform for enzyme stabilization due to its curvature, which, in turn, leads to more power generation and much longer lifetimes (Table 10.5).

TABLE 10.5 Recent Developments of CNTs as Biofuel Cells.

Material	Function	Comments	Refs.
Carbon fiber microelectrodes (CFMEs) modified with (SWNTs)	$Glucose/O_2$ biofuel cells	SWNTs were used as a support for steadily confining the methylene green (MG) electrocatalyst for the oxidation of NADH, for the oxidation of glucose (as anodic biocatalyst, that is, NAD^+-dependent glucose dehydrogenase, GDH) and for the reduction of O2 (cathodic biocatalyst (i.e., laccase)). They showed good bioelectrocatalytic activity toward the oxidation of glucose and reduction of oxygen, respectively.	[58]

TABLE 10.5 *(Continued)*

Material	Function	Comments	Refs.
CNTs	Glucose/ oxygen-CNT biofuel cell	Comparison of conventional carbon fiber fuel cell with CNT-microwire fuel cell was done and the later was found to have tenfold higher performances. Also, the power density of CNT biofuel cellare is found to be much higher than the conventional carbon-fiber biofuel cells under physiological conditions.	[59]
(SWCNTs) Covalently phenylated, naphthylated or terphenylated	Glucose oxidation	It was reported that the fully enzymatic fuel cell had power density of 40 μW/ cm^2 at 20 kΩ loading and the open-circuit potential for the biofuel cell was found to be 0.4 V.	[60]
CNTs)	Glucose oxidase (GOx)	The authors of this research had found that the good CNT dispersion and enzyme immobilization occurred simultaneously through sequential enzyme adsorption, precipitation, and crosslinking (EAPC) by directly adding MWCNTs onto an aqueous solution of GOx. They also established that EAPC-driven GOx activity was 4.5 and 11-times better than covalently attached GOx (CA) on acid-treated CNTs and simply adsorbed GOx (ADS) on intact CNTs, respectively.	[61]
Au-NPs/ functionalized-MWCNTs	Air-glucose fuel cell	It was concluded that better catalytic activities and stability for Au-NPs/f-MWCNTs compared to Au sheet with three-fold increment in the power density of air-glucose fuel cell having an open circuit voltage of ~1.3 V.	[62]

10.5 CARBON NANOTUBES FOR RENEWABLE ENERGY APPLICATIONS

Conventional energy sources like oil and natural gas are not enough to meet the long-term energy requirements. Instead we have to utilize renewable energy sources such as solar, wind, or biological resources. Of these renewable resources, the implications of CNTs are mostly achieved in solar cell applications. The current major issue with the currently available solar cells such as silicon-based and semiconducting material-based ones are its low performance ability, less material flexibility, and device cost. The electronic properties of CNTs are the best choice to address the current issues with the solar cell performance. Organic solar cells, organic–inorganic hybrid solar cells, and dye sensitized solar cells (DSSC) are now using CNTs as one of the component to improve its performance.

Photovoltaic devices harvest electricity through photons absorbed from the light energy of sun. CNTs allow high-photon absorption and the charge transfer phenomenon due to the presence of delocalized Π electron system (Table 10.6).

TABLE 10.6 Recent Developments of CNTs for Solar Cell Applications.

Material	Function	Comments	Refs.
$CH_3NH_3PbI_3$/ CNTs	Hole collector for perovskite solar cells	The fabricated semitransparent $CH_3NH_3PbI_3$/CNTs solar cells, due to their transparency, produce photovoltaic output with dual-side illuminations.	[63]
		They added spiro-OMeTAD to the CNT network and the composite electrode enhanced the efficiency to 9.90% owing to the superior hole extraction and reduced recombination in solar cells.	
Aligned CNT fibers	Catalytic counter electrode as conductive materials in DSSCs	The synthesized CNT-fibers solar cell have a power conversion efficiency of 2.94%, which is independent of incident light angle and cell length. They proposed that it can be woven into textiles by a suitable weaving technology.	[64]
CNT-PEDOT: PSS /n-Si	Hole-selective layers	The prepared CNT-PEDOT: PSS /n-Si is an excellent hole-selective layer in flexible solar cells in which CNT network will function as a carrier transport bridge between the PEDOT:PSS domains and also as stress centers preventing the breakage of transport path of carriers.	[65]
		The designed solar cell with such composite has a power conversion efficiency of 9.24% having an open circuit voltage of 576 mV.	

TABLE 10.6 *(Continued)*

Material	Function	Comments	Refs.
3D (CNT) sponge	Counter electrode for a DSSC	The developed CNT sponge with high flexibility and superior mechanical strength, showed high catalytic activity to the counter electrode reduction reactions of I^-/I_3^- redox couple and had a photoelectric conversion efficiency of 6.21%	[66]
CNT-Si	To construct heterojunction solar cells	The fabricated CNT-Si solar cells did gas doping of ozone without damaging the CNT structure. They found that the cell efficiency increased from 5.29% to 12.70%. The authors proposed that the ozone treatment donated holes to CNTs, which increased its work function and consequently its buit-in-potential and schottky barrier at the heterojunction for more-efficient charge separation.	[67]

10.6 CARBON NANOTUBES IN ELECTRONIC DEVICES

Supercapacitors are effective replacement for conventional batteries, which turn out mass amount of waste to the environment. Supercapacitors also have the advantage of less size, high-power capacity, long cycle life, and high-energy capacity compared to the traditional batteries. Studies are reported for CNTs as efficient electrode material for the supercapacitors (Table 10.7).

TABLE 10.7 Recent Developments of CNTs for Electronic Applications.

Material	Function	Highlights	Refs.
RuO_2/MWNT, TiO_2/MWNT, and SnO_2/MWNT	Electrodes	Comparison was done for pure MWNT electrode with TiO_2, and SnO_2-dispersed MWNT electrodes and found the later to have higher specific capacitance attributed to the pseudo capacitance of the nanocrystalline metal oxides dispersed on the functionalized MWNT. This investigation showed that TiO_2/MWNT was more cost-effective and novel nanocrystalline composite material for electrochemical double-layer capacitor.	[68]

TABLE 10.7 *(Continued)*

Material	Function	Highlights	Refs.
MWCNT	Electrodes	The authors had compared activated carbon with functionalized MWCNTs and later they were found to be a better supercapacitors due to their accessible mesopores and moderate specific-surface area. They varied the functionality of MWCNTs and found to have pure electrostatic attraction of ions and quick pseudofaradaic reactions. They fabricated super capacitors of varying capacitance from 4 to 135 F/g, depending on the type of nanotubes and their post treatments.	[69]
Sprayed networks of SWCNTs	As electrodes and charge collectors	The demonstrated printable thin-film super capacitors using aqueous gel electrolyte and organic liquid electrolyte with high energy and power densities (6 W h/kg for both electrolytes and 23 and 70 kW/kg for aqueous gel electrolyte and organic electrolyte, respectively) and having comparable performance with respect to other SWCNT-based supercapacitor devices made-up using altered methods.	[70]
RuO_2 NP/ carboxylated CNT	Electrodes	Well-dispersed RuO_2 nanoparticles were produced on carbon nanotubes with increased capacitance since the inner part of RuO_2 are highly accessed by the protons due to its decreased size and also due to its good dispersion.	[71]
(PANI/sMWCNTs)	Electrodes	The authors have compared PANI nanoroads and PANI/sMWCNT composite and the later was found to have a good cycling stability (below 10% capacity loss after 1000 cycles) and the specific capacitance of the PANI/sMWCNT composite was found to be a maximum value of 515.2 F/g.	[72]

10.7 CARBON NANOTUBES FOR POLYMER REINFORCEMENT

Use of green nanocomposites using biodegradable polymers is an efficient approach for waste management due to its degradability in the natural environment. The inconvenience associated with the use of biodegradable polymers is their poor mechanical performance and less thermal stability. CNTs are used as a better reinforcement material for these types of biodegradable polymers, which enhances their mechanical properties and thermal stabilities. The properties of the composites mainly depend on the good dispersion of CNTs in the polymer matrix, surface interactions between polymer and CNTs, and alignment of CNTs in the matrix (Table 10.8).

TABLE 10.8 Recent Developments of CNTs for Biopolymer Reinforcement.

Material	Function	Comments	Refs.
SWNTs/ poly(propylene fumarate) (PPF)	Bone tissue engineering scaffolds	The authors compared pure PPF and SWNT/PPF and found that the later have higher compressive modulus, flexural modulus, compressive offset yield strength, and flexural strength.	[73]
MWCNT-g-(PLLA) Poly (L-lactide)	Thermal stability	It has been found that the MWCNT-g-PLLAs were more thermally stable than those of neat PLLA by studying crystallinity of the composites through XRD analysis.	[74]
MWNT/ polylactic acid (PLA)	Flame retardancy	PLA ternary system based on sepiolite and MWNTs showed that the heat release capacity (HRC) reduces by 58%, which is an indicator of a materials fire hazard. The authors also showed that 45% reduction in peak heat release (PHR).	[75]
MWCNT/ poly(L-lactide-*co*-ε-caprolactone) (PLACL) composites	Physical and thermomechanical property enhancement	The surface functionalization of MWCNTs improved the dispersion and adhesion, which acted as reinforcing filler in the PLACL polymer matrix and improved the physical and thermomechanical properties of the nanocomposites.	[76]
Poly(lactic acid) (PLA)/(CNTs)/ poly(propylene carbonate) (PPC)	Mechanical and conducting properties	The authos had proved that the prepared PPC/PLA/CNT composites exhibited with better biodegradability and excellent conductivity compared to petroleum-based polymers such as low-density polyethylene, high density polyethylene, and polypropylene (PP); hence a better alternative to petroleum-based polymers, especially at room temperatures.	[77]

TABLE 10.8 *(Continued)*

Material	Function	Comments	Refs.
CNT/ Poly(lactide-co-glycolide) (PLGA)	Bone tissue engineering	The authors of this work had found that PLGA scaffolds containing water-dispersible MWCNTs exhibited high mechanical strength and good cellular and tissue compatibility and are possible candidates for bone tissue engineering.	[78]

10.8 CARBON NANOTUBES FOR LARGE-SCALE ENVIRONMENTAL APPLICATIONS

Large-scale productions of CNTs are limited by its production cost and compromised purity levels compared to laboratory synthesis. Nevertheless, recent literatures are available for the production of CNTs with minimal cost especially through catalytic chemical vapor deposition (CCVD) techniques. The low-cost production technique envisaged the interest toward large-scale production and commercialization of CNTs.[79]

Other than the cost and purity, the major concern is regarding its toxicity toward human health and environment. Large scale use of CNTs can cause an accumulation of CNTs in the environment and which may enter into the food cycle of humans. In-vivo studies show the cytotoxic effect of CNTs on the animal cells and organs, due to its chemical and physical nature.[80] Many researches are going on to minimize the cytotoxic effect of CNTs by various functionalization and dispersion methods.[81,82] We are optimistic that the best ways to avoid the toxic effect of CNTs will be explored soon and much extensive use of CNTs can be done in coming days.

10.9 CONCLUSIONS AND FUTURE OUTLOOK

The prominent green applications of CNTs are discussed in the chapter. Especially the usage of CNTs for water and air pollution treatment and monitoring by using CNTs as sorbents, filters, antimicrobial and antiviral agents, and gas-sensors are elucidated. The energy applications of CNTs by using them in batteries, supercapacitors, solar cells, hydrogen fuel cells, and biofuel cells have made them extremely significant. However, their industrial-level production is a challenge due to their cost, purity, and toxicity issues. The environmental implications are also a major exigent issue due to their toxic effect on human body and researches are going on to deal with this concern and we are sanguine that it will be addressed in near future, which would enable the large-scale production and usage of CNTs for various technological and environmental applications.

In future, for the commercialization of CNT products,[83] it is important to integrate the lab-scale production methods toward the existing manufacturing techniques without compromising its intrinsic nanoscale properties. Lithographic techniques can be a breakthrough in this direction, especially for electronic devices using CNTs. Many companies are investing on the production and sales on the CNT and CNT-based materials, although their manufacturing techniques are hidden. For the up-and-coming CNT applications, their production and sales are of much vital. Furthermore, the health and safety standards are of prime importance for the manufacturing and use of CNTs. The National Institute of Standards and Technology, IEEE, The Chinese government, the Bayer are the some institutes, which established some standards for the processing and use of CNTs.[84,85] It is of prime important to find the disposal and reuse/recycle measures of CNT when large-scale production and use were implemented. Industry, academia, and government should go hand-in-hand to resolve the environmental and social impact of CNTs during their lifecycle. Finally, we believe the properties and application of CNTs from surface chemistry to large-scale commercialization will contribute to new frontiers of nanotechnology in the years to come.

KEYWORDS

- **carbon nanotubes**
- **surface area**
- **functionalization**
- **physical properties**
- **green applications**

REFERENCES

1. Iijima, S. Helical Microtubules of Graphitic Carbon. *Nature* 1991, *354*, 56–58. doi:10.1038/354056a0.
2. Yaragalla, S.; Anilkumar, G.; Kalarikkal, N.; Thomas, S. Structural and Optical Properties of Functionalized Multi-Walled Carbon Nanotubes. *Mater. Sci. Semicond. Process*. 2016. doi:10.1016/j.mssp.2015.10.022.
3. Yaragalla, S.; Mishra, R.; Thomas, S.; Kalarikkal, N.; Maria, H. J;. Carbon-Based Nanofillers and Their Rubber Nanocomposites. 2018. doi:10.1016/C2016-0-03648-3.
4. Yaragalla, S.; Rajendran, R.; Jose, J.; Almaadeed, M. A.; Kalarikkal, N.; Thomas, S. Preparation and Characterization of Green Graphene Using Grape Seed Extract for Bioapplications. *Mater. Sci. Eng. C*. 2016. doi:10.1016/j.msec.2016.04.050.
5. Yaragalla, S.; Anilkumar, G.; Vineeshkumar, T. V.; Kalarikkal, N.; Thomas, S. Preparation of Epoxy Graphene and Its Structural and Optical Properties. *Adv. Mater. Lett*. 2015. doi:10.5185/amlett.2015.5914.
6. Yaragalla, S.; Rajendran, R.; AlMaadeed, M. A.; Kalarikkal, N.; Thomas, S. Chemical Modification of Graphene with Grape Seed Extract: Its Structural, Optical and Antimicrobial Properties. *Mater. Sci. Eng. C*. 2019. doi:10.1016/j.msec.2019.04.061.
7. Yaragalla, S.; Thomas, S.; Maria, H. J.; Mishra, R. K.; Kalarikkal, N. Carbon-Based Nanofiller and Their Rubber Nanocomposites. 2019. doi:10.1016/C2018-0-02522-0.
8. Yaragalla, S.; Meera, A. P.; Kalarikkal, N.; Thomas, S. Chemistry Associated with Natural Rubber-Graphene Nanocomposites and Its Effect on Physical and Structural Properties. *Indust. Crops Prod*. 2015. doi:10.1016/j.indcrop.2015.05.079.
9. Yaragalla, S.; Sarath Chandran, C.; Kalarikkal, N.; Subban, R. H. Y.; Chan, C. H.; Thomas, S. Effect of Reinforcement on the Barrier and Dielectric Properties of Epoxidized Natural Rubber-Graphene Nanocomposites. *Polym. Eng. Sci*. 2015. doi:10.1002/pen.24131.
10. Yaragalla, S.; Sindam, B.; Abraham, J.; Raju, K. C. J.; Kalarikkal, N.; Thomas, S. Fabrication of Graphite-Graphene-Ionic Liquid Modified Carbon Nanotubes Filled Natural Rubber Thin Films for Microwave and Energy Storage Applications. *J. Polym. Res*. 2015. doi:10.1007/s10965-015-0776-5.
11. Srinivasarao, Y.; Ri Hanum, Y. S.; Chan, C. H.; Nandakumar, K.; Sabu, T. Electrical Properties of Graphene Filled Natural Rubber Composites. *Adv. Mater. Res.*. 2013. doi:10.4028/www.scientific.net/AMR.812.263.
12. Ponnamma, D.; Jose Chirayil, C.; Sadasivuni, K. K.; Somasekharan, L.; Yaragalla, S.; Abraham, J.; Thomas, S. Special Purpose Elastomers: Synthesis, Structure-Property Relationship, Compounding, Processing and Applications. 2013. doi:10.1007/978-3-642-20925-3_3.
13. Yaragalla, S.; Chandran, N.; Chan, C. H.; Kalarikkal, N.; Thomas, S. A Review on Mechanical Properties of Semicrystalline/Amorphous Polymer Blends. In *Functional Polymeric Composites: Macro to Nanoscales*. 2017. doi:10.1201/9781315207452.
14. Yaragalla, S.; Bhavitha, K. B.; Kalarikkal, N.; Thomas, S. Novel Graphene-Based Polymer Nanocomposites. In *Handbook of Polymer and Ceramic Nanotechnology*. 2020. doi:10.1007/978-3-030-10614-0_29-1.

15. Ghasempour, R.; Narei, H. CNT Basics and Characteristics. In *Carbon Nanotube-Reinforced Polymers*; Elsevier, 2018; pp 1–24. doi:10.1016/B978-0-323-48221-9.00001-7.
16. Basu-Dutt, S.; Minus, M. L.; Jain, R.; Nepal, D.; Kumar, S. Chemistry of Carbon Nanotubes for Everyone. *J. Chem. Educ.* 2012, *89*, 221–229. doi:10.1021/ed1005163.
17. Su, M.; Zheng, B.; Liu, J. A Scalable CVD Method for the Synthesis of Single-Walled Carbon Nanotubes with High Catalyst Productivity. *Chem. Phys. Lett.* 2000, *322*, 321–326. doi:10.1016/S0009-2614(00)00422-X.
18. Ohashi, T. Carbon Nanotubes. In *Carbon Nanomaterials for Advanced Energy Systems*; John Wiley & Sons, Inc: Hoboken, NJ, 2015; pp 47–84. doi:10.1002/9781118980989.ch2.
19. Ying, L. S.; bin Mohd Salleh, M. A.; Mohamed Yusoff, H. b.; Abdul Rashid, S. B.; Abd. Razak, J. b. Continuous Production of Carbon Nanotubes—A Review. *J. Indust. Eng. Chem.* 2011, *17*, 367–376. doi:10.1016/j.jiec.2011.05.007.
20. Eschemann, T. O.; Lamme, W. S.; Manchester, R. L.; Parmentier, T. E.; Cognigni, A.; Rønning, M.; De Jong, K. P. Effect of Support Surface Treatment on the Synthesis, Structure, and Performance of Co/CNT Fischer-Tropsch Catalysts. *J. Catalysis.* 2015. doi:10.1016/j.jcat.2014.12.010.
21. Han, Y.; Li, S.; Chen, F.; Zhao, T. Multi-Scale Alignment Construction for Strong and Conductive Carbon Nanotube/Carbon Composites. *Mater. Today Commun.* 2016. doi:10.1016/j.mtcomm.2015.12.002.
22. Boccaccini, A. R.; Cho, J.; Roether, J. A.; Thomas, B. J. C.; Jane Minay, E.; Shaffer, M. S. P. Electrophoretic Deposition of Carbon Nanotubes, *Carbon.* 2006. doi:10.1016/j.carbon.2006.06.021.
23. Dervishi, E.; Li, Z.; Xu, Y.; Saini, V.; Biris, A. R.; Lupu, D.; Biris, A. S. Carbon Nanotubes Synthesis, Properties, and Applications. *Particulate Sci. Technol.* 2009, *27*, 107–125. doi:10.1080/02726350902775962.
24. Liu, P. Modifications of Carbon Nanotubes with Polymers. *Eur. Polym. J.* 2005, *41*, 2693–2703. doi:10.1016/j.eurpolymj.2005.05.017.
25. Battigelli, A.; Ménard-Moyon, C.; Da Ros, T.; Prato, M.; Bianco, A. Endowing Carbon Nanotubes with Biological and Biomedical Properties by Chemical Modifications. *Adv. Drug Deliv. Rev.* 2013, *65*, 1899–1920. doi:10.1016/j.addr.2013.07.006.
26. Liu, C. H.; Fan, S. S. Effects of Chemical Modifications on the Thermal Conductivity of Carbon Nanotube Composites. *Appl. Phys. Lett.* 2005, *86*, 123106. doi:10.1063/1.1887839.
27. Schnorr, J. M.; Swager, T. M. Emerging Applications of Carbon Nanotubes †. *Chem. Mater.* 2011, *23*, 646–657. doi:10.1021/cm102406h.
28. Singh, S.; Vardharajula, S.; Tiwari, P.; Eroğlu, E.; Vig, K.; Dennis, V.; Ali. Functionalized Carbon Nanotubes: Biomedical Applications. *Int. J. Nanomed.* 2012, 5361. doi:10.2147/IJN.S35832.
29. Pan, B.; Xing, B. Adsorption Mechanisms of Organic Chemicals on Carbon Nanotubes. *Environ. Sci. Technol.* 2008, *42*, 9005–9013. doi:10.1021/es801777n.
30. Sarkar, B.; Mandal, S.; Tsang, Y. F.; Kumar, P.; Kim, K.-H.; Ok, Y. S. Designer Carbon Nanotubes for Contaminant Removal in Water and Wastewater: A Critical Review. *Sci. Total Environ.* 2018, *612*, 561–581. doi:10.1016/j.scitotenv.2017.08.132.

31. Liu, C.; Cheng, H. M. Carbon Nanotubes for Clean Energy Applications. *J. Phys. D Appl. Phys*. 2005. doi:10.1088/0022-3727/38/14/R01.
32. Elhissi, A. M. A.; Ahmed, W.; Hassan, I. U.; Dhanak, V. R.; D'Emanuele, A. Carbon Nanotubes in Cancer Therapy and Drug Delivery. *J. Drug Deliv*. 2012, *2012*, 1–10. doi:10.1155/2012/837327.
33. Li, Y.-H.; Wang, S.; Cao, A.; Zhao, D.; Zhang, X.; Xu, C.; Luan, Z.; Ruan, D.; Liang, J.; Wu, D.; Wei, B. Adsorption of Fluoride from Water by amorphous alumina supported on Carbon Nanotubes. *Chem. Phys. Lett*. 2001, *350*, 412–416. doi:10.1016/S0009-2614(01)01351-3.
34. Lu, C.; Chiu, H. Adsorption of Zinc(II) from Water with Purified Carbon Nanotubes. *Chem. Eng. Sci*. 2006, *61*, 1138–1145. doi:10.1016/j.ces.2005.08.007.
35. Cantalini, C.; Valentini, L.; Armentano, I.; Kenny, J. M.; Lozzi, L.; Santucci, S. Carbon Nanotubes as New Materials for Gas Sensing Applications. *J. Eur. Ceram. Soc*. 2004, *24*, 1405–1408. doi:10.1016/S0955-221903,00441-2.
36. Xie, X. L.; Mai, Y. W.; Zhou, X. P. Dispersion and Alignment of Carbon Nanotubes in Polymer Matrix: A Review. *Mater. Sci. Eng. R Rep*. 2005. doi:10.1016/j.mser.2005.04.002.
37. Ong, Y. T.; Ahmad, A. L.; Zein, S. H. S.; Tan, S. H. A Review on Carbon Nanotubes in an Environmental Protection and Green Engineering Perspective. *Braz. J. Chem. Eng*. 2010, *27*, 227–242. doi:10.1590/S0104-66322010000200002.
38. Latrous El Atrache, L.; Hachani, M.; Kefi, B. B. Carbon Nanotubes as Solid-Phase Extraction Sorbents for the Extraction of Carbamate Insecticides from Environmental Waters. *Int. J. Environ. Sci. Technol*. 2016, *13*, 201–208. doi:10.1007/s13762-015-0861-7.
39. Elshafiee, C.; abd elsalam, M.; Moalla, S.; Ali, H.; Osman, D.; Abdalla, R.; Moustafa, Y. Carbon Nanotubes as Superior Sorbent for Removal of Phenol from Industrial Waste Water. *Egypt. J. Chem*. 2017, 0–0. doi:10.21608/ejchem.2017.1756.1149.
40. Peter, K. T.; Vargo, J. D.; Rupasinghe, T. P.; De Jesus, A.; Tivanski, A. V.; Sander, E. A.; Myung, N. V.; Cwiertny, D. M. Synthesis, Optimization, and Performance Demonstration of Electrospun Carbon Nanofiber–Carbon Nanotube Composite Sorbents for Point-of-Use Water Treatment. *ACS Appl. Mater. Interf*. 2016, *8*, 11431–11440. doi:10.1021/acsami.6b01253.
41. Wu, J.; An, A. K.; Guo, J.; Lee, E.-J.; Farid, M. U.; S. Jeong, S. CNTs Reinforced Super-Hydrophobic-Oleophilic Electrospun Polystyrene Oil Sorbent for Enhanced Sorption Capacity and Reusability. *Chem. Eng. J*. 2017, *314*, 526–536. doi:10.1016/j.cej.2016.12.010.
42. Shawky, H. A.; El-Aassar, A. H. M.; Abo-Zeid, D. E. Chitosan/Carbon Nanotube Composite Beads: Preparation, Characterization, and Cost Evaluation for Mercury Removal from Wastewater of Some Industrial Cities in Egypt. *J. Appl. Polym. Sci*. 2012, *125*, E93–E101. doi:10.1002/app.35628.
43. Inyang, M.; Gao, B.; Zimmerman, A.; Zhang, M.; Chen, H. Synthesis, Characterization, and Dye Sorption Ability of Carbon Nanotube–Biochar Nanocomposites. *Chem. Eng. J*. 2014, *236*, 39–46. doi:10.1016/j.cej.2013.09.074.
44. Addo Ntim, S.; Mitra, S. Adsorption of Arsenic on Multiwall Carbon Nanotube-Zirconia Nanohybrid for Potential Drinking Water Purification. *J. Colloid Interf. Sci*. 2012. doi:10.1016/j.jcis.2012.01.063.

45. Kanel, S. R.; Misak, H.; Nepal, D.; Mall, S.; Brittle, W.; Sizemore, I.; Kempisty, D. M.; Goltz, M. N. The Use of Carbon Nanotube Yarn as a Filter Medium to Treat Nitroaromatic-Contaminated Water. *N. Carbon Mater*. 2016, *31*, 415–423. doi:10.1016/S1872-580516,60021-5.
46. Mauter, M. S.; Elimelech, M. Environmental Applications of Carbon-Based Nanomaterials. *Environ. Sci. Technol*. 2008, *42*, 5843–5859. doi:10.1021/es8006904.
47. Li, H.; Gui, X.; Zhang, L.; Wang, S.; Ji, C.; Wei, J.; Wang, K.; Zhu, H.; Wu, D.; Cao, A. Carbon Nanotube Sponge Filters for Trapping Nanoparticles and Dye Molecules from Water. *Chem. Commun*. 2010, *46*, 7966. doi:10.1039/c0cc03290e.
48. Oza, G.; Pandey, S.; Gupta, A.; Shinde, S.; Mewada, A.; Jagadale, P.; Sharon, M.; Sharon, M. Photocatalysis-Assisted Water Filtration: Using TiO_2-Coated Vertically Aligned Multi-Walled Carbon Nanotube Array for Removal of *Escherichia coli* O157:H7. *Mater. Sci. Eng.* C. 2013.
49. Vecitis, C. D.; Schnoor, M. H.; Rahaman, M. S.; Schiffman, J. D.; Elimelech, M. Electrochemical Multiwalled Carbon Nanotube Filter for Viral and Bacterial Removal and Inactivation. *Environ. Scie. Technol*. 2011, *45*, 3672–3679. doi:10.1021/es2000062.
50. Brady-Estévez, A. S.; Schnoor, M. H.; Vecitis, C. D.; Saleh, N. B.; Elimelech, M. Multiwalled Carbon Nanotube Filter: Improving Viral Removal at Low Pressure. *Langmuir*. 2010, *26*, 14975–14982. doi:10.1021/la102783v.
51. Lee, C. H.; Johnson, N.; Drelich, J.; Yap, Y. K. The Performance of Superhydrophobic and Superoleophilic Carbon Nanotube Meshes in Water–Oil Filtration. *Carbon* 2011, *49*, 669–676. doi:10.1016/j.carbon.2010.10.016.
52. Star, A.; Joshi, V.; Skarupo, S.; Thomas, D.; Gabriel, J.-C. P. Gas Sensor Array Based on Metal-Decorated Carbon Nanotubes. *J. Phys. Chem B* 2006, *110*, 21014–21020. doi:10.1021/jp064371z.
53. Liu, J.; Guo, Z.; Meng, F.; Jia, Y.; Liu, J. A Novel Antimony−Carbon Nanotube−Tin Oxide Thin Film: Carbon Nanotubes as Growth Guider and Energy Buffer. Application for Indoor Air Pollutants Gas Sensor. *J. Phys. Chem. C* 2008, *112*, 6119–6125. doi:10.1021/jp712034w.
54. Park, Y.; Dong, K.-Y.; Lee, J.; Choi, J.; Bae, G.-N.; Ju, B.-K. Development of an Ozone Gas Sensor Using Single-Walled Carbon Nanotubes. *Sens. Actuat. B Chem* 2009, *140*, 407–411. doi:10.1016/j.snb.2009.04.055.
55. Penza, M.; Rossi, R.; Alvisi, M.; Cassano, G.; Signore, M. A.; Serra, E.; Giorgi, R. Pt- and Pd-Nanoclusters Functionalized Carbon Nanotubes Networked Films for Sub-Ppm Gas Sensors. *Sens. Actuat. B Chem*. 2008, *135*, 289–297. doi:10.1016/j.snb.2008.08.024.
56. Wei, B.-Y.; Hsu, M.-C.; Su, P.-G.; Lin, H.-M.; Wu, R.-J.; Lai, H.-J. A Novel SnO_2 Gas Sensor Doped with Carbon Nanotubes Operating at Room Temperature. *Sens. Actuat. B Chem*. 2004, *101*, 81–89. doi:10.1016/j.snb.2004.02.028.
57. Rigoni, F.; Drera, G.; Pagliara, S.; Goldoni, A.; Sangaletti, L. High Sensitivity, Moisture Selective, Ammonia Gas Sensors Based on Single-Walled Carbon Nanotubes Functionalized with Indium Tin Oxide Nanoparticles. *Carbon* 2014, *80*, 356–363. doi:10.1016/j.carbon.2014.08.074.
58. Li, X.; Zhou, H.; Yu, P.; Su, L.; Ohsaka, T.; Mao, L. A Miniature Glucose/O_2 Biofuel Cell with Single-Walled Carbon Nanotubes-Modified Carbon Fiber Microelectrodes

as the Substrate. *Electrochem. Commun.* 2008, *10*, 851–854. doi:10.1016/j.elecom.2008.03.019.

59. Gao, F.; Viry, L.; Maugey, M.; Poulin, P.; Mano, N. Engineering Hybrid Nanotube Wires for High-Power Biofuel Cells. *Nat. Commun.* 2010, *1*, 2. doi:10.1038/ncomms1000.
60. Stolarczyk, K.; Łyp, D.; Żelechowska, K.; Biernat, J. F.; Rogalski, J.; Bilewicz, R. Arylated Carbon Nanotubes for Biobatteries and Biofuel Cells. *Electrochim. Acta* 2012, *79*, 74–81. doi:10.1016/j.electacta.2012.06.050.
61. Kim, B. C.; Lee, I.; Kwon, S.-J.; Wee, Y.; Kwon, K. Y.; Jeon, C.; An, H. J.; Jung, H.-T.; Ha, S.; Dordick, J. S.; Kim, J. Fabrication of Enzyme-Based Coatings on Intact Multi-Walled Carbon Nanotubes as Highly Effective Electrodes in Biofuel Cells. *Sci. Rep.* 2017, *7*, 40202. doi:10.1038/srep40202.
62. Naruse, J.; Hoa, L. Q.; Sugano, Y.; Ikeuchi, T.; Yoshikawa, H.; Saito, M.; Tamiya, E. Development of Biofuel Cells Based on Gold Nanoparticle Decorated Multi-Walled Carbon Nanotubes. *Biosens. Bioelectron.* 2011, *30*, 204–210. doi:10.1016/j.bios.2011.09.012.
63. Li, Z.; Kulkarni, S. A.; Boix, P. P.; Shi, E.; Cao, A.; Fu, K.; Batabyal, S. K.; Zhang, J.; Xiong, Q.; Wong, L. H.; Mathews, N.; Mhaisalkar, S. G. Laminated Carbon Nanotube Networks for Metal Electrode-Free Efficient Perovskite Solar Cells. *ACS Nano* 2014, *8*, 6797–6804. doi:10.1021/nn501096h.
64. Chen, T.; Qiu, L.; Cai, Z.; Gong, F.; Yang, Z.; Wang, Z.; Peng, H. Intertwined Aligned Carbon Nanotube Fiber Based Dye-Sensitized Solar Cells. *Nano Lett.* 2012, *12*, 2568–2572. doi:10.1021/nl300799d.
65. Kurias, K. M.; Jasna, M.; Menon, M. R. R.; Antony, A.; Jayaraj, M. K. Fabrication of CNT-PEDOT:PSS/Si Heterojunction Carrier Selective Solar Cell. In *AIP Conference Proceedings*, 2019; p 050008. doi:10.1063/1.5093868.
66. Chen, J.; Meng, F.; Gui, X.; Sun, H.; Zeng, Z.; Li, Z.; Zhou, Y.; Tang, Z. The Application of a Three Dimensional CNT-Sponge as the Counter Electrode for Dye-Sensitized Solar Cells. *Carbon* 2012, *50*, 5624–5627. doi:10.1016/j.carbon.2012.08.002.
67. Zhao, X.; Wu, H.; Yang, L.; Wu, Y.; Sun, Y.; Shang, Y.; Cao, A. High Efficiency CNT-Si Heterojunction Solar Cells by Dry Gas Doping. *Carbon* 2019, *147*, 164–171. doi:10.1016/j.carbon.2019.02.078.
68. Reddy, A. L. M.; Ramaprabhu, S. Nanocrystalline Metal Oxides Dispersed Multiwalled Carbon Nanotubes as Supercapacitor Electrodes. *J. Phys. Chem. C* 2007, 111, 7727–7734. doi:10.1021/jp069006m.
69. Frackowiak, E.; Metenier, K.; Bertagna, V.; Beguin, F. Supercapacitor Electrodes from Multiwalled Carbon Nanotubes. *Appl. Phys. Lett.* 2000, *77*, 2421–2423. doi:10.1063/1.1290146.
70. Kaempgen, M.; Chan, C. K.; Ma, J.; Cui, Y.; Gruner, G. Printable Thin Film Supercapacitors Using Single-Walled Carbon Nanotubes. *Nano Lett.* 2009, *9*, 1872–1876. doi:10.1021/nl8038579.
71. Kim, Y.-T.; Tadai, K.; Mitani, T. Highly Dispersed Ruthenium Oxide Nanoparticles on Carboxylated Carbon Nanotubes for Supercapacitor Electrode Materials. *J. Mater. Chem.* 2005, *15*, 4914. doi:10.1039/b511869g.
72. Zhu, Z.; Wang, G.; Sun, M.; Li, X.; Li, C. Fabrication and Electrochemical Characterization of Polyaniline Nanorods Modified with Sulfonated Carbon

Nanotubes for Supercapacitor Applications. *Electrochimica Acta.* 2011, *56*, 1366–1372. doi:10.1016/j.electacta.2010.10.070.
73. Shi, X.; Hudson, J. L.; Spicer, P. P.; Tour, J. M.; Krishnamoorti, R.; Mikos, A. G. Injectable Nanocomposites of Single-Walled Carbon Nanotubes and Biodegradable Polymers for Bone Tissue Engineering. *Biomacromolecules* 2006, *7*, 2237–2242. doi:10.1021/bm060391v.
74. Amirian, M.; Chakoli, A. N.; Cai, W.; Sui, J. Effect of Functionalized Multiwalled Carbon Nanotubes on Thermal Stability of Poly L-Lactide, Biodegradable Polymer. *Sci. Iranica.* 2013. doi:10.1016/j.scient.2013.05.019.
75. Hapuarachchi, T. D.; Peijs, T. Multiwalled Carbon Nanotubes and Sepiolite Nanoclays as Flame Retardants for Polylactide and Its Natural Fibre Reinforced Composites. *Compos. Part A Appl. Sci. Manuf.* 2010, *41*, 954–963. doi:10.1016/j.compositesa.2010.03.004.
76. Chakoli, A. N.; Sui, J.; Amirian, M.; Cai, W. Crystallinity of Biodegradable Polymers Reinforced with Functionalized Carbon Nanotubes. *J. Polym. Res.* 2011, *18*, 1249–1259. doi:10.1007/s10965-010-9527-9.
77. Yang, G.; Geng, C.; Su, J.; Yao, W.; Zhang, Q.; Fu, Q. Property Reinforcement of Poly(Propylene Carbonate) by Simultaneous Incorporation of Poly(Lactic Acid) and Multiwalled Carbon Nanotubes. *Compos. Sci. Technol.* 2013, *87*, 196–203. doi:10.1016/j.compscitech.2013.08.010.
78. Mikael, P. E.; Amini, A. R.; Basu, J.; Josefina Arellano-Jimenez, M.; Laurencin, C. T.; Sanders, M. M.; Barry Carter, C.; Nukavarapu, S. P. Functionalized Carbon Nanotube Reinforced Scaffolds for Bone Regenerative Engineering: Fabrication, in Vitro and in Vivo Evaluation. *Biomed. Mater.* 2014, *9*, 035001. doi:10.1088/1748-6041/9/3/035001.
79. Colomer, J.-F.; Stephan, C.; Lefrant, S.; Van Tendeloo, G.; Willems, I.; Kónya, Z.; Fonseca, A.; Laurent, C.; Nagy, J. Large-Scale Synthesis of Single-Wall Carbon Nanotubes by Catalytic Chemical Vapor Deposition (CCVD) Method. *Chem. Phys. Lett.* 2000, *317*, 83–89. doi:10.1016/S0009-261499,01338-X.
80. Lam, C.-W. Pulmonary Toxicity of Single-Wall Carbon Nanotubes in Mice 7 and 90 Days After Intratracheal Instillation. *Toxicol. Sci.* 2003, *77*, 126–134. doi:10.1093/toxsci/kfg243.
81. Wick, P.; Manser, P.; Limbach, P.; Dettlaffweglikowska, U.; Krumeich, F.; Roth, S.; Stark, W.; Bruinink, A. The Degree and Kind of Agglomeration Affect Carbon Nanotube Cytotoxicity. *Toxicol. Lett.* 2007, *168*, 121–131. doi:10.1016/j.toxlet.2006.08.019.
82. Reijnders, L. Cleaner Nanotechnology and Hazard Reduction of Manufactured Nanoparticles. *J. Cleaner Prod.* 2006, *14*, 124–133. doi:10.1016/j.jclepro.2005.03.018.
83. De Volder, M. F. L.; Tawfick, S. H.; Baughman, R. H.; Hart, A. J. Carbon Nanotubes: Present and Future Commercial Applications, *Science* 2013, *339*, 535–539. doi:10.1126/science.1222453.
84. Zhang, Q.; Huang, J.; Zhao, M.; Qian, W.; Wei, F. Carbon Nanotube Mass Production: Principles and Processes. *ChemSusChem* 2011, *4*, 864–889.
85. Pauluhn, J. Multi-Walled Carbon Nanotubes (Baytubes®): Approach for Derivation of Occupational Exposure Limit. *Regulat. Toxicol. Pharm.* 2010, *57*, 78–89.

CHAPTER 11

CARBON NANOTUBE COMPOSITES FOR AEROSPACE APPLICATIONS

ARUNIMA REGHUNADHAN[1,*], ARUNI SHAJKUMAR[2], JIJI ABRAHAM[3], and NIMITHA K.C.[3]

[1]*TKM College of Engineering, Karicode, Kollam, Kerala-691005*

[2]*Laboratory for Advanced Research in Polymer Materials, Bhubaneswar, India*

[3]*Department of Chemistry, Vimala College, Thrissur, Kerala, India*

[*]*Corresponding author. E-mail: arunimarenjith02@gmail.com*

ABSTRACT

The wonder material carbon nanotubes (CNTs) have been utilized in many fields of application which include conducting materials, polymer composites, sensing applications, space applications, etc. These materials have been employed in various types of materials for the aerospace industry and research. The lightweight nature, versatility in properties, especially the high mechanical strength and thermal properties make them included in rotorcrafts, in propellants, in sensing applications, and in the manufacturing of components. This chapter is a brief discussion on the applications of CNT and the composites based on them for various aerospace applications.

11.1 CARBON NANOTUBES AND THEIR PROPERTIES: A BRIEF INTRODUCTION

11.1.1 INTRODUCTION

Carbon nanotube (CNT), one of the allotrope of carbon was discovered by the Japanese Scientist Iijima in 1991. It is a one-dimensional nanomaterial with tubular structure in which carbon atom is sp^2 hybridized with honeycomb atomic arrangement. The length of the CNT ranges to several micrometers and the dimeter is in the nanometer range. So the length-to-dimeter ratio or aspect ratio is high for CNT. CNTs can be classified into single-walled CNT (SWCNT), double-walled CNT (DWCNT) and multiwalled CNT (MWCNT) depending on the graphene sheet rolled-up into seamless tube. The spacing between adjacent layers is about ~0.34 nm.[16] The main routes for the synthesis of CNT include chemical vapor deposition (CVD), arc discharge, laser ablation, catalytic growth, and gas phase processes, such as carbon monoxide disproportionation, etc.[8]

11.1.2 PROPERTIES

CNTs have got unique properties. These materials have inherent extra ordinary electrical, optical, electronic, thermal and mechanical (high young's modulus, 1 TPa and tensile strength, 100 GPa) characteristics. Chemical stability and ultralightweight have made CNTs as an attractive candidate for various applications since from its discovery in 1991. The chirality and diameter of CNT have a significant role in determining their metallic or semiconducting behavior. MWNTs are usually metallic and can carry currents of up to 10^9 A/cm^2 and their thermal conductivity is about 3500 W/m/K.[24]

11.1.3 PURIFICATION AND FUNCTIONALIZATION OF CNT

CNTs possess many excellent properties. But still, there are some problems associated with CNTs which include impurities in synthesized CNT, processing difficulty, and formation of agglomeration. Several methods including physical separation, chemical oxidation, and combinations of chemical and physical techniques have been adopted for achieving CNTs with preferred purity.[18] Processing difficulty and formation of agglomeration can be rectified by the proper surface functionalization of CNT either by a covalent or by a noncovalent approach. In covalent approach, a covalent bond is formed between incoming group and the CNT surface. A change occurs in the carbon atoms in the hybridization of CNT from sp^2 to sp^3. This disrupts the inherent electrical conductivity of CNT. On the other hand, in noncovalent approach, there exists only secondary interaction between CNT and the incoming group. So the inherent electrical conductivity of CNT is preserved. Covalent methods include thermally activated chemical functionalization, chemical oxidation, electrochemical modification of nanotubes and photochemical functionalization.[4] The noncovalent route for the functionalization of CNT is by the use of polymers, surfactants, ionic liquid, etc.

11.1.4 GENERAL APPLICATIONS

CNTs offer a plethora of applications in many different divisions of basic and applied science. Major fields of application include composite materials, field emitters, catalyst and catalyst supports, sensors and actuators, thermal conductors, thermal interface materials, tips for scanning probe microscopy, conductive films, bionanomaterials, energy storage and conversion materials (battery, supercapacitor, and solar cell) and nanoelectronic devices (transistors), nanofiltration, artificial muscles, drug and gene delivery, Tissue engineering.[9]

11.2 AERONAUTICS AND AEROSPACE APPLICATIONS OF CARBON MATERIALS: AN OVERVIEW

Carbon-based composite materials are widely used in air craft industry because of their unique properties, such as lightweight, high mechanical stability, excellent shock absorbing fracture toughness and impact resistance, easy processability and shaping options, and resistance to numerous aggressive conditions (corrosion, flame, moisture, etc.). Carbon fiber reinforced polymer (CFRP) composites cane be used for the production of structural component of aircraft especially fuselage components and panels of airplane.[14]

11.3 POTENTIAL ROLE OF CARBON NANOTUBES IN AEROSPACE APPLICATIONS

Requirements of the next-generation aircraft, rotorcraft, unmanned aerial vehicles, and missiles include lightweight, visual and thermal signature, increased speed, and man euverability. These requirements can be fulfilled using some advanced functional materials especially nanomaterials. Among various nanomaterials, CNTs are found to be an ideal candidate to meet these requirements. CNTs have so many additional advantages over traditional aerospace materials. Table 11.1 presents an outline of the properties of CNTs in comparison with traditional aerospace materials .[6] CNTs can be used in several components of the airplane. CNT incorporated polymer composites were employed as the primary airframe material on four major aircraft structures, such as the Boeing 747-400, Boeing 757-200, Airbus A320, and Embraer E145.[34]

CNTs can be used as a substituent for copper wiring thereby reducing the weight dramatically. Scientists observed a 69% weight saving using CNT instead of copper wire. CNTs are identified as propellant additives for the next-generation chemical propulsion systems.[28] Several Studies have been undertaken by researchers where CNTs have been utilized for lightning strike protection.[48] Another important application of CNT in aircraft industry is its use as a sensor.[19]

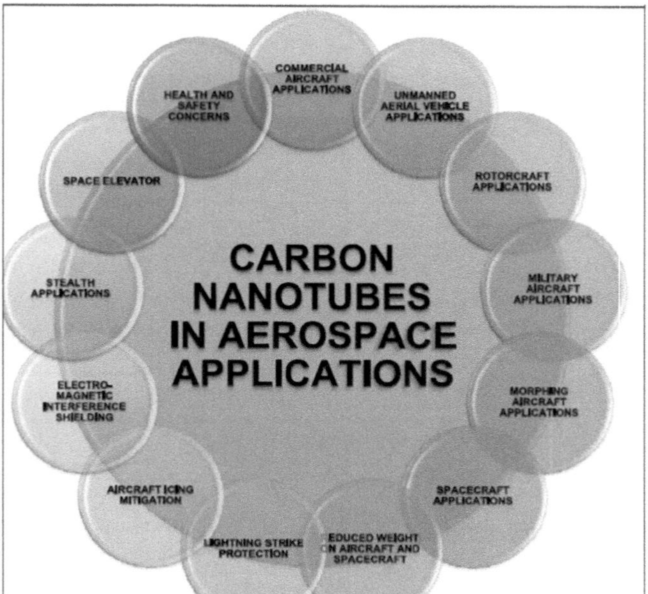

FIGURE 11.1 An overview of aerospace applications of CNTs.

Source: Reproduced with permission from Ref. [15]. © 2014 Elsevier.

CNTs-reinforced epoxy composites were used for structural health monitoring in aeronautic structural parts. Both AC and DC conductivities are measured for various levels of strain loaded in both axial tension and flexural mode. It was found that the electrical resistance or impedance of the nanocomposites samples is intensely affected by the applied mechanical stress.[46] Figure 11.1 represents an overview of aerospace applications of CNTs.[13] Expected CNT employment in aeronautics and astronautics are identified for commercial aircraft, military aircraft, rotorcraft, unmanned aerial vehicles, satellites, and space launch vehicles.

TABLE 11.1 A Comparison of CNT Properties with Traditional Aerospace Materials. The Symbols *and # Refer to Theoretical and Measured Values Respectively.

Material	Specific gravity (g/cm³)	Yield strength (GPa)	Elastic modulus (GPa)	Thermal conductivity (Wm⁻¹/K)	Electrical resistivity (μΩ cm)	Normalized strength-to-mass ratio
SWCNT*	1.4	65	1000	6000	30–100	225
SWCNT#	1.4	1.8	80	150	150	7
Conventional carbon fiber,M55J)	2.2	4	550	70	800	9
IM7 carbon composites	1.6	2.1	152	30	2000	7
Titanium	4.5	0.9	103	12	127	1
Aluminum	2.7	0.5	69	180	4.3	1

11.3.1 CARBON NANOTUBES AND THEIR COMPOSITES IN AIRCRAFTS

Composite structure owing to its lightweight, high strength-to-weight ratio, high-shock absorbance etc. is used in aircrafts to reduce weight which in turn reduces CO_2 emission and improved engine performance. Composites based on carbon are widely used in the production of aircraft parts. Incorporation of nanofillers like CNT into a polymer matrix enhances the electrical conductivity of the composite which can be used as laminates in aircrafts.[31] Another use of CNT-based composites in aircraft is to use it as a lightning strike protector (LSP). To improve the lightning strike resistance, Xia and co-workers modified the CNT film composites with silver.[49] This composite assembly exhibited enhanced electrical conductivity. In a similar work done by Chu and his co-workers[7] involved the development of a silver nanoparticle with sesame-cookie topography-modified carbon nanotube paper (SMCNP). This structural assembly has demonstrated enhanced LSP efficiency.

Another approach is to prepare a multilayer structure where the first layer is composed of a composite of CNTs and silicone.[2] The second layer composed of a PTC heater which is a composite of carbon black and polymer. This assembly not only possess the improved power output efficiency but also the power to sustain different temperature variations on the wing surface thereby preventing overheating.

CNT/carbon (CNT/C) composites have a lot of applications in military and aerospace due to their lightweight, high strength, and excellent conductivity. Zhang et al. modified the composite using pyrolyzed polydopamine to act as an interface between CNTs and carbon matrix. This modification has improved its thermal properties, load transfer, and electron transport.[57]

11.3.2 CARBON NANOTUBES AND THEIR COMPOSITES IN EMI SHIELDING

Electromagnetic interference or EMI is an unwanted issue that mainly happens in remotely piloted aircraft controls, particularly in the radio frequency band emitted by the motor and power supplies. This causes negative impact on human health as well as on the functionality of electronic devices. Therefore, this interference need to be shielded (EMI shielding) to avoid interruptions in communication signals. Initially, conductive paints, rubbers with dispersed metallic or carbon black particles widely categorized as radar absorbing materials (RAM) was used for EMI shielding particularly in military aircrafts which enabled them to operate in stealth mode, that is, it can operate with less detectability metallic structures are commonly used for EMI shielding in passenger aircraft and remotely piloted aircraft system (RPAS). However, the metallic structures possess certain drawbacks such as heavy weight, corrosion susceptibility, weak flexibility, etc. Thus, a need for lightweight, corrosion resistant, and flexible materials are increasing in demand.

Although metallic foams were one solution to this problem, its lack of efficiency resulted in the development of composites which composed of lightweight polymers and effective fillers, such as nanofibers, flakes, sheets and films of carbon nanostructures, and CNTs.

Wang and co-workers[47] developed a 3D nanostructure based on CNTs. In this work, carbon nanospheres were grown on a 3D CNT hybrid architecture resulting in a CNT-necklace-like hybrid 3D structure. This structure possessed lightweight and conductivity, and demonstrated an excellent magnetic characteristic which can be used for electromagnetic shielding in aircrafts. Segregated conductive structures in a polymer matrix can enhance the electrical and EMI shielding of the composite. Feng and co-workers[11] have studied a polyetherimide/CNT (PEI/CNT) composite with a segregated structure for its EMI shielding capability. They have synthesized conductive CNT-coated PEI granules via a microwave-assisted synthesis. The coated CNT layer under irradiation exhibited localized heating. This structure has demonstrated excellent performance for EMI-shielding applications. The same group has also studied the EMI-shielding effect of segregated thermoplastic polyurethane/CNT composite synthesized via microwave selective sintering.[12]

Combining electrical conductive performance and lightweight by the preparation of porous CNT and reduced graphene oxide foam composite exhibited an excellent performance in EMI-shielding effectiveness (547 dB cm^3/g).[27] The EMI-shielding properties of CNTs modified with ferromagnetic materials has been explored. They encapsulated Fe@CNT composites in silicon carbide (SiC) which exhibited adequate EMI shielding at high temperature resistance.[32] CNT and Fe nanoparticle decorated porous carbon/graphene foam composite has been studied for its electromagnetic interference shielding performance.[52] This composite has also exhibited ultraviolet shielding. Zeng and co-workers synthesized a flexible PVDF/CNT/Ni@CNT composites with chain-like Ni@CNTs which possess excellent electrical conductivity and electromagnetic interference shielding.[54] Joseph et al. prepared composites of PMMA/Multilayered Graphene/ MWCNT and PVC/multilayered graphene/MWCNT hybrid nanocomposites. The multilayered structure prepared from this hybrid nanocomposite exhibited good EMI-shielding capacity.[21]

CNTs incorporated in ultrahigh performance concrete have exhibited good mechanical properties and electromagnetic shielding capacity.[23] Bagotia and co-workers prepared a composite based on polycarbonate/ethylene methyl acrylate with a graphene:MWCNT hybrid filler in different ratios.[3] The composite with a combination of graphene and MWCNT in the ratio 1:3 exhibited good mechanical properties, electrical conductivity, and EMI shielding. A thin film structure made of ultrathin CNT networks and CNT–epoxy nanocomposite exhibited high conductivity, excellent adherence to both polymer and metals and a tunable EMI-shielding efficacy.[25] This structural composition is a potential application in EMI shielding. A similar three-layered sandwich structure prepared using poly(vinyledene fluoride) (PVDF), graphene nanoplatelets, nickel, and CNT showed excellent EMI-shielding capability due to the presence of multiple interfaces of the multilayered structure.[35]

Nanofiller reinforced carbon–matrix nanocomposites exhibit lightweight, high-specific strength, and electromagnetic interference shielding efficacy at extreme temperature conditions. Feng and co-workers prepared a 3D core-shell structure with CNT@graphene nanofiller incorporated to polycarbon matrix.[13] This structure has promising applications in EMI shielding.

11.3.3 CARBON NANOTUBE-BASED PROPELLANTS

Porous carbon-based materials possess excellent catalytic properties due to their high surface area and therefore can act as an excellent propellant. MWCNTs possess high surface area and good catalytic properties, hence they can be used as a carrier for energy components. Elbausney and co-workers coated MWCNTs with copper particles (CuO) and encapsulated into ammonium perchlorates (APC) oxidizer to be used as an excellent propellant. Bao et al. incorporated MWCNTs and nano/micro particles of aluminum powder to electrically controlled solid propellant such as hydroxyl ammonium nitrate.[5,10] This assembly enhanced the electrical conductivity and energy release. Hassan et al. demonstrated the preparation of a stable and cost-effective solid-contact ion-selective electrode for perchlorate determination. SWCNTs were used as contact material and indium (III) 5, 10, 15, 20-(tetraphenyl) porphyrin chloride was utilized as ion carrier in this study.. This assembly can be effectively used in the determination of ClO_4^- in propellants.[17]

Kim and coworkers[26] developed a highly flexible and patternable MWCNT paper electrode to use as a heating platform to fire Al/CuO nanoparticle-based nanoenergetic materials. To demonstrate its real-time application, this assembly was used to ignite solid propellants enough to launch a small rocket. Another application of CNTs in the area of propellant is their use to stabilize highly energetic green explosive, such as polymeric nitrogen. This was achieved through the electrochemical functionalization of CNTs with clusters of nitrogen.[29,39]

11.3.4 CNTS AND THEIR COMPOSITES IN ROTORCRAFTS

The rotary-wing aircrafts often abbreviated as rotorcrafts are heavy aircrafts that use their heavy rotor blades or wings to lift, with rotor blades revolve around a mast. The helicopters, cyclogyro or cyclocopter, auto gyro, gyrodyne, rotor kite, etc. are examples of types of rotorcrafts. The CNTs can be embedded in various systems, such as metals and polymers to alter the properties like damping. There are some reports that CNTs are employed for the rotorcraft applications. CNTs or their composite structures can be used to fabricate rotor blades. CNTs have the capacity to dissipate energy efficiently. The chief researchers in this field are army scientists of different countries. The driving force for the focus of research is the fact that the helicopters generally need more cost for the maintenance of their components. If CNTs can be incorporated in larger amounts in the fabricating materials, then the disadvantages of present materials can be overcome. The effect of CNT inclusion on the beams and blades are studied by army research laboratories.

Secondly, CNT composites are utilized as anti-icing agents in the rotor blades. When a layer of ice is formed on the rotor blades then the aerodynamic properties get deteriorated as temperatures falls below zero range under freezing storm conditions. When such formation is increased, it will lead to the breakdown of the blades. These icy layers can be removed with the coatings of CNT composite materials. Since, the accumulation of ice on aircraft has posed serious issues of safety and performance of the aircraft. Aircrafts usually use an electrothermal heating technology in their wings to protect the surface from icing. Owing to the properties of CNTs such as a high strength-to-weight ratio and increased power density, heaters based on CNTs can be a potential choice for this purpose. Zangrossi et al.[53] has demonstrated the electrothermal and mechanical performance of MWCNTs buckypapers embedded in fiber-reinforced polymer composites and their potential application in ice-protection.

11.3.5 CNT-BASED ADHESIVES FOR AEROSPACE APPLICATIONS

Adhesives are an integral part in the aerospace industry and research. Adhesives in aerospace applications are used to join metal to metal, metal to composite, and composite to composite parts. Aerospace applications also require increased strength-to-weigh ratio, heat dissipation and static charge dissipation.[30] The most common adhesive material used in this area is the well-known epoxies. Epoxies possess good thermal stability, chemical stability, and dimensional stability. Despite all these advantages, they are brittle in nature, which limit their application in many fields. Suitable modifiers can enhance the strength and reduce the brittleness of the epoxy adhesives. For aerospace applications, CNTs are often selected as a filler material to fabricate the nanocomposites of epoxies. The addition of CNTs in the epoxies makes them conductive. Epoxy-based conductive CNT adhesives display an exceptionally high strength. The CNT-based conductive adhesives are selected in the place of conventional metallic materials because of the lightweight nature, lead-free environment, and avoids the metallic corrosion, etc. The metal parts are usually joined using conductive adhesives, but they will have high contact resistivity. When the epoxy adhesives are modified with CNT as a filler, it will enhance the resistance due to high contact resistivity. The resistivity can be reduced by forcing a controlled amount of electric current through the bonded joint. Hoa and Rosca reported a decrease of 10 times in resistivity by applying the 0.5 A/cm^2 of current densities for 30 s. It was observed that perhaps the oxide layer surrounding many of the metals is responsible for high contact resistivity of the adhesive joints. The wide gradient of electric field created around the nanotubes during the current treatment will breakdown the oxide layer creating permanent conductive channels.This can be done by producing large potential gradient along the surface which breaks the oxide layers and hence enhance the fatigue resistance and mechanical strength (Fig. 11.2).[37]

Sometimes even smaller amount such as 1 wt.% of the CNT can enhance the bonding strength of the adhesives.[20] When large amount of the SWCNTs are added, agglomerates formed will lead to decreased mechanical properties. Zhao et al. reported the synthesis of a dry adhesive which was thermally and electronically conducting. In order to enhance the adhesion strength over cycles, they have added molybdenum catalyst.[22,58]

FIGURE 11.2 Preparation of adhesive-bonded panels using SWCNT–adhesive composite.

Source: Reproduced with permission from [20]. © 2014 Elsevier.

11.3.6 CNT-BASED COATINGS FOR AEROSPACE APPLICATIONS

Coating materials for aerospace applications can be derived from CNTs and graphene. These coatings can be used in de-icing, self-cleaning, anti-fouling, wear and corrosion resistance, flame retardancy, etc. The conventional coatings used for anti-icing will switch a fraction of hot air from the engines and blow it to the wing's inner surface. The heat transports to the surface by thermal conduction, prevents ice buildup. An innovative thought has been put forward by researchers from Battelle Memorial Institute, a technology R&D organization based in Columbus by creating a CNT web, in which the individual CNTs are aligned in the draw direction, and stacking between 10 and 40 web layers horizontally, at different orientations to achieve the desired heating characteristics. Scientists have suggested a heat coat made up of layers of CNTs for unmanned space vehicles. They can also be potentially useful in larger aircraft, the Heat

Coat technology consists of a series of layers that can be sprayed onto the aircraft surface like paint or applied as a laminate sheet. The first layer is a primer coating, followed by the heater coating that is made up of CNTs. This is topped off by a barrier coating and an outer top coating.

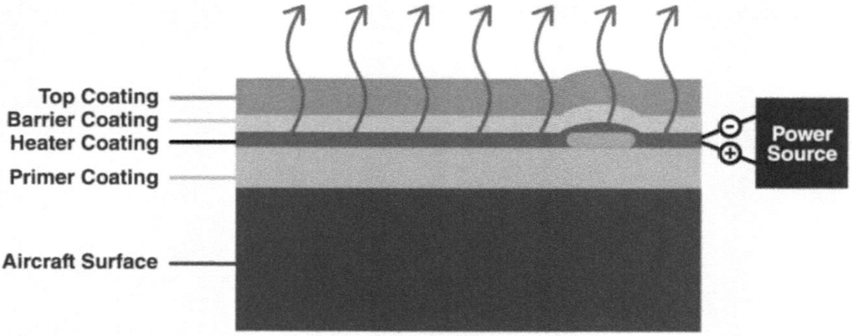

FIGURE 11.3 Battelle's Heat Coat for aircrafts with CNT as a component.
Source: Adapted from the website https://www.battelle.org

Heat is created to prevent ice from forming when power is applied to the heater layer. The aircraft's intelligent controller tracks the heater layer's output so that the power level and the heat produced are dynamically altered to ensure the minimum power level for the current conditions.

Similar to the above-mentioned invention, the chemical vapor deposition method was employed to produce CNT webs. The CNT web is a continuous sheet or layer of CNTs aligned horizontally, which are obtained from specially grown CNT forests. Different layers of the CNT web of size 70 × 50 mm were included in imparting de-icing property to the epoxy composite laminates. Within an environmental chamber, the de-icing efficiency of the laminate with various heating elements was studied. To track the de-icing process, a camera was mounted inside the room, capturing an image every 10 s. The ice was formed beforehand on one side of the samples, the samples were fixed upright, connecting a thermocouple to the middle of the surface on the other side. The result suggested that the GF laminate could remove the ice within 25–50 s with 30–40 layers of CNT web as the heating element. Samples with 20 CNT web layers required 150 s, while approximately 300 s were required with 10 CNT web layers or four CF layers.[50]

11.3.7 CNTS IN SENSING

The sensing ability of the CNTs are well known. The carbon nanotubes have been utilized in the sensing applications along with different polymers in composites. CNT sensors can work at high temperatures, they can be used as physical and chemical sensors, sensors that can conduct safety inspections cost-effectively, easily, and reliably than the existing procedures.[55,56] CNT added fiber-reinforced laminates with damage sensing has been extensively studied, but still this field lacks advancement.[48]

11.4 ADVANCES IN THE FIELD OF CNTs FOR AEROSPACE APPLICATIONS

CNT-based materials like carbon fiber-reinforced polymer composites provide a better way to construct lightweight aerospace structures. The superior mechanical properties of CNTs provide them a good mass advantage over conventionally used metallic materials in commercial aircraft. Because of this reason CNT-based alternatives have been widely using in this field for the past 50 years. CNTs are having high aspect ratio cylinders of graphene. The specific strength of CNTs are found to be 10 times greater than that of conventionally used in polyacrylonitrile (PAN) in aerospace applications and the specific modulus of CNT is five-fold higher than that of PAN-based fiber.[41] Figure 11.3 depicts the mile stone events of application of CNT-based materials in aerospace applications. The timeline gives a clear idea of implementation of CNT-based materials in aircrafts even before the official discovery of CNTs. Application of CNT-based materials provide several advantages like reduction of vehicle mass, improvement of functionality, damage tolerance, durability, and self-healing capacity. It also improves the thermal protection, control and energy generation, and distribution[13] (Fig. 11.4).

Various studies shows that the wings, body, and cryogenic propellant tanks can be replaced with CNT-reinforced functionally graded composites. Low temperature applications of CNTs like for the nacelle, fan, blades, case, and ducts involve the reinforcement of CNT polymer composites. High temperature applications, such as for combustor, turbine components, and nozzles demand the use of functionally graded ceramic matrix CNT composites. Second- and third-generation of propulsion system also replace the conventionally used polymer materials with CNT-based materials. It is found that CNT ceramic matrix composites can effectively be used in the production of reusable launch vehicles (RLV).[45]

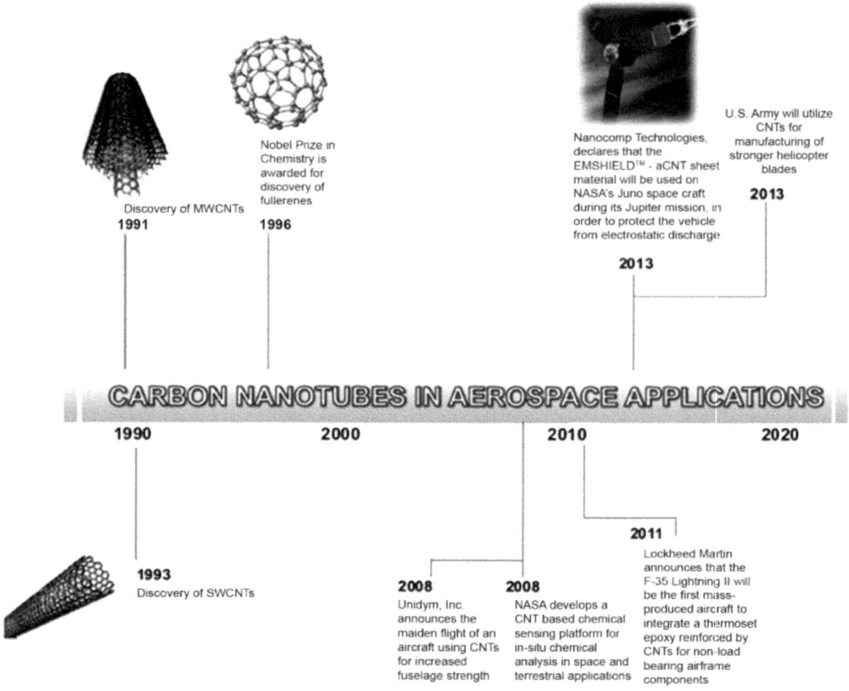

FIGURE 11.4 Milestones of CNTs and their application in aerospace sciences.
Source: Reproduced with permission from [13]. © 2014 Elsevier.

Dispersion of CNT in a medium like epoxy resin by methods like ultrasonication, followed by curing produces polymer–CNT nanocomposites. These nanocomposites show excellent thermo mechanical properties. –COOH functionalized epoxy–MWCT composites show a tensile strength of 121.8 MPa. As the concentration of CNT increases, the modulus of elasticity also increases. But a decrease in the tensile strength is observed when the concentration of CNT exceeds a limit. Therefore, the addition of CNTs is limited to 2%.[36]

One of the most efficient applications of CNT-based materials in aircrafts is the replacement of aluminum-based air frames with

CNT-reinforced polymer. This can reduce the weight by almost 14.1% and a decrease of fuel consumption by 9.8%. Copper wiring inside the aircraft can also be effectively replaced by wires made of CNTs. This will also reduce the weight of wire conductors by about 60%. It also prevents the premature failures and overheating of copper wires by vibrational fatigue and completely eliminates the chances of corrosion. CNT-based nanocomposites can also be used to control the phenomenon of aircraft icing. It is a phenomenon of the growth of water molecules as supercooled droplets in the atmosphere when they come in contact with the surfaces of aircraft during flight. This will adversely affect the efficiency of aircraft and will cause profile drag. Several studies have been carried out to find out icephobic materials. Development of meso/micropore CNT paper composite by Chu et al. shows exceptional de-icing properties. CNTs are also used as propellant additives. This will completely avoid the possibilities of agglomeration, earlier ignition, and higher rate of release of heat. Carbon fiber-reinforced polymers exhibit low electrical conductivity. This will enable them to use in devices used for lighting protection. CNTs and buckypaper which is a microscopic assembly of entangled CNTs can be utilized in the field of high current carrying capacity upon lightning. This will help to protect the electric circuits and devices on board aircraft.[1,33,40,51]

Self-sensing is another promising field where the use of CNT-based materials can be employed. Here, sensing is the measurement of electrical resistance at the surface of aircraft. The property of CNTs to undergo mechanical deformation, real-time strain monitoring by means of change in conductivity can be helpful to use them as sensors in aircrafts.[43]

11.4.1 RECENT DEVELOPMENTS IN THE AEROSPACE APPLICATIONS OF CNTS

Recent advances in the field of CNTs for aerospace applications involve the use of CNTs as space elevators, morphine devices or wings and structural health monitoring systems to provide efficient damage detection. Surface treatment with CNTs will increase observational efficiencies. CNTs can be used in a regenerable system for the removal of carbon dioxide from the cabin atmosphere in space flights. This will provide more efficient oxygen supply in flights. Multi-layer arrangement of CNTs can act as radar absorbing materials. Substitution of conventional materials with CNTs in patch antennas will definitely reduce weight and provide a planar structure for aircraft.

CNTs offer wide applications in unmanned aerial vehicles (UAVs) in which the pilot no longer is onboard the aircraft. Implementation of CNTs will increase the efficiency of UAVs by increasing the operational hours, reducing the weight and resistance toward icing. Use of CNTs in electrical UAVs increases its survival time.[44] Reduction of weight and increase in the battery energy can be done by using CNTs in UAVs.

11.4.1.1 SATELLITES AND SPACECRAFT

Use of CNTs in satellite applications can reduce the weight of a satellite in an effective way. Patch antennas fabricated from CNTs exhibit various features like planar structure, low profile and bandwidth of 5–20%. Studies show that usage of silane-grafted MWCNTs improved thermal stability and mechanical properties than unmodified CNTs. Exceptionally good thermal conductivity of MWCNTs offer greater potential for distribution of heat flow in satellite and thermal protection systems. Street et al. shows that a long wear life and low friction can be attained for CNTs in aerospace applications and can even replace the most widely used solid lubricants, graphite.[42] Number of patents has been filed on the basis of usage of CNTs and derivatives in aerospace applications recently. This confirms the importance of CNT-based materials in aerospace applications.

11.4.1.2 SOLAR SAIL

Solar sails are nonrocket spacecraft that uses sunlight for propulsion. The dimension of solar sail is several meters wide and is made up of reflective ultrathin materials. Efficiency of solar sails is generally stated in terms of spacecraft acceleration. It is inversely proportional to the sail areal density. Conventional sails are made up of aluminized materials and consist of a reflecting layer to absorb sunlight and another layer to emit heat. CNT membranes can effectively replace these layers so that the single layer of CNT can itself reflect sunlight and emit heat. Very thin nanotube sheets weigh only few kilograms. Reflectivity of CNTs can be enhanced by doping and found special applications. Studies show that doped CNT made sail could reach Pluto in days and the nearest star in a few decades.[38]

11.4.1.3 WHIPPLE SHIELD

Whipple shield is a protective layer used in spacecraft to overcome MMOD impact. MMOD can make impacts on spacecraft at hypervelocities. Usage of CNTs in whipple shield contributes enhanced resistance.

11.4.1.4 SURFACE COATINGS

CNT coating can be employed for electromagnetic shielding and to increase lateral thermal conductivity. Surface coating with CNTs also helps to improve satellite thermal control and electrical conductivity, radiation hardness of microprocessors used. CNTs can attain good conductivity without any doping. This will protect the spacecraft from severe electromagnetic events.

11.4.1.5 THERMAL RADIATORS

Improper functioning of thermal radiators in aircrafts will lead to shortened life and degradation of performance. CNTs can replace traditionally used radiators made of aluminum or copper. In order to avoid the limitations, CNTs are needed to design in such a way to consider the anisotropic thermal conductivity. CNT-based hybrids are then used in this field to increase the efficiency.

11.5 CHALLENGES AND FUTURE SCOPE

CNT-embedded polymeric matrix faces several challenges. Most important challenge is the homogeneous dispersion of CNTs in polymer matrix. Another problem is the lack of proper processing methods. Sometimes, theoretically predicted thermal conductivity values cannot match with the experimental results due to the ineffective transport of phonons at the end of the tubes. In order to avoid this problem, reinforced interface gaskets are to be used in future. Partial or full dissipation of energy to a failure mode makes CNTs a wrong choice for lightning strike protection in aircrafts. But this problem can be resolved by not endangering the aircraft. Another important challenge is the alignment of nanoparticles. Studies show that SWCNTs have fewer defects compared with MWCNTs. Generally, the flexibility and superlative resilience observed at the nanoscale is not observed at the macroscale. Several researches are going on nowadays so as to meet the environmental concerns related to the usage of CNTs. Chances of fibrosis and lung inflammation on exposure to CNTs needed to be considered. For future spacecraft, release of CNTs from the components of aircraft will become much more prominent. These materials can persist in the environment for a very long period due to their exceptionally high resistance toward chemical, biological, and physical responses. Therefore, health and safety concerns regarding the use of CNTs in aircrafts are to be addressed as soon as possible.

Usage of CNT-based materials with reduced environmental impact is of paramount importance. CNTs will have a significant role in the aerospace industry in future. Once the proper characterization and large-scale production of CNTs can be done by addressing the safety concerns, recyclability, and cost reduction, a breakthrough can be observed in this field in future. Tremendous decrease in weight of aerospace vehicles paves way for the potential applications of nanotechnology in aerospace science. For future space programs, CNTs need to undergo desirable functionalization to give materials with higher electrical, magnetic, mechanical and thermal properties. Along with these, self-cleaning and self-repairing properties need to be considered.

11.6 CONCLUSIONS

The importance or significance of the CNTs and their composites is never ending. The advance applications such as aerospace lead them to a new horizon. In the aerospace field, CNTs and their composites have been tested and found effective than the conventional materials for applications like rotor blades and their coatings, sensing, in propellants, for EMI-shielding devices, and so on. Even though there are so many proposed applications for the nanotube material, advancement is still required for the enhancement of properties and manufacturing. The chapter discussed simply about these applications.

KEYWORDS

- **CNTs**
- **aerospace**
- **sensing**
- **propellants**
- **composites**

REFERENCES

1. Agee, F. J. et al. Nanotechnology Research for Aerospace Applications. In *Independent Component Analyses, Wavelets, Neural Networks, Biosystems, and Nanoengineering VII.* 2009; p 73430U.
2. Anon, scholar (1).
3. Bagotia, N.; Choudhary, V.; Sharma, D. K. Synergistic Effect of Graphene/ Multiwalled Carbon Nanotube Hybrid Fillers on Mechanical, Electrical and EMI Shielding Properties of Polycarbonate/Ethylene Methyl Acrylate Nanocomposites. *Compos. Part B Eng.* **2019,** *159*, 378–388.
4. Balasubramanian, K.; Burghard, M. Chemically Functionalized Carbon Nanotubes. *Small* **2005**
5. Bao, L. et al. Impact of MWCNT/Al on the Combustion Behavior of Hydroxyl Ammonium Nitrate (HAN)-Based Electrically Controlled Solid Propellant. *Combust. Flame* **2020,** *218*, 218–228. http://www.sciencedirect.com/science/article/pii/S0010218020301942.
6. Baur, J.; Silverman, E. Challenges and Opportunities in Multifunctional Nanocomposite Structures for Aerospace Applications. *MRS Bull.* **2007**.
7. Chu, H. et al. Sesame-Cookie Topography Silver Nanoparticles Modified Carbon Nanotube Paper for Enhancing Lightning Strike Protection. *Carbon* **2019,** *143*, 204–214.
8. Dai, H. Carbon Nanotubes: Synthesis, Integration, and Properties. *Acc. Chem. Res* **2002**.
9. De Volder, M. F. L. et al. Carbon Nanotubes: Present and Future Commercial Applications. *Science* **2013**.
10. Elbasuney, S. et al. MWNTs Coated with CuO Particles: A Novel Nano-Catalyst for Solid Propellants. *J. Inorg. Organometal. Polym. Mater.* **2019,** *29* (6), 2064–2071.
11. Feng, D.; Wang, Q.; et al. Microwave Assisted Sinter Molding of Polyetherimide/ Carbon Nanotubes Composites with Segregated Structure for High-Performance EMI Shielding Applications. *Compos. Sci. Technol.* **2019,** *182*, 107753. http://www.sciencedirect.com/science/article/pii/S0266353819315404.
12. Feng, D.; Xu, D.; et al. Highly Stretchable Electromagnetic Interference (EMI) Shielding Segregated Polyurethane/Carbon Nanotube Composites Fabricated by

Microwave Selective Sintering. *J. Mater. Chem. C* **2019,** *7* (26), 7938–7946. http://dx.doi.org/10.1039/C9TC02311A.
13. Feng, L. et al. Development of Light Cellular Carbon Nanotube@Graphene/Carbon Nanocomposites with Effective Mechanical and EMI Shielding Performance. *Carbon* **2020,** *168*, 719–731. http://www.sciencedirect.com/science/article/pii/S0008622320306928.
14. Gigliotti, M.; Lafarie-Frenot, M. C.; Grandidier, J. C. Development of Experimental and Modelling Tools for the Characterisation of the Thermo-Electro-Mechanical Behaviour of Composite Materials for Aircraft Applications. *Mecanique et Indust.* **2011**.
15. Gohardani, O.; Elola, M. C.; Elizetxea, C. Potential and Prospective Implementation of Carbon Nanotubes on Next Generation Aircraft and Space Vehicles: A Review of Current and Expected Applications in Aerospace Sciences. *Progr. Aerospace Sci.* **2014,** *70*, 42–68.
16. Haddon, R. C. Carbon Nanotubes. *Acc. Chem. Res.* **2002**.
17. Hassan, S. S. M. et al. Single-Walled Carbon Nanotubes (SWCNTs) as Solid-Contact in All-Solid-State Perchlorate ISEs: Applications to Fireworks and Propellants Analysis. *Sensors (Switzerland)* **2019,** *19* (12).
18. Hou, P.-X.; Liu, C.; Cheng, H.-M. Purification of Carbon Nanotubes. *Carbon* **2008,** *46* (15), 2003–2025. http://www.sciencedirect.com/science/article/pii/S000862230800448X.
19. Jaillet, C.; Alexopoulos, N. D.; Poulin, P. Carbon Nanotube Structures with Sensing and Actuating Capabilities. 2013; pp. 57–97.
20. Jakubinek, M. B. et al. Single-Walled Carbon Nanotube–Epoxy Composites for Structural and Conductive Aerospace Adhesives. *Compos. Part B Eng.* **2015,** *69*, 87–93.
21. Joseph, J. et al. Graphene and CNT Filled Hybrid Thermoplastic Composites for Enhanced EMI Shielding Effectiveness. *Mater. Res. Express* **2019,** *6* (8).
22. Joshi, M.; Chatterjee, U. *8—Polymer Nanocomposite: An Advanced Material for Aerospace Applications*; Elsevier Ltd., 2016.
23. Jung, M. et al. Carbon Nanotubes (CNTs) in Ultra-High Performance Concrete (UHPC): Dispersion, Mechanical Properties, and Electromagnetic Interference (EMI) Shielding Effectiveness (SE). *Cement Concr. Res.* **2020,** *131*, 106017. http://www.sciencedirect.com/science/article/pii/S0008884619311172.
24. Khanna, S. Carbon Nanotubes: Properties and Applications. In *Carbon Nanotubes and Nanoparticles*; 2019.
25. Kim, H. et al. Multifunctional Primer Film Made from Percolation Enhanced CNT/Epoxy Nanocomposite and Ultrathin CNT Network. *Compos. Part B Eng.* **2019,** *175*, 107107. http://www.sciencedirect.com/science/article/pii/S135983681932164X.
26. Kim, H. et al. Highly Flexible and Patternable Multiwalled-Carbon Nanotube/Nitrocellulose Hybrid Conducting Paper Electrodes as Heating Platforms for Effective Ignition of Nanoenergetic Materials. *ACS Appl. Mater. Interf.* **2020,** *12* (25), 28586–28595. https://doi.org/10.1021/acsami.0c02226.
27. Kong, L. et al. Powerful Absorbing and Lightweight Electromagnetic Shielding CNTs/RGO Composite. *Carbon* **2019,** *145*, 61–66. http://www.sciencedirect.com/science/article/pii/S0008622319300090.

28. Law, C. K. Fuel Options for Next-Generation Chemical Propulsion. *AIAA J.* **2012,** *50* (1), 19–36.
29. Li, F. et al. Application of Nanometer Materials for Solid Propellants. *Int. J. Energetic Mater. Chem. Propulsion* **2011,** *10* (1), 67–83.
30. Li, J.; Lumpp, J. K.; Grulke, E. Carbon Nanotube Filled Conductive Adhesives. *2003 Int. Symp. Microelectron.* **2003,** 5288, 35–38.
31. Lin, Y. et al. Experimental Study to Assess the Effect of Carbon Nanotube Addition on the through-Thickness Electrical Conductivity of CFRP Laminates for Aircraft Applications. *Compos. Part B: Eng.* **2015,** *76*, 31–37.
32. Mei, H. et al. SiC encapsulated Fe@CNT Ultra-High Absorptive Shielding Material for High Temperature Resistant EMI Shielding. *Ceram. Int.* **2019,** *45* (14), 17144–17151. http://www.sciencedirect.com/science/article/pii/S0272884219313811.
33. Murday, J. S. et al. Nanotechnology in Aerospace Applications. *Adv. Mater. Process.* **2005,** *163* (12), 21–23.
34. O'Donnell, S. E.; Sprong, K. R.; Haltli, B. M. Potential Impact of Carbon Nanotube Reinforced Polymer Composite on Commercial Aircraft Performance and Economics. In *Collection of Technical Papers—AIAA 4th Aviation Technology, Integration, and Operations Forum, ATIO*; 2004; pp 693–702.
35. Qi, Q. et al. An Effective Design Strategy for the Sandwich Structure of PVDF/GNP-Ni-CNT Composites with Remarkable Electromagnetic Interference Shielding Effectiveness. *ACS Appl. Mater. Interf.* **2020,** *12* (32), 36568–36577. https://doi.org/10.1021/acsami.0c10600.
36. Rathod, V. T.; Kumar, J. S.; Jain, A.; 2017. Polymer and Ceramic Nanocomposites for Aerospace Applications. *Appl. Nanosci. (Switzerland)* **2017,** *7* (8), 519–548.
37. Rosca, I. D.; Hoa, S. V. Method for Reducing Contact Resistivity of Carbon Nanotube-Containing Epoxy Adhesives for Aerospace Applications. *Compos. Sci. Technol.* **2011,** *71* (2), 95–100.
38. Samareh, J. A.; Siochi, E. J. Systems Analysis of Carbon Nanotubes: Opportunities and Challenges for Space Applications. *Nanotechnology* **2017,** *28* (37).
39. Shao, Y.; Wang, Z.; Xu, L. The Application of Carbon Nanotubes in Propellants and Explosives Abroad. *Huaxue Shijie* **2013,** *54* (5), 311–314.
40. Sharon, M. Applications of Nanotechnology in Aerospace. In *Nanotechnology in the Defense Industry*; 2019; pp 289–307.
41. Siochi, E. J.; Harrison, J. S. Structural Nanocomposites for Aerospace Applications. *MRS Bull.* **2015,** *40* (10), 829–835.
42. Street, K. W.; Miyoshi, K.; Vander Wal, R. L. Application of Carbon Based Nano-Materials to Aeronautics and Space Lubrication. In *Superlubricity*; 2007; pp 311–340.
43. Theodosiou, T. C.; Saravanos, D. A. Mechanical and Electrical Response Models of Carbon Nanotubes. *Solid Mechan. Its App.* **2013,** *188*, 219–266.
44. Thomas, J. P.; Qidwai, M. A. The Design and Application of Multifunctional Structure-Battery Materials Systems. *JOM* **2005,** *57* (3), 18–24.
45. Udupa, G.; Shrikantha Rao, S.; Gangadharan, K. V. Future Applications of Carbon Nanotube Reinforced Functionally Graded Composite Materials. In *IEEE-International Conference on Advances in Engineering, Science and Management, ICAESM-2012*. 2012; pp 399–404.

46. Vertuccio, L. et al. Piezoresistive Properties of Resin Reinforced with Carbon Nanotubes for Health-Monitoring of Aircraft Primary Structures. *Compos. Part B Eng.* **2016,** *107*, 192–202.
47. Wang, H. et al. Bead Nano-Necklace Spheres on 3D Carbon Nanotube Scaffolds for High-Performance Electromagnetic-Interference Shielding. *Chem. Eng. Journal* **2019,** *360*, 1241–1246.
48. Wu, C. et al. Study of Carbon Nanotubes/Short Carbon Fiber Nanocomposites for Lightning Strike Protection. In *Proc.SPIE*, 2010. https://doi.org/10.1117/12.847491.
49. Xia, Q. et al. Fabrication of the Silver Modified Carbon Nanotube Film/Carbon Fiber Reinforced Polymer Composite for the Lightning Strike Protection Application. *Compos. Part B: Eng.* **2020,** *180*, 107563. http://www.sciencedirect.com/science/article/pii/S1359836819329750.
50. Yao, X.; Falzon, B. G.; Hawkins, S. C. Carbon Nanotube Web: A Promising Heating Element for Anti-Icing/De-Icing Applications. *ICCM International Conferences on Composite Materials*, 2017–Augus(August).
51. Yowell, L. L.; Moloney, P. G. Applications of Nanotechnology in Aerospace. In *Nanotechnology for Sustainable Manufacturing*; 2014; pp 93–111.
52. Yu, C. et al. Fe Nanoparticles and CNTs Co-Decorated Porous Carbon/Graphene Foam Composite for Excellent Electromagnetic Interference Shielding Performance. *J. Alloys Compounds* **2020,** *820*, 153108. http://www.sciencedirect.com/science/article/pii/S0925838819343543.
53. Zangrossi, F. et al. Electro-Thermal and Mechanical Performance of Multi-Wall Carbon Nanotubes Buckypapers Embedded in Fibre Reinforced Polymer Composites for Ice Protection Applications. *J. Compos. Mater.* **2020,** *54* (23), 3457–3469.
54. Zeng, S. et al. Flexible PVDF/CNTs/Ni@CNTs Composite Films Possessing Excellent Electromagnetic Interference Shielding and Mechanical Properties under Heat Treatment. *Carbon* **2019,** *155*, 34–43. http://www.sciencedirect.com/science/article/pii/S000862231930822X.
55. Zhang, H.; Bilotti, E.; Peijs, T. The Use of Carbon Nanotubes for Damage Sensing and Structural Health Monitoring in Laminated Composites: A Review. *Nanocomposites* **2015,** *1* (4), 167–184. http://dx.doi.org/10.1080/20550324.2015.1113639.
56. Zhang, H.; Liu, Y.; et al. Improved Fracture Toughness and Integrated Damage Sensing Capability by Spray Coated CNTs on Carbon Fibre Prepreg. *Compos. Part A Appl. Sci. Manuf.* **2015,** *70*, 102–110. http://dx.doi.org/10.1016/j.compositesa.2014.11.029.
57. Zhang, S. et al. Carbon Nanotube/Carbon Composite Fiber with Improved Strength and Electrical Conductivity via Interface Engineering. *Carbon* **2019,** *144*, 628–638. http://www.sciencedirect.com/science/article/pii/S0008622318312260.
58. Zhao, Y. et al. Interfacial Energy and Strength of Multiwalled-Carbon-Nanotube-Based Dry Adhesive. *J. Vacuum Sci. Technol. B Microelectron. Nanometer Struct. Process. Measurement, Phenomena* **2006,** *24* (1), 331–335.

CHAPTER 12

CARBON NANOTUBES FOR SOLAR ENERGY CONVERSION APPLICATIONS

JIJI ABRAHAM[1,*], ARUNIMA REGHUNADHAN[2], NIMITHA K.C.,[1], SONEY C. GEORGE[3], and SABU THOMAS[4]

[1]*Vimala College (Autonomous) Thrissur 680009, Kerala, India*

[2]*Department of Chemistry, TKM College of Engineering, Karicode, Kollam 691005, Kerala, India*

[3]*Centre for Nanoscience and Nanotechnology, Amal Jyothi College of Engineering, Kottayam, Kerala, India*

[4]*International and Inter University Centre for Nanoscience and Nanotechnology, Mahatma Gandhi University, P.D. Hills, Kottayam 686560, Kerala, India*

[*]*Corresponding author. E-mail: jijiabraham02@gmail.com*

ABSTRACT

Solar cells are the most familiar energy storage devices and they are considered as the storage devices for future, when the conventional energy sources will expire. This chapter deals with the applications of carbon nanotubes in solar energy conversion. The characteristics of the carbon nanotubes which make these materials suitable for the solar cell applications are wide range of direct bandgaps, strong absorptions in the visible spectrum, and high charge carrier mobility, and chemical stability. In the photovoltaic technology, carbon-based materials, especially CNTs are being tried to enhance the parameters, such as energy efficiency and lightweight nature. This chapter summarizes the use of CNT as electrode material, electrolytes, antireflective, and self-cleaning coating composites in photovoltaics and in thermal energy conversions.

12.1 INTRODUCTION

Fullerenes are allotropes of carbon. They are molecules consisting of only carbon. Cylindrical fullerenes are called carbon nanotubes (CNTs). They are also called buckytubes. Carbon nanotubes are generally having nanosized diameter and micrometer-sized length. Their length-to-diameter ratio is in the range of 1000 or more. In CNT, the atoms are arranged in hexagons. CNT consists of cylindrical graphitic sheet, that is, graphite rolled into a cylinder with diameter in the nano range. CNTs are classified into two categories on the basis of their structure. They are single-walled (SWCNT) and multiwalled (MWCNT) CNTs. Double-walled CNTs (DWCNTs) are perfect examples of multiwalled nanotubes. SWCNT consists of a rolled single layer of graphite into seamless cylinder. Multiwalled carbon nanotubes can be considered as a collection of several SWCNTs with different diameters. The properties of SWCNTs are entirely different from that of MWCNTs. The first scanning image of carbon nanotube was done in 1993. SWCNTs are much narrower than MWCNTs. There are other rare forms of carbon nanotubes other than SWCNT and MWCNT. They are fullerite, torus, and nanoknot.[1] Bonding in carbon nanotubes is sp^2 and consists of honeycomb lattices. Here, each atom is joined to three neighbors as in graphite. Carbon nanotubes are considered as the wondering materials of the 21st century. They are found enormous applications in diverse fields, like electronics, biomedical engineering, drug delivery, neuroengineering, and gene therapy.[2] Figure 12.1 shows the schematic representation of single-walled, double-walled, and multiwalled carbon nanotubes.

12.1.1 SYNTHESIS AND PURIFICATION OF CARBON NANOTUBES

Carbon nanotubes can be generally synthesized by three methods. They are arc discharge, laser ablation, and chemical vapor deposition. In arc-discharge technique, two carbon electrodes are taken and an electric arc is supplied. Carbon electrodes get evaporated, and from these carbon vapors,

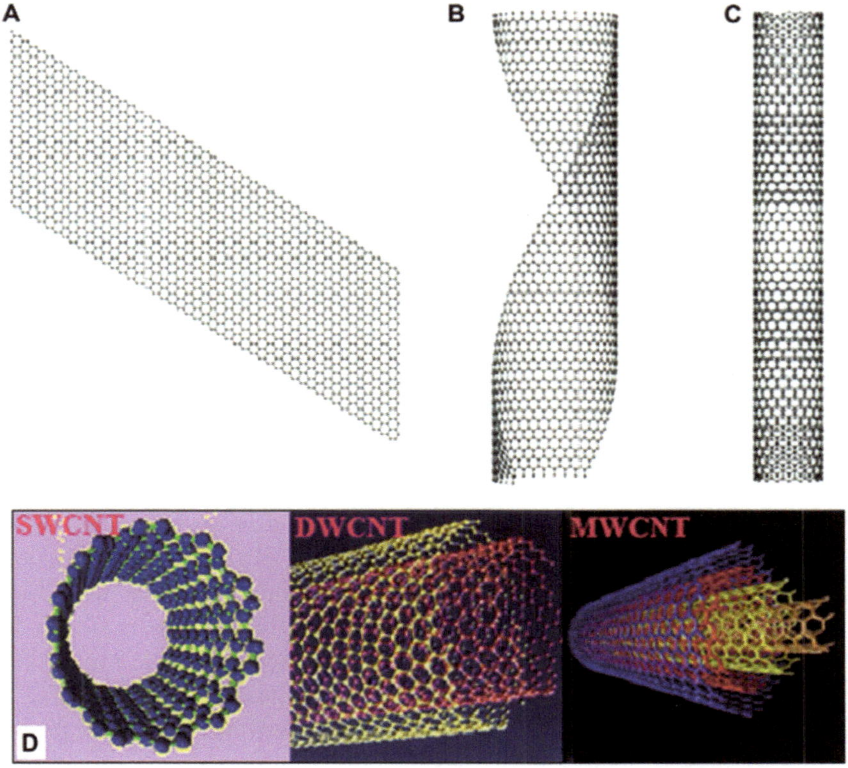

FIGURE 12.1 Schematic diagrams showing different types of CNTs and other carbon structures: (A) Flat sheet of Graphite. (B) Partially rolled sheet of graphite. (C) SWCNT. (D) Structures of the three CNT types; SWCNT, DWCNT, and MWCNT, respectively.

---

nanotubes self-assemble. Laser ablation method consist of a carbon feed stock gas, such as methane or carbon monoxide and powerful laser beams are impinged on it. As a result of this, carbon nanotubes are produced with high purity. One of the most adapted methods for the synthesis of multi-walled carbon nanotubes over single-walled carbon nanotubes is chemical vapor deposition (CVD). It always offers homogeneous product outcome. CVD involves the reaction of carbon-containing gas, such as acetylene, ethylene, etc. with a metal catalyst like Mo, Co, Ni, and Fe at an elevated temperature.[1] Purification of carbon nanotubes is of paramount importance because the electronic structure of CNT highly depends on the diameter of the CNT. Even advanced methods could not be able to produce CNT with high electronic type selectivity. Purification of CNT can be further done by using several chemical and electrochemical methods, such as electrophoresis, chromatography, etc. Centrifugal sorting is one of the most efficient and flexible method used now as a purification method. This method is more specifically called density-gradient ultracentrifugation (DGU). This helps the separation by diameter by varying the concentration of surfactant used. DGU is an efficient way to sort out DWCNT and SWCNT. SWCNT with varying chiralities and SWCNT enantiomers can also be prepared by this method using ionic and nonionic surfactants.[4]

12.2 CARBON NANOTUBES: THE WONDER MATERIALS IN ELECTRONICS

Electronic properties of carbon nanotubes are exceptionally great and can be used as a substitute for silicon in several fields, such as low-cost thin film transistors and high performance logic devices.[3] Semiconducting nanotubes are direct bandgap materials. Carbon nanotubes have highly delocalized electronic structure and are helpful for their use in high mobility electronic materials. Electronic and optoelctronic properties of carbon nanotubes can be modified by tuning the bandgap of CNTs. This can be done by controlling the diameter.[4] The most important property of SWCNTs is their ability to show the semiconducting and metallic properties. Nanotubes are characterized by a set of electrical properties like electrical conductivity, capacitance, and inductance. It is due to the intrinsic structure of nanotubes and their interaction with the objects. Electrical conductance offered by carbon nanotubes is affected by scattering by tube imperfections.[5]

12.2.1 ELECTRONIC STRUCTURE OF CARBON NANOTUBES

Electronic structure of CNTs can be explained on the basis of band structure of graphene. The rolling process of graphene into CNT is specified by a pair of integers (n, m) that defines the chiral vector that determines the circumference of the nanotube. CNT is defined by the chiral vector. Geometry of SWCNT can be specified by the relative position of $C = na_1 +$

ma_2 where a_1 and a_2 are the unit vector of honeycomb lattice (Fig. 12.2).[6] From C, diameter of the nanotube can be calculated using eq 12.1.

$$d = \frac{C}{\pi}\sqrt{n^2 + nm + m^2} \tag{12.1}$$

Here, "a" is the lattice constant of the honeycomb network. It is equal to 1.42 A° for C-C bond length in carbon nanotubes. Chiral angle θ is another parameter that defines the geometry of the tube.[7]

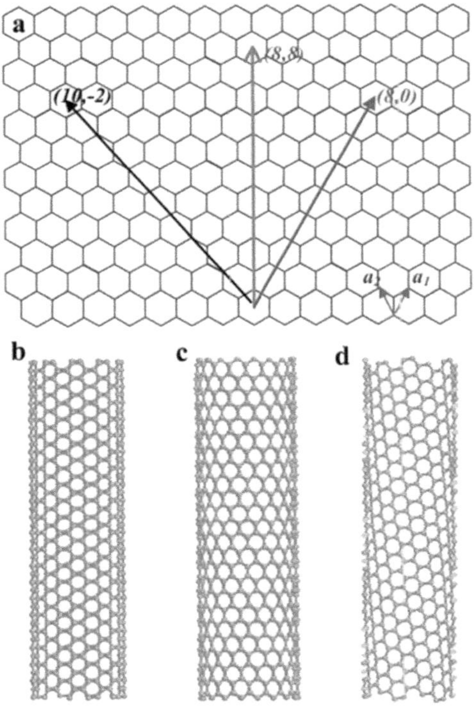

FIGURE 12.2 Schematic honeycomb structure of a graphene sheet. Single-walled carbon nanotubes can be formed by folding the sheet along lattice vectors. The two basis vectors **a**1 and **a**2 are shown. Folding of the (8,8), (8,0), and (10,-2) vectors leads to armchair (b), zigzag (c), and chiral (d) tubes, respectively.

Source: Reprinted with permission from Ref. 7. © 2002 American Chemical Society.

The very distinct and important geometries of carbon nanotubes are of three types and represented in Figure 12.2. They are as follows:

1. Armchair: Nanotubes of the type (n, n). The name is due to their structure.
2. Zigzag nanotubes: Nanotubes of the type (n, 0) perpendicular to the tube axis.
3. All the other types are commonly called chiral nanotubes.

Armchair and zigzag structures of CNTs have high degree of symmetry. These are arrangement of hexagons around the circumference. CNT endings are generally made up of pentagons and hexagons. Generally, CNTs have the diameter in the range of 0.4 nm. CNT endings are capped with C60 hemisphere and sometimes with C20 dodecahedron. Nature of bonding in CNT is based on the quantum chemistry more specifically orbital hybridization. sp^2 bonds in CNT are stronger than those in diamond (sp^3). CNTs can align themselves in the form of ropes by van der Waals forces. CNTs merged under high pressure and ensure possibilities for producing strong wires of unlimited length.[1]

12.2.2 CONDUCTIVITY AND OTHER ELECTRICAL PROPERTIES OF CARBON NANOTUBES

Armchair nanotubes are generally metallic in nature. So they offer a tremendous conductivity in their applications. Most CNTs are semiconductors. Semiconducting CNTs are major channel materials in field effect transistors (FETs). Metallic CNT films are useful as good transparent conductors. CNT-based field effect transistors is a three terminal switch in which current is passed through the CNT connected to two electrodes called a source (S) and drain(D). Switching can be done by controlling the carrier density in CNT by a system called gate which is an electrically isolated electrode.[4] The carriers move from source to drain and offers an energy barrier in the semiconductor. The electrical field generated by the gate is responsible for the change in conductivity. SWCNTs act as perfect material for replacing metal oxides as transparent conducting materials due to their high conductivity, flexibility, and low refractive index. The extraordinary electrical conductivity of SWCNTs arises due to the high charge carrier mobility in the quantum channel of SWCNT. This property will allow them to carry current densities 1000 times higher than the metal conductors like copper. Conductivity of SWCNT depends on the resistance of tubes, barriers, and junction resistances, etc.[8] For metallic SWCNT, there is a finite value of density of states (DOS) at the Fermi level. The energy separation between the pair of peaks represents the bandgap of SWCNT.[2] SWCNTs are obtained as a mixture of metallic and semi-conducting materials during their preparation. Therefore, the interband transitions are arising from both. There are several factors that influence the conductivity of CNTs. One of the factors is pressure. It is found that the conductivity of MWCNTs was found to be decreased while applying pressure. When pressure is applied, MWCNTs and graphene show less conductivity than graphite. Figure 12.3A illustrates the effect of pressure on the conductivity of different materials, such as carbon black, graphite, graphene, and MWCNT.[9] Another important factor is density. It is found that the conductance of MWCNTs increase with density. Figure 12.3B represents the bilinear relationship between density and conductance of MWCNTs.

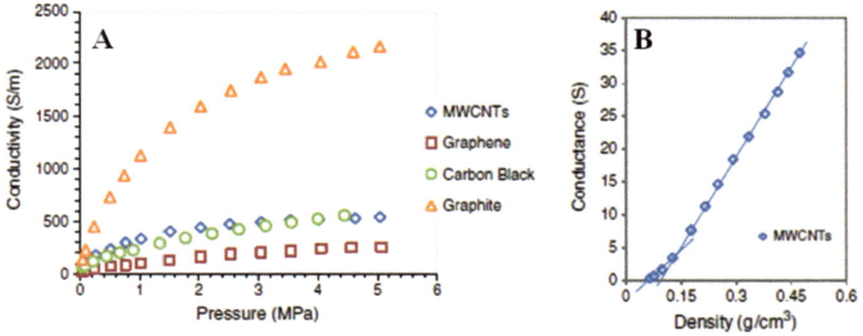

FIGURE 12.3 (A) Electrical conductivity behavior of the different carbon powders as a function of pressure. For each material, the data points represent an average of at least three identical assays differing not more than 3%. (B) Conductance of MWCNTs as a function of density.[9]

Source: Reprinted with permission from Ref. 9. © 2012 Elsevier.

12.2.3 OPTOELECTRONIC PROPERTIES OF CARBON NANOTUBES

Several studies have been carried out to study the optical and optoelectronic properties of carbon nanotubes. The discovery of bandgap photo-luminescence from individual SWCNTs in aqueous solutions was in the year 2002. Semiconductors consist of free electron–hole pair excitations as well as electron–hole pairs called excitons. Various studies, such as nanospectroscopy, resonant Raman scattering, and magnetic optical spectroscopy of SWCNTs are being carried out currently. These methods help to understand the fundamental properties of one-dimensional excitons and phonons. The exciton binding energies in CNTs are much higher than that of the conventional semiconductors, such as GaAs due to strong electrostatic interactions. The large binding energies result in large radiative lifetimes and fluorescence lifetimes at ordinary temperature. Radiative recombination of excitons results in photoluminescence and electroluminescence. Electroluminescence has been observed in ambipolar FETs.[4] Photoconductivity of carbon nanotubes is a promising field of investigation which finds applications of CNTs in photodetecting materials and in solar cells.[10] Nanotubes have been recently identified as building blocks in solar cells. Reduced manufacturing cost, unlimited material availability unlike silicon, and good quantum efficiency, etc. are the advantages of nanotube-incorporated solar cells. Bandgap of CNTs can vary from 0.5 to 0.6 V. This is inversely proportional to tube diameter.

Carbon nanotube diode can also exhibit a photovoltaic effect and was observed primarily by Lee. The device consists of a semiconducting CNT suspended between a source and drain. Electrostatic doping forms a p–n junction at room temperature. Potential voltage shows the typical behavior of CNT diode as an ideal p–n junction diode, and when light hits on it, photovoltaic effect is observed. This will increase the photocurrent. Infrared laser excited photoconductivity from CNTs has also been observed.[6]

Photochromic molecular switches are systems exhibiting two or more stable configurations with peculiar chemophysical features. When these molecular systems are irradiated with photons, they change their configuration and properties. Spiropyran (SP) is a typical example for this. On irradiation with UV light, spiropyran is converted into an open form called merocyanine (MC).When CNTs are functionalized with SP, then every 100 carbon atoms in SWCNTs carried 1 SP group. When the material is irradiated with UV, SP-functionalized SWCNTs are converted into MC-functionalized SWCNTs. This can be confirmed by a change in the absorption band. Figure 12.8 represents the SP-functionalized carbon nanotubes and absorption and emission spectra of SP- and MC-functionalized SWCNTs on UV irradiation.[11]

SWCNTs can absorb light in the visible spectral range due to their diameter-dependent optical transitions. Intraband absorption in the infrared region by metallic and semiconductor nanotubes is a promising field for ultrabroadband detection. SWCNT films show properties like photodesorption, bolometric effects, and photothermoelectric effects.[9] Ambipolar CNT–FETs can be used to produce a light source. The shape of electroluminescence peak depends on the length of the CNT. Long CNTs will give narrow and symmetric peaks. Short CNTs produce broad and asymmetric spectra. In LEDs, chemical doping defines the location of light emission, whereas in CNTs, gate will determine the same. The point at which electron and hole current meet will be the pint of emission of light. The shape of emitting spot allows us to find the electron–hole recombination lengths.[12]

12.3 SOLAR CELL OR PHOTOVOLTAICS BASICS

Solar cells are the most familiar energy storage devices and they are considered as the storage devices for future, when the conventional energy sources will expire. Apart from the electricity generation from geothermal sources, such as tides, wind, etc. solar energy will be the chief source of energy in the future. Hence, the devices to harvest solar energy makes their own place in the energy storage technology. They have been used in small-scale purposes, such as in calculators, toys, torches, etc. to large-scale purposes such as in satellites. Solar cells are scientifically termed as photovoltaics (in this photo means light, voltaic means electricity), which are defined as the devices that convert light energy into electrical energy. They are the materials which are able to absorb radiations from the visible region.

12.3.1 MECHANISM OF LIGHT ENERGY CONVERSION

The principle acting behind the solar cells are photoelectric effect. In photoelectric effect, a potential difference is produced at the junction of two materials by the action of electromagnetic radiation. The key composition of the photovoltaics or simply solar cells is the semiconductor material. The photo voltaic effect takes place in three stages.

a. Generation of charge carriers by the absorption of photons in the materials that can form a junction.
b. Separation of photo-generated charge carriers at the junction.
c. Collection of the charge carriers from the terminals of the junction.

When the sunlight strikes the cell, a portion of it is absorbed within the semiconductor material and the energy of the absorbed light is transferred to the semiconductor. The transferred energy then knocks electrons loose, allowing them to flow freely. The semiconductor material is basically a p-n junction diode. It is made up of p-type materials which are very rich in holes and n-type materials rich in electrons. At the junction, excess electrons from n-type material try to diffuse to p-side and vice-versa. Migration of electrons to the p-side exposes positive ion centers in n-side. Similarly, migration of holes to the n-side exposes negative ion centers in the p-side. This may result in the production of an electric field at the junction and forms a depletion layer. The whole process is schematically represented in Figure 12.4.

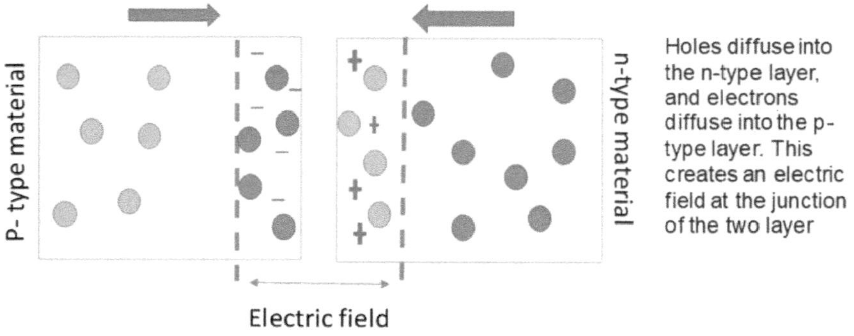

FIGURE 12.4 Mechanism of energy conversion in solar cells

When sunlight falls on the solar cell, photons with energy greater than the bandgap of the semiconductor are absorbed by the cell and create electron–hole pair. These pairs migrate respectively to n- and p- side of the p–n junction due to electrostatic force across the junction. In this way, a potential difference is recognized between two sides of the cell. The schematic representation of the processes in a solar cell is given in Figure 12.5.

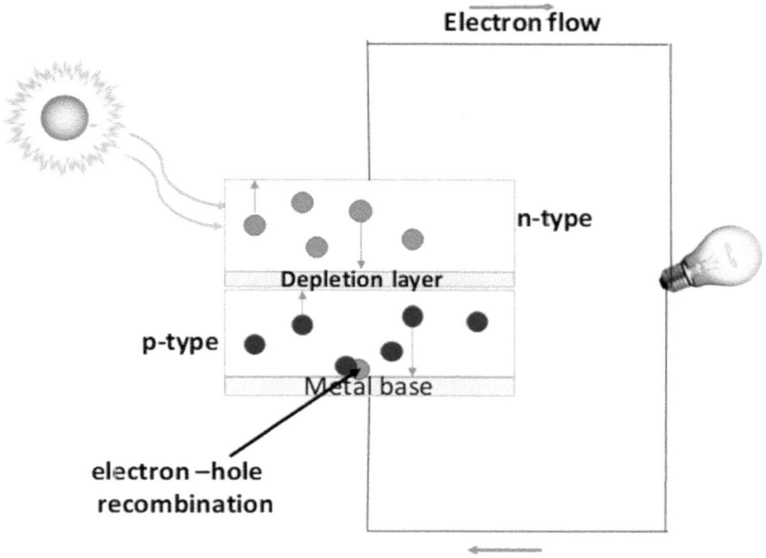

FIGURE 12.5 Schematic representation of the processes in a solar cell

12.3.2 IMPORTANT CHARACTERISTICS OF A SOLAR CELL

12.3.2.1 OPEN-CIRCUIT VOLTAGE (V_{OC})

When potential is generated by the photovoltaic cell in the open circuit, it is denoted by V_{OC} (OC means-open circuit). It is measured by noting the voltage across the terminals of the cell when no load is connected to the cell.

12.3.2.2 SHORT-CIRCUIT CURRENT OR SATURATION CURRENT (I_{SC})

The maximum current that a solar cell can deliver without harming its own constriction. It is measured by short-circuiting the terminals of the cell at most optimized condition of the cell for producing maximum output. When a high resistance is applied in the external circuit, a minor current flows through it and the voltage falls. The voltage goes on falling and the current goes on increasing when the resistance in the external circuit is reduced. So when the resistance is reduced to zero and the current increases to its maximum value which is known as saturation current (I_{SC}), then the voltage becomes zero. Short-circuit current density (J_{SC}) is the ratio of maximum or short-circuit current (I_{SC}) to exposed surface area (A) of the cell.

$$J_{sc} = \frac{I_{sc}}{A}$$

12.3.2.3 FILL FACTOR (FF)

The product of open current and saturation current is called the ideal power. Similarly, for any system, there will be a maximum possible current and maximum voltage, and the product of these two is called the maximum useful power. The fill factor is the ratio of useful power to ideal power.

Fill factor (FF) = maximum useful power/ideal power

12.3.2.4 CONVERSION EFFICIENCY (η)

Efficiency is defined as the ratio of the amount of energy drawn from the solar cell to input energy from the sun. It is often expressed in terms of fill factor, open-circuit voltage, saturation current, and input power.

$$\eta = \frac{V_{oc} I_{sc} FF}{P_{in}}$$

12.3.3 SOLAR CELLS: CLASSIFICATION AND STRUCTURE

Solar cells can be made of only single-junction or use multijunctions of light-absorbing material to take advantage of various absorption and charge separation mechanisms. Depending on the development stages of the solar cells, they are divided into three generations. In each generation, the materials used to construct the solar cell changed. The first-generation cells (also called conventional, traditional, or wafer-based cells) are made up of crystalline silicon. Second-generation solar cells are thin film photovoltaics that include amorphous silicon-based, CdTe-based, GaAs-based solar cells, etc. and are commercially significant. The third-generation solar cells include thin-film technologies and are often described as emerging photovoltaics—most of them have not yet been commercially applied and are still in the research and or in development phase. The classification based on the material of each generation is given in Figure 12.6.

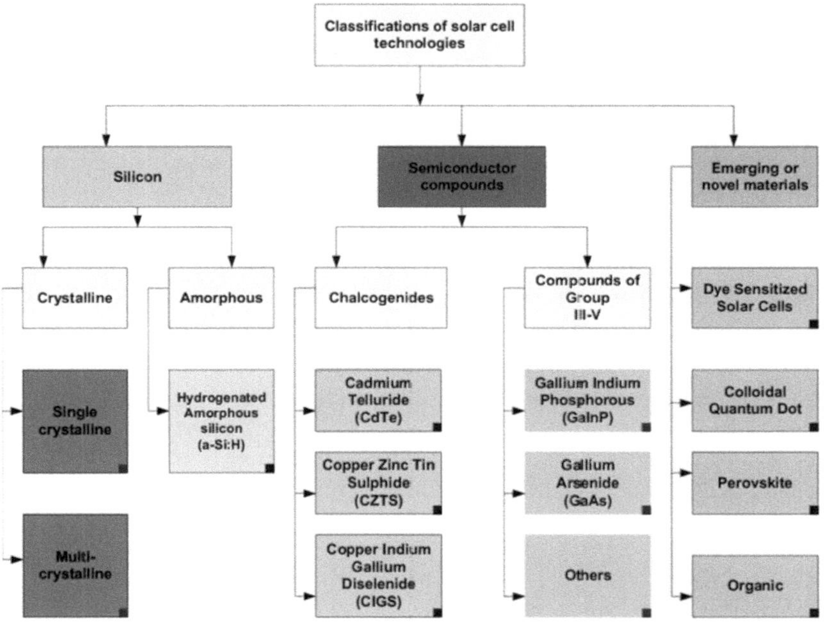

FIGURE 12.6 The classification based on the material of each generation

12.3.3.1 SILICON-BASED SOLAR CELLS

Silicon was the first material chosen for the making of solar cells. They are the first-generation photovoltaics which are made up of monocrystalline silicon material. In the advanced stage of these materials, amorphous silicon-based thin films were used, were expensive and the thickness of the solar cell was just 1 μ, or about 1/300th the size of monocrystalline silicon solar cell. But the efficiency was low compared with the ancestor.

12.3.3.2 ORGANIC SOLAR CELLS

The term organic solar cells are used for materials in which the light-harvesting layer or materials is made up of a polymer. The organic solar cells had the advantage of easiness in the fabrication and lightweight nature. But they were far behind the monocrystalline solar cells, because of their low efficiency. Conjugated polymers with ability to be doped and conducting polymers are employed in the organic solar cells.

12.3.3.3 DYE-SENSITIZED SOLAR CELLS

The dye-sensitized solar cells (DSSCs) are included in the third-generation and are mimicking the natural photosynthetic phenomena. They can absorb any light falling on them and were invented in 1991 by Professor Michael Graetzel and Dr Brian O'Regan. DSSCs can produce electrical energy by placing them indoor or outdoor. DSSCs are based on a semiconductor formed between a photosensitized anode and an electrolyte which is a photo electrochemical system. They have the advantages of being cost-effective, lightweight, thin and have the ability to convert both natural and artificial light, but lacks the high-energy conversion efficiency. The schematic structure of DSSC is given in Figure 12.7.[13]

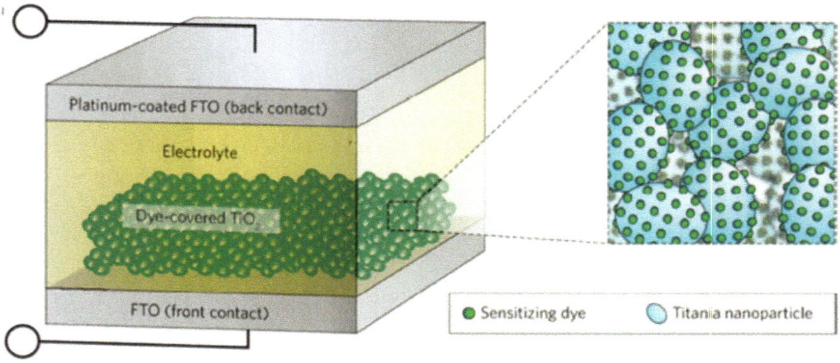

FIGURE 12.7 Schematic representation of a dye-sensitized solar cell.

12.3.3.4 PEROVSKITE SOLAR CELLS

Perovskite materials have the structural similarity of Ca TiO_3. They are well known for their light absorption capacity.[14] The perovskite materials are the favorite candidates in the solar cell industry because of their high-energy efficiency.[15] These are having a layered structure in which the perovskite materials are usually sandwiched as shown in the schematic Figure 12.8.

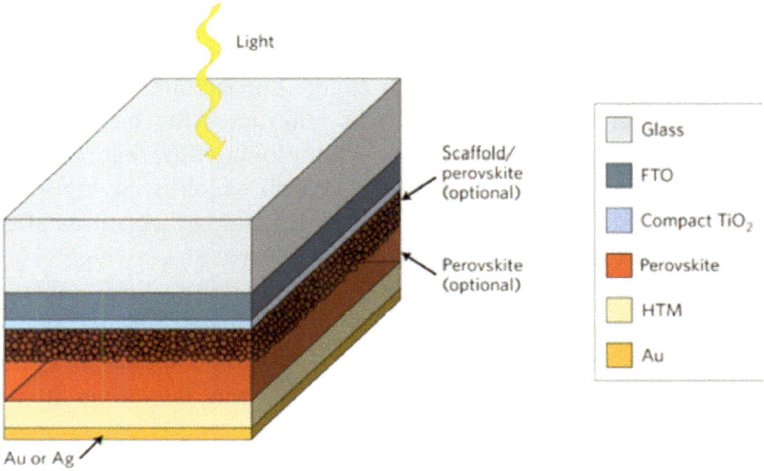

FIGURE 12.8 The layered structure of the perovskite materials used in solar cells

12.4 CARBON NANOTUBES AS CANDIDATES FOR COMPONENTS IN SOLAR CELLS

Carbon nanotubes are the materials with exceptional mechanical and electrical properties. Especially their conductive nature, surface area, and aspect ratio make them favorite among all the nanomaterials used in electronic industry. The single-walled and multiwalled nanotubes are used in a large number of electronic applications. In the photovoltaic technology, nowadays, carbon-based materials are being employed to enhance the parameters, such as energy efficiency and lightweight nature. Carbon nanotubes are being utilized in combination with organic, inorganic, or polymeric materials. They have been tested as the anodic materials instead of indium–titanium oxide. When looking into the role of CNTs in the solar cell or photovoltaic applications, they are selected as light-harvesting materials, transparent electrode, support in dye-sensitized cells, composites of CNTs with polymers act as photoactive layer, etc.

12.4.1 POSSIBILITIES OF SOLVING PROBLEMS IN ENERGY CONVERSION USING CNTS

The characteristics of the SWCNTs which make them suitable for the solar cell applications are wide range of direct bandgaps, strong absorptions in the visible spectrum, and high charge carrier mobility and chemical stability. When compared with their allotropes, the bandgap of CNTs may be adjusted over a wide range by employing different radii and functionalities.[16] In the case of organic solar cell which is having low carrier mobility and stability, incorporation of CNTs impart a positive effect. The addition of CNTs results in the electron mobility to a very high value equal to 10000 $m^2/V/s$ for SWCNT which exceeds those of any other organic semiconductor at room temperature.[17]

The CNTs can form a p–n junction very effectively and which in turn gives rise to photovoltaic effect easily. So, they have very high power conversion efficiency. The carbon nanotubes are generally incorporated with a polymer matrix in the photovoltaic materials. This combination act as the donor/acceptor complex (D/A), where the polymers are the donors and the CNTs the acceptor. Poly(3-hexylthiophene) (P3HT) and poly(3-octylthiophene) (P3OT) are the most common polymers used as donors in combination with SWCNTs. It is found that the CNTs can alter the HOMO levels of P3HT which make the conduction process faster.[18]

The addition of suitable surface modifiers or color enhancers, such as dye molecules into the polymer–CNT solar cells improves light absorption in the UV and red region, which in turn leads to an increase in the short-circuit current (Isc) by more than five times. SWCNTs can improve the crystallinity of conducting polymers. The crystalline structure of the polymer is a more efficient intermedium for the interchain transport.[18] In addition to the above-mentioned advantages, CNTs have the ability to absorb the infrared rays including visible range of sunlight, so the CNT-based solar cells can work at night times. Their thermal stability is very high and is generally about 1600°F. The amount of the CNT material to be used for the construction of solar cells is very less.

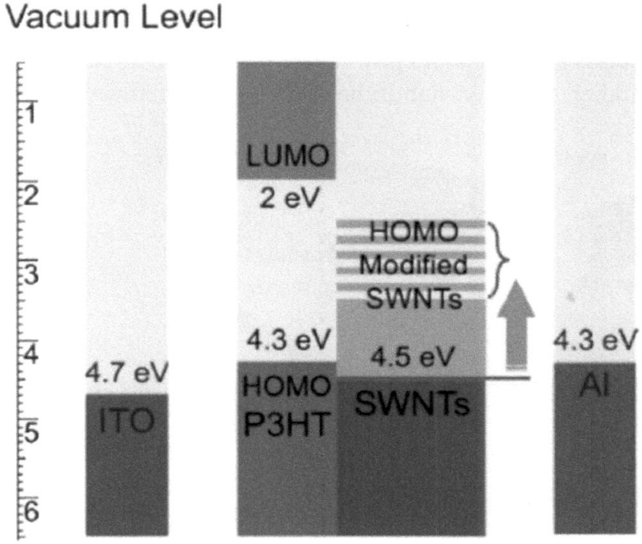

FIGURE 12.9 Electronic band line-up of the P3HT:SWCNTs. The arrow indicates the up shift of the nanotube Fermi level.[18]

Source: Reprinted with permission from Ref. 18. © 2006 American Chemical Society.

12.5 CNT-BASED SOLAR CELLS

SWCNTs offer an ideal light-harvesting medium that has a wide range of direct bandgaps, strong absorptions within the solar spectrum, and high carrier mobility with excellent chemical stability. An individual SWCNT can form a p–n junction diode, giving rise to the photovoltaic effect. So SWCNTs display high power conversion efficiency (PCE) under illumination. SWCNTs contain semiconducting and metallic forms, normally in a 2:1 ratio. Semiconducting SWCNTs form Schottky contacts with metallic objects, which are responsible for the ideal diode behavior, whereas metallic SWCNTs easily recombine electrons and holes. CNT acts as electron acceptors/transporters or light absorber and electron donor.[19]

Wei et al. constructed double-walled carbon nanotubes as energy conversion materials to assemble thin-film solar cells and to remove the use of polymers. The DWCNTs assist as photogeneration sites as well as a ballistic transport path for charge carriers (holes), and a piece of n-type silicon wafer was used to extract electrons. The various heterojunctions created between p-DWCNTs and n-Si improve the efficiency of solar cell. The large carrier density and high mobility of DWCNTs confirm much improved current density and power efficiency of solar cells compared with broadly studied polymer–nanotube composite structures (Fig. 12.10).[20]

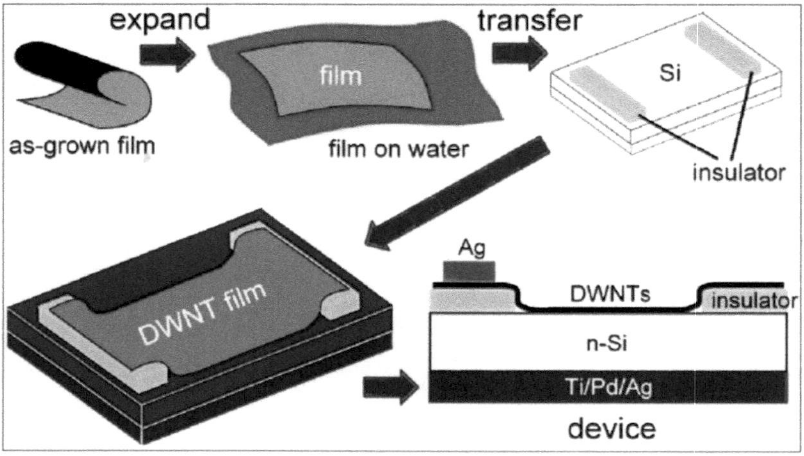

FIGURE 12.10 Illustration of the fabrication process of DWCNT/n-Si solar cell. The as-grown DWCNT films were fully expanded in a water solution and then conformally transferred to a patterned Si substrate.[20]

Source: Reprinted with permission from Ref. 20. © 2007 American Chemical Society.

12.6 CNT-BASED ELECTRODE MATERIALS AND ELECTROLYTES

CNTs have good optical transparency over a broad range from the visible to the near-infrared as well as high electrical conductivity. SWCNTs are effective for hole collection because their work function is in the range 4.8–5.0 eV, which is higher than that of ITO (usually less than 4.8 eV). Biris et al. reported the first SWCNT-based silicon solar cell with PCE of 1.3%. They formed this SWCNT film by spray-coating from a dimethylformamide suspension.[21] Gruner et al. reported the use of SWCNTs as a transparent anode and demonstrated efficient, flexible organic solar cells with a structure of polyethylene terephthalate (PET)/SWCNT/poly(3,4-ethylenedioxythiophene) (PEDOT)/P3HT:PCBM/Al.[22]

Seo et al. reported a remarkable conversion efficiency of ~7% employing CNT-based counter electrodes.[23] Cha et al. achieved an efficiency higher than 80% for a Pt counter electrode-based DSSC using CNT microballs.[24] Lee et al. demonstrated that vertically aligned nitrogen-doped CNT (N-CNT) counter electrodes can be employed in DSSCs via a "growth-detachment-transfer" process. This novel methodology has several advantages which include (1) surface area of vertically aligned CNT arrays is extremely large with a highly ordered geometry, (2) both electrical conductivity and catalytic activity of nitrogen-doped CNT is excellent. Doping of CNT with electron-rich nitrogen led to the reduction in the bandgap of the CNT to the metallic level CNT, and (3) this approach is substrate-independent. After growing the vertical N-CNTs via general catalytic CVD, the N-CNTs are readily transferrable to an arbitrary substrate.[25] In another study, Yang et al. reported the use of penetrated and aligned carbon nanotubes as counter electrodes of highly efficient dye-sensitized solar cells. Penetrated and aligned CNT films with high electrical conductivity and high catalytic activity and the resulting cells presented high-energy conversion efficiencies up to 8.46%. Figure 12.11a schematically shows the preparation of penetrated and aligned CNT film. A CNT array was first synthesized by a chemical vapor deposition process. A typical scanning electron microscopy (SEM) image of CNT film by side view is provided in Figure 12.11b. The resulting CNT film can be easily peeled off from the silicon wafer or stabilized on conductive glass after the further pressing treatment. Figure 12.11c compares J–V curves of the cells based on CNT films with different thicknesses. The open-circuit photovoltage (V_{OC}) remained almost unchanged with the increasing thickness from ~10 to ~40 lm, while the FF increased with the film thickness slightly increased from ~10 to ~20 lm and stabilized with the further increase of film thickness to ~40 lm. The short-circuit photocurrent density (J_{SC}) increased with the increasing thickness from ~10 to ~30 lm and then stabilized with the further increase of film thickness to ~40 lm. As a result, the energy conversion efficiency increased with the increasing film thickness from ~10 to ~30 lm and then stabilized with the further increase of film thickness. J_{SC} was increased with the increasing thickness from ~10 to ~30 lm as more CNTs increased contact areas with the electrolyte under the same CNT number density. As a comparison, the dye-sensitized solar cell based on the conventional Pt as counter electrode exhibited V_{OC} of 0.735 V, J_{SC} of 16.66 mA/cm^2, and FF of 0.60 (Figure 12.11c). The efficiency is calculated to be 7.32%. Therefore, such CNT films show promising applications as counter electrodes in dye-sensitized solar cells. Electrochemical impedance spectroscopy was further used to understand the high performance of penetrated and aligned CNT film as counter electrode. Two semicircles are typically observed in the Nyquist plot (Figure 12.11d). The first semicircle represents the electrochemical reaction at counter electrode in the high-frequency region (i.e., R_{CT1}), while the second semicircle reflects the sum of the charge transfer at TiO_2/dye/electrode interfaces in the middle-frequency region (i.e., R_{CT2}) and the Warburg diffusion process of I^- and I_3^- ions in the electrolyte in the low-frequency region.[26]

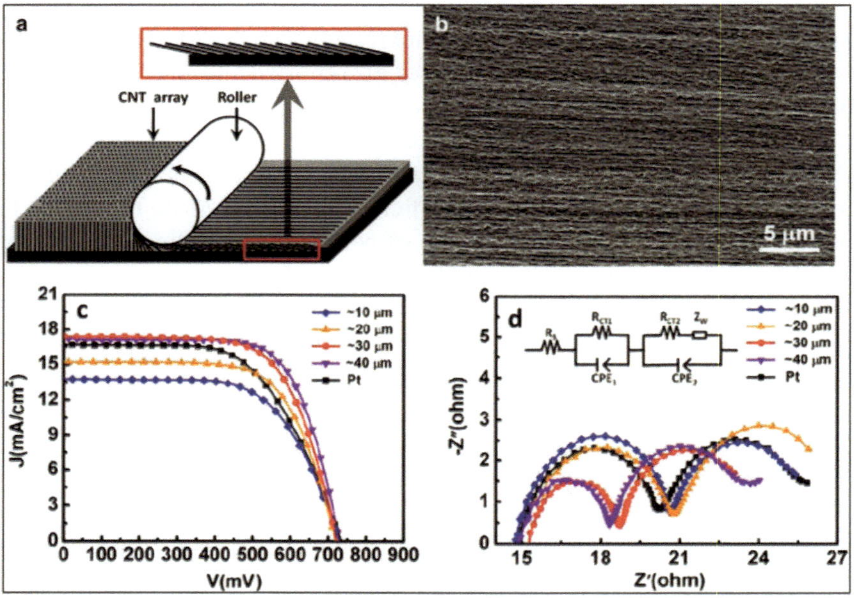

FIGURE 12.11 (a) Schematic illustration for the preparation of penetrated and aligned CNT film. (b) Scanning electron microscopy (SEM) image of the CNT film from a side view. (a) J–V curves of dye-sensitized solar cells by using penetrated and aligned CNT films with different thicknesses and platinum (Pt) film as counter electrodes under AM1.5 illumination. (b) Nyquist plots of the cells at (a) (the inserted image showing an equivalent circuit). The frequencies ranged from 0.1 to 100 kHz with an applied voltage of −0.8 V. Here, R_S, R_{CT1}, R_{CT2}, Z_W, and CPE stand for serial resistance, charge-transfer resistance at the counter electrode, charge-transfer resistance at TiO_2/dye/ electrode interfaces, diffusion impedance, and constant phase element, respectively.[26]

Source: Reprinted with permission from Ref. 26. © 2012 Elsevier.

Mehmood et al. used TiO_2–MWCNTs photoanodes for efficient DSSCs. Outstanding electrical conductivity of CNTs enhances Photoconversion efficiency (PCE) of DSSCs by decreasing the charge recombination. The integration of MWCNTs in the TiO_2 network considerably improves the efficiency of a DSSC because of the improvement in the surface area of the hybrid anode, causing in a higher dye loading. MWCNTs also improve the injection efficiency of the electrons due to the augmented positive potential. Here, experimental results are correlated with simulation results of a quantum modeling technique based on the density functional theory. This concept is used to analyze the thermodynamic aspects of the charge transport processes in DSSCs.[27] High surface area of carbon nanotube has begun to attract considerable interest for its use as counter electrodes in solar cell. An economic alternative for high cost Pt electrode is necessary for the low-cost production of solar cell. Studies have been reported on the use of CNT as a coating material for counter electrode in dye-sensitized solar cell. Performance of randomly dispersed multiwall CNT and relatively well-aligned CNTs directly grown by catalytic CVD was compared in one study. It is found that conversion efficiency of well-aligned CNT-coated counter electrode is much higher than that of the electrode coated by randomly oriented CNT.[28] Stable, ultra-strong, and highly flexible aligned carbon nanotube fibers can be used not only as catalytic counter electrodes but also as conductive materials to support dye-loaded TiO_2 nanoparticles in DSSCs (Fig. 12.12). Among two CNT fibers, one CNT fiber integrated with dye-loaded titanium oxide (TiO_2) nanoparticles is used as a working electrode, while another bare CNT fiber is used as a counter electrode. Power conversion efficiency of fiber-shaped DSSC is 2.94%. Since the PCE is independent of both incident light and cell length, this fiber solar cell may be easily scaled up for a practical application. Moreover, the flexible CNT fiber solar cell can act as woven fabrics to make textiles by a conventional weaving technology, which enables a wide variety of applications in portable devices, for example, power regeneration for space aircraft and clothing integrated photovoltaics.[29]

Surface-modified carbon nanotube by a polymer pyrene-functionalized poly(maleic acid-co-phydroxystyrene)-block-poly(p-hydroxystyrene) (HSPM) can be used as counter electrodes for dye-sensitized solar cells. HSPM-modified CNT displayed enhanced and stable dispersion in a-terpineol, which resulted in more facile and uniform CNT coatings on electrodes as compared with unmodified CNT. Surface modification led to improvement in fill factor and solar energy conversion efficiency of dye-sensitized solar cell.[30]

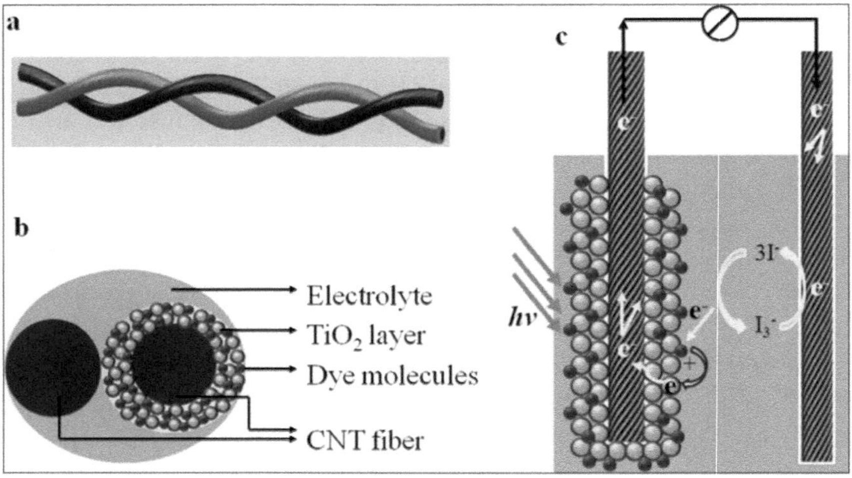

FIGURE 12.12 Schematic illustration of a wire-shaped DSSC fabricated from two CNT fibers. (a) Two CNT fibers twined into a cell. (b) Top view of a cell. (c) Working mechanism.[29]

Source: Reprinted with permission from Ref. 29. © 2012 American Chemical Society.

MWCNT/mesoporous carbon nanofibers composite prepared by electrospinning and silica template as counter electrodes for dye-sensitized solar cells has been explored by Park et al. The preparation technique involves three major steps, such as electrospinning, stabilization, and carbonization. Excellent mesoporous structure, excellent distribution of CNT, and good electrical conductivity of prepared material made it as an excellent component of DSSC. Its characteristics promote the electron and ion transfer, decreases the resistance of charge transfer, and increases the contact area between liquid electrolyte and CNT/Meso-ACNF. The overall conversion efficiency of CNT/Meso-ACNF augmented to 6.35% which is slightly greater than Pt counter electrode.[31]

Transparent films of CNT with exceptional mechanical strength have been used in solar cells as electrodes. Researchers have created three types of solar cells to investigate the efficiency of free-standing CNT film as electrodes. Three types of anodes include: (1) ITO on glass, (2) t-CNT on glass, and (3) a hybrid anode with t-CNT on the top of ITO on glass. It is found that the combination of 3-D t-CNT network with planar ITO leads to twice-increased photocurrent. The role of thin PEDOT–PSS coating on t-CNT for planarizing CNT network and avoiding shorts and for blocking electrons from recombination on t-CNTs has also been investigated.[32]

12.7 CNT AS ANTIREFLECTIVE AND SELF-CLEANING COATINGS

Incorporating CNTs, graphene, or conducting polymers into conventional silicon wafers leads to promising solar cell architectures with significantly enhanced power conversion efficiency.[33] Transparent surface coatings with suitable optical path differences can reduce the reflection of substrates and these coatings are called antireflection coating (ARC). It is well understood that smooth surfaces shine more than rough ones. So roughness is essential to decrease the reflectivity of surfaces. Materials which are used to construct ARCs, include silicon, silica, titania, zirconia, zinc oxide, cobalt oxide, tin oxide, carbon, and PET, polystyrene (PS), and gallium nitride.[34] ARCs arrays also exhibit self-cleaning ability because of the high fraction of air trapped between arrays. Self-cleaning surface is created by making the surface super hydrophobic. Normally, carbon nanotubes can be used to make the surface hydrophobic.

Zhao et al. reported silica-coated MWCNT composites (SiO_2/MWCNTs) synthesized by sol–gel method. Vinyl triethoxy modification of SiO_2/MWCNTs leads to a decrease in surface energy of the prepared material. Therefore, a solar cell incorporating CNTs has a super hydrophobic surface with maximum contact angle (CA) of 156.[35] Jia et al. reported a metal separator semiconductor heterojunction solar cell generated by placing a carbon nanotube film onto a silicon substrate. The nanotube–oxide–Si solar cells with polymer embodiment exhibited stable efficiencies of over 10%, due to the enhanced photon absorption, controlled charge recombination, and reduced internal resistance.[36] Kim et al. reported a super hydrophobic surface created through spraying with PS and multiwalled MWCNTs. Spraying is one of the easy and cheap systems for manufacturing super hydrophobic surfaces.[37]

12.8 CARBON NANOTUBE–POLYMER COMPOSITES IN PHOTOVOLTAICS

Fan et al. developed an effective approach by combining both highly transparent and conductive CNT network and poly (3, 4-ethylene dioxythiophene):poly (styrene sulfonate) (PEDOT:PSS) film to co-form coordinate heterojunctions with silicon to develop a hybrid PEDOT:PSS-CNT/n-Si solar cell. The PCE of polymer–CNT is improved upto 10.2%, which is much higher than the PCE values of both PEDOT:PSS/n-Si (5.5%) and CNT/n-Si (6.1%) solar cells fabricated using the same materials. Both the PEDOT:PSS patches and CNT "highway network" can co-form coordinate heterojunctions with the silicon substrate and fully contribute their advantages.[38] The effects of integrating different types of carbon nanotubes (SWCNT; DWCNT; and MWCNT) in organic solar cells were explored by Arranz-Andre's et al. The incorporation of CNTs to the polymer surges the power conversion efficiency by three orders of magnitude compared with the device without CNTs. However, there is no clear connection with the number of walls or diameter of the nanotubes, suggesting that the purity and defect density of the CNTs, and therefore, the metallic to semiconducting ratios are variables to consider.[39] Akhtar et al. developed a solid-state dye-sensitized solar cell in which carbon nanotubes–polyethylene oxide (PEO) composite electrolyte is used as electrolyte.[40] The incorporation of CNTs into PEO matrix considerably enhanced the electrolyte features of DSSC, such as roughness, amorphicity, and ionic conductivity. The solid-state DSSC fabricated with the optimum composite electrolyte (added 1% CNTs in PEO matrix) attained maximum conversion efficiency of 3.5%, an open-circuit voltage (V_{OC}) of 0.589 V, short-circuit current density (J_{SC}) of 10.64 mA/cm^2 and FF of 56%. Same group developed solid-state dye-sensitized solar cells which are based on CNTs–polyacrylonitrile (PAN) composite electrolyte. This solar cell has conversion efficiency of 3.9%.[41]

12.9 ROLE OF CNTs AND ITS COMPOSITES IN THERMAL ENERGY CONVERSIONS

Novel single-walled nanotube/phase change material composites were designed and synthesized by Wang et al. Composites displayed numerous distinctive features, such as a wider absorption range for sunlight, high light to heat conversion and energy storage efficiencies. These composites can be used as smart clothing or leather through fabric blending or wire drawing. Moreover, composites exhibited strong near-infrared absorbance characteristics and high-energy storage density with minor temperature change during heat storage and release processes, which have a potential application in military stealth.[42] Hu et al. used carbon nanotube-based thermoelectrochemical cell to harvest waste thermal energy. CNT electrodes offer high electrochemically accessible surface areas and fast redox-mediated electron transfer, which considerably improves thermocell current generation capacity and overall effectiveness. Thermocell efficiency is further upgraded by directly synthesizing MWCNTs as vertical forests that reduce electrical and thermal resistance at electrode/substrate junctions.[43] Im et al. fabricated carbon nanotube aerogel-based thermoelectrochemical cells as high-efficiency electrochemical thermal energy harvester.[44]

12.10 ADVANCES AND FUTURE PERSPECTIVES IN CNT-BASED ENERGY CONVERSION MATERIALS

Carbon nanotubes found enormous applications in the field of energy conversion. In DSSCs, carbon nanomaterials are being used. These are capable of reducing the use of electrocatalytic triiodide. First use of carbon nanomaterials in flat cells was done in 1996. CNTs are incorporated in energy conversion systems since 1999. 3D photovoltaic cells with CNTs as counter electrodes are highly flexible in nature. Extraordinary behavior of CNTs like lightweight, flexibility, excellent mechanical strength, and good electrocatalytic property enrich the performance of solar cells. CNTs also help to enhance the charge separation and electron transportation which is very important in solar cell applications. CNTs can also be used in the form of CNT sheets and CNT-dispersed films which will improve the efficiency of solar cells and their flexibility. CNTs are having high electrochemical activity with interactive surface and this is beneficial to exciton dissociation and charge carrier transport. This will definitely improve the power conversion efficiency. Platinized CNTs can act as good conducting electrodes in solar cells. Ti-TiO_2-CNTs can be merged into DSSC which shows extraordinary energy conversion efficiency.[45] Figure 12.13 shows the SEM images of Ti-coated TiO_2 working electrode with CNTs and a schematic representation of Ti–TiO_2–CNT film-structured DSSC.

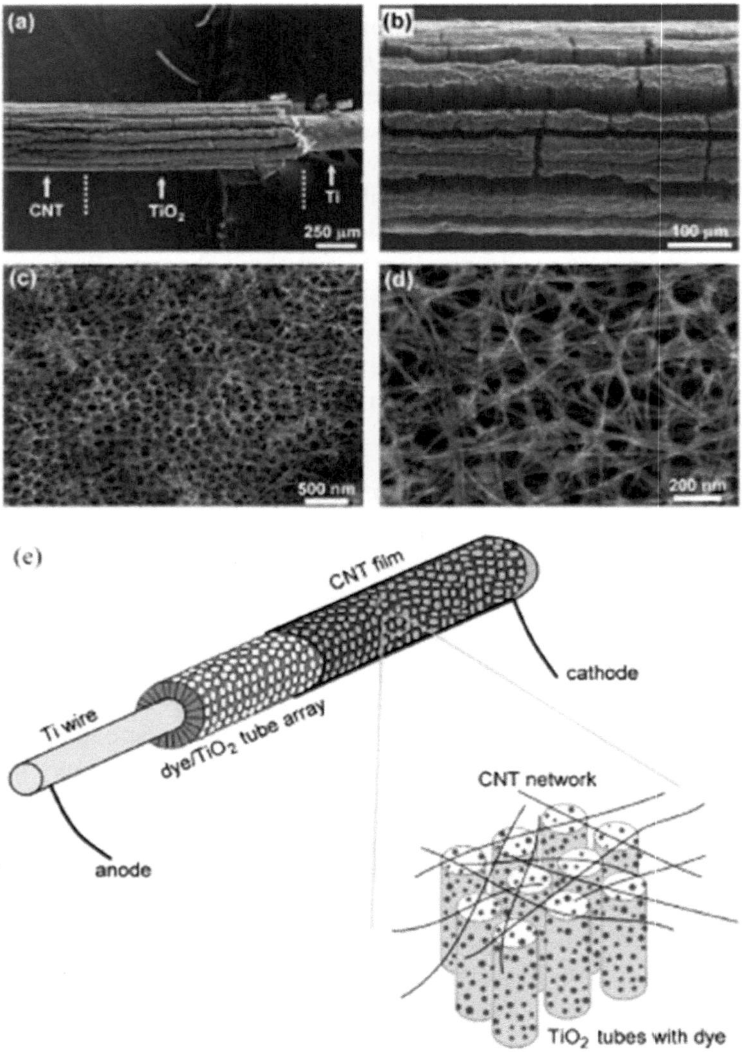

FIGURE 12.13 (a) SEM image of TiO_2-coated Ti wire (WE) wrapped with thin CNT film (CE). (b) Enlarged view of (a). (c) SEM image of CNT network deposited on the surface of WE. (d) Enlarged view of CNT network. (e) Schema of Ti–TiO_2–CNT film-structured DSSC.[45]

Source: Reprinted with permission from Ref. 45. © 2013 Elsevier.

CNTs can be used in low-temperature fuel cells. It can be used for the oxygen reduction reaction (ORR) in cathode of the fuel cell. Platinum materials are being used as catalysts for ORR reactions nowadays. But high cost and nonavailability of platinum-based catalysts offer a challenge in the field of fuel cells. Commercialization of low-cost catalysts offers a broad way for manufacturing fuel cells. High surface area and chemical stability of CNTs replaces Pt catalysts in fuel cells. Limited ORR activities of CNTs are further compensated by doping CNTs with heteroatoms like nitrogen. The electron donation process from carbon to oxygen in CNT network is facilitated by the electrons provided by heteroatoms. This will lead to improved ORR activity. [46]

12.11 CONCLUSION

This chapter discussed the electrical and optoelectronic properties of NT which make them as an ideal candidate for solar energy conversion application. Carbon nanotubes are being utilized in combination with organic, inorganic, or polymeric materials. These are selected as light-harvesting materials, transparent electrode, support in dye-sensitized cells, composites of CNTs with polymers act as photoactive layer, etc. Extraordinary behavior of CNTs like lightweight, flexibility, excellent mechanical strength and good electrocatalytic property enrich the performance of solar cells. CNTs also help to enhance the charge separation and electron transportation which is very important in solar cell applications. CNTs can also be used in the form of CNT sheets and CNT-dispersed films which will improve the efficiency of solar cells and their flexibility.

KEYWORDS

- **Solar cells**
- **carbon nanotubes**
- **energy**
- **perovskite**
- **conversion**

REFERENCES

1. Ahmad, A.; Kholoud, M. M.; Abou, E-N.; Reda, A. A.; Ammar, A. A-W. Carbon Nanotubes, Science and Technology Part (I) Structure, Synthesis and Characterization. *Arab. J. Chem.* **2012,** *5*, 1–23.
2. MORADI, O.; YARI, M.; ZARE, K.; MIRZA, B.; NAJAFI, F. Carbon Nanotubes: A Review of Chemistry Principles and Reactions. *Fullerenes Nanotubes Carbon Nanostruct.* **2012,** *20*, 138–151.
3. Cao, Q.; Han, S-j.; Tulevski, G. S.; Zhu, Y.; Lu, D. D.; Haensch, W. Arrays of Single-Walled Carbon Nanotubes with Full Surface Coverage for High-Performance Electronics. *Nat. Nanotechnol.* Mar **2013,** *8*.
4. Jariwala, D.; Sangwan, V. K.; Lauhon, L. J.; Marks, T. J.; Hersam, M. C. Carbon Nanomaterials for Electronics, Optoelectronics, Photovoltaics, and Sensing. *Chem. Soc. Rev.* **2013,** *42*, 2824.
5. Avouris, P.; Chen, Z.; Perebeinos, V. Carbon-Based Electronics. *Nat. Nanotechnol.* Oct **2007,** *2*.
6. Dai, H. Carbon Nanotubes: Synthesis, Integration, and Properties. *Acc. Chem. Res.* Dec 17, **2002,***35* (12), 1035–1044.
7. Scarselli, M.; Castrucci, P.; De Crescenzi, M. Electronic and Optoelectronic Nano-Devices Based on Carbon Nanotubes. *J. Phys. Condens. Matter* **2012,** *24*.
8. Mustonen, K.; Laiho, P.; Kaskela, A.; Susi, T.; Nasibulin, A. G.; Kauppinen, E. I. Uncovering the Ultimate Performance of Single-Walled Carbon Nanotube Films as Transparent Conductors. *Appl. Phys. Lett.* **2015,** *107*, 143113.
9. Marinho, B.; Ghislandi, M.; Tkalya, E.; Koning, C. E.; Gijsbertus de With. Electrical Conductivity of Compacts of Graphene, Multi-Wall Carbon Nanotubes, Carbon Black, and Graphite Powder. *Powder Technol.* **2012,** *221*, 351–358.
10. Nanot, S.; Hároz, E. H.; Kim, J-H.; Hauge, R. H.; Kono, J. Optoelectronic Properties of Single-Wall Carbon Nanotubes. *Adv. Mater.* **2012,** *24*, 4977–4994.
11. Setaro, A.; Adeli, M.; Glaeske, M.; Przyrembel, D.; Bisswanger, T.; Gordeev, G.; Maschietto, F.; Faghani, A.; Paulus, B.; Weinelt, M.; Arenal, R.; Haag. R.; Reich, S. Preserving p-Conjugation in Covalently Functionalized Carbon Nanotubes for Optoelectronic Applications. *Nat. Commun. 8*, **14281**. doi: 10.1038/ncomms14281.
12. Avouris, P.; Radosavljevic, M.; Shalom, J. Carbon Nanotube Electronics and Optoelectronics. doi: 10.1007/3-540-28075-8_9.
13. Bagher, A. M.; Vahid, M. M. A.; Mohsen, M. Types of Solar Cells and Application. *Am. J. Opt. Photon.* **2015,** *3* (5), 94–113.
14. Liu, M.; Johnston, M. B.; Snaith, H. J. Efficient Planar Heterojunction Perovskite Solar Cells by Vapour Deposition. *Nature* 2013 *501* (7467), 395–398.
15. Snaith, H. J. Perovskites: The Emergence of a New Era for Low-Cost, High-Efficiency Solar Cells. *J. Phys. Chem. Lett.* **2013,** *4* (21), 3623–630.
16. Cataldo, S.; Salice, P.; Menna, E.; Pignataro, B. Carbon Nanotubes and Organic Solar Cells. *Energy Environ. Sci.* **2012,** *5*, 5919–5940.
17. Zhou, X. J.; Park, J. Y.; Huang, S. M.; Liu, J.; McEuen, P. L. *Phys. Rev. Lett.* **2005,** *95*, 146805.

18. Geng, J.; Zeng, T. Influence of Single-Walled Carbon Nanotubes Induced Crystallinity Enhancement and Morphology Change on Polymer Photovoltaic Devices. *J. Am. Chem. Soc.* **2006,** *128*, 16827.
19. Jeon, I.; Matsuo, Y.; Maruyama, S. Single-Walled Carbon Nanotubes in Solar Cells. *Top Curr. Chem. (Z)* **2018,** *376*, 4. https://doi.org/10.1007/s41061-017-0181-0
20. Wei, J.; Jia, Y.; Shu, Q.; Gu, Z.; Wang, K.; Zhuang, D.; Zhang, G.; Wang, Z.; Luo, J.; Cao, A.; Wu, D. Double-Walled Carbon Nanotube Solar Cells. *Nano Lett.* **2007,** *7* (8).
21. Li, Z.; Kunets, V. P.; Saini, V.; Xu, Y.; Dervishi, E.; Salamo, G. J.; Biris, A. R.; Biris, A. S. $SOCl_2$ Enhanced Photovoltaic Conversion of Single Wall Carbon Nanotube/n-Silicon Heterojunctions. *Appl. Phys. Lett.* **2008,** *93* (24), 243117.
22. Rowell, M. W.; Topinka, M. A.; McGehee, M. D.; Prall, H-J.; Dennler, G.; Sariciftci, N. S.; Hu, L.; Gruner, G. Organic Solar Cells with Carbon Nanotube Network Electrodes. *Appl. Phys. Lett.* **2006,** *88* (23), 233506.
23. Seo, S. H.; Kim, S. Y.; Koo, B. K.; Cha, S. I.; Lee, D. Y. *Langmuir* **2010,** *26*, 10341.
24. Cha, S. I.; Koo, B. K.; Seo, S. H.; Lee, D. Y. *J. Mater. Chem.* **2010,** *20*, 659.
25. Lee, K. S.; Lee, W. J.; Park, N-G.; Kim, S. O.; Park, J. H. Transferred Vertically Aligned N-Doped Carbon Nanotube Arrays: Use in Dye-Sensitized Solar Cells as Counter Electrodes. *Chem. Commun.* **2011,** *47*, 4264–4266.
26. Yang, Z.; Li, L.; Lin, H.; Luo, Y.; He, R.; Qiu, L.; Ren, J.; Peng, H. Penetrated and Aligned Carbon Nanotubes for Counter Electrodes of Highly Efficient Dye-Sensitized Solar Cells. *Chem. Phys. Lett.* **2012,** *549*, 82–85.
27. Mehmood, U.; Hussein, I. A.; Harrabi, K.; Mekki, M. B.; Ahmed, S.; Tabet, N. Hybrid TiO_2–Multiwall Carbon Nanotube (MWCNTs) Photoanodes for Efficient Dye Sensitized Solar Cells (DSSCs). *Solar Energy Mater. Solar Cells* **2015,** *140*, 174–179.
28. Nam, J. G.; Park, Y. J.; Kimc, B. S.; Lee, J. S. Enhancement of the Efficiency Of Dye-Sensitized Solar Cell by Utilizing Carbon Nanotube Counter Electrode. *Scripta Materialia* **2010,** *62*, 148–150.
29. Chen, T.; Qiu, L.; Cai, Z.; Gong, F.; Yang, Z.; Wang, Z.; Peng, H. Intertwined Aligned Carbon Nanotube Fiber Based Dye-Sensitized Solar Cells, dx.doi.org/10.1021/nl300799d | *Nano Lett.* **2012,** *12*, 5, 2568–2572.
30. Choi, H. J.; Shin, J. E.; Lee, G-W.; Park, N-G.; Kim, K.; Hong, S. C. Effect of Surface Modification of Multi-Walled Carbon Nanotubes on the Fabrication and Performance of Carbon Nanotube Based Counter Electrodes for Dye-Sensitized Solar Cells. *Curr. Appl. Phys.* **2010,** *10*, S165–S167.
31. Park, S-H.; Jung, H-R.; Kim, B-K.; Lee, W-J.; MWCNT/Mesoporous Carbon Nanofibers Composites Prepared by Electrospinning and Silica Template as Counter Electrodes for Dye-Sensitized Solar Cells. *J. Photochem Photobiol. A Chem.* **2012,** *246*, 45–49.
32. Ulbricht, R.; Lee, S. B.; Jiang, X.; Inoue, K.; Zhang, M.; Fang, S.; Baughman, R. H.; Zakhidov, A. A. Transparent Carbon Nanotube Sheets as 3-D Charge Collectors in Organic Solar Cells. *Solar Energy Mater. Solar Cells* **2007,** *91*, 416–419.
33. Shi, E. et al. TiO_2-Coated Carbon Nanotube-Silicon Solar Cells with Efficiency of 15%. *Sci. Rep.* **2012,** *2*, 884.

34. Hanaei, H.; Assadi, M. K.; Saidur, R. Highly Efficient Antireflective and Self-Cleaning Coatings That Incorporate Carbon Nanotubes (CNTs) into Solar Cells: A Review. *Renew. Sustain. Energy Rev.* **2016,** *59*, 620–635.
35. Zhao, L. et al. Super Hydrophobic Surface Fabricated by Modifying Silica Coated Multiwalled Carbon Nanotubes Composites. *J. Sol–Gel Sci. Technol.* **2013,** *69* (1), 107.
36. Jia, Y. et al. Encapsulated Carbon Nanotube–Oxide–Silicon Solar Cells with Stable 10% Efficiency. *Appl. Phys. Lett.* **2011,** *98* (13), 133115.
37. Kim, H-K.; Cho, Y-S. Fabrication of a Super Hydrophobic Surface via Spraying with Polystyrene and Multi-Walled Carbon Nanotubes. *Colloids Surf A: Physicochem. Eng. Asp.* **2015,** *465*, 77.
38. Fan, Q.; Zhang, Q.; Zhou, W.; Xia, X.; Yang, F.; Zhang, N.; Xiao, S.; Li, K.; Gu, X.; Xiao, Z.; Chen, H.; Wang, Y.; Liu, H.; Zhou, W.; Xie, S. Novel Approach to Enhance Efficiency of Hybrid Silicon-Based Solar Cells via Synergistic Effects of Polymer and Carbon Nanotube Composite Film. http://dx.doi.org/10.1016/j.nanoen.2017.02.003
39. Arranz-Andre's, J.; Blau, W. J. Enhanced Device Performance Using Different Carbon Nanotube Types in Polymer Photovoltaic Devices. Carbon **2008,** *46*, 2067–2075.
40. Akhtar, M. S.; Park, J-G.; Lee, H-C.; Lee, S.-K.; Yang, S.-K. Carbon Nanotubes–Polyethylene Oxide Composite Electrolyte for Solid-State Dye-Sensitized Solar Cells. *Electrochim. Acta* **2010,** *55*, 2418–2423.
41. Akhtara, M. S.; Li, Z. Y.; Park, D. M.; Oh, D. W.; Kwak, D-H.; Yanga, O-B. A New Carbon Nanotubes (CNTs)–Poly Acrylonitrile (PAN) Composite Electrolyte for Solid State Dye Sensitized Solar Cells. *Electrochimica Acta* **2011,** *56*, 9973–9979.
42. Wang, Y.; Tang, B.; Zhang, S. Single-Walled Carbon Nanotube/Phase Change Material Composites: Sunlight-Driven, Reversible, Form-Stable Phase Transitions for Solar Thermal Energy Storage. *Adv. Funct. Mater.* **2013**. doi: 10.1002/adfm.201203728.
43. Hu, R.; Cola, B. A.; Haram, N.; Barisci, J. N.; Lee, S.; Stoughton, S.; Wallace, G.; Too, C.; Thomas, M.; Gestos, A.; dela Cruz, M. E.; Ferraris, J. P.; Zakhidov, A. A.; Baughman, R. H. Harvesting Waste Thermal Energy Using a Carbon-Nanotube-Based Thermo- Electrochemical Cell. *Nano Lett.* **2010,** *10*, 838–846. doi: 10.1021/nl903267n.
44. Im, H.; Kim, T.; Song, H.; Choi, J.; Park, J. S. Ovalle-Robles, R.; Yang, H. D.; Kihm, K. D.; Baughman, R. H.; Lee, H. H.; Kang, T. J.; Kim, Y. H. High-Efficiency Electrochemical Thermal Energy Harvester Using Carbon Nanotube Aerogel Sheet Electrodes. doi: 10.1038/ncomms10600 |www.nature.com/nature communications.
45. Yan, J.; Uddin, M. J.; Dickens, T.J.; Okoli, O. I. Carbon Nanotubes (CNTs) Enrich the Solar Cells. *Solar Energy* **2013,** *96*, 239–252.
46. Chen, Z.; Higgins, D.; Chen, Z. Nitrogen Doped Carbon Nanotubes and Their Impact on the Oxygen Reduction Reaction in Fuel Cells. Carbon **2010,** *48*, 3057–3065.

INDEX

D

E

F

G

H

I

L

M

N

O

P

Q

S